Mensuration of Areas and Volumes

Parallelogram

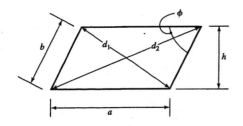

$$p = 2(a + b)$$

$$d_1 = \sqrt{a^2 + b^2 - 2ab(\cos\phi)}$$

$$d_2 = \sqrt{a^2 + b^2 + 2ab(\cos\phi)}$$

$$d_1^2 + d_2^2 = 2(a^2 + b^2)$$

$$A = ah = ab(\sin\phi)$$

If $a = b$, the parallelogram is a rhombus.

Regular Polygon (n equal sides)

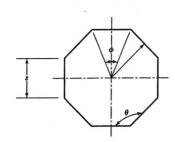

$$\phi = 2\pi/n$$

$$\theta = [\pi(n - 2)]/n = \pi - \phi$$

$$p = ns$$

$$s = 2r[\tan(\phi/2)]$$

$$A = (nsr)/2$$

Right Circular Cone

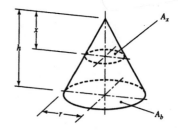

$$V = (\pi r^2 h)/3$$

$$A = \text{side area} + \text{base area}$$

$$= \pi r(r + \sqrt{r^2 + h^2})$$

$$A_s : A_h = x^2 : h^2$$

Right Circular Cylinder

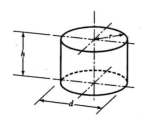

$$V = \pi r^2 h = \pi d^2 h/4$$

$$A = \text{side area} + \text{end areas}$$

$$= 2\pi r(h + r)$$

Paraboloid of Revolution

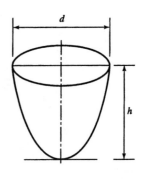

$$V = \pi d^2 h/8$$

Construction Cost
Analysis and Estimating

PRENTICE HALL INTERNATIONAL SERIES
IN CIVIL ENGINEERING AND ENGINEERING MECHANICS

William J. Hall, Editor

Au and Christiano, *Structural Analysis*
Bathe, *Finite Elements Procedures*
Biggs, *Introduction to Structural Engineering*
Chopra, *Dynamics of Structures: Theory and Applications to Earthquake Engineering*
Cooper and Chen, *Designing Steel Structures*
Cording et al., *The Art and Science of Geotechnical Engineering*
Hendrickson and Au, *Project Management for Construction, 2/e*
Higdon et al., *Engineering Meachanics, 2nd Vector Edition*
Holtz and Kovacs, *Introduction in Geotechnical Engineering*
Johnston, Lin, and Galambos, *Basic Steel Design, 3/e*
Kelkar and Sewell, *Fundamentals of Analysis and Design of Shell Structures*
Kramer, *Geotechnical Earthquake Engineering*
MacGregor, *Reinforced Concrete: Mechanics and Design, 3/e*
Melosh, *Structural Engineering Analysis by Finite Elements*
Nawy, *Prestressed Concrete: A Fundamental Approach, 3/e*
Nawy, *Reinforced Concrete: A Fundamental Approach, 4/e*
Ostwald, *Construction Cost Analysis and Estimating*
Pfeffer, *Solid Waste Management*
Popov, *Engineering Mechanics of Solids, 2/e*
Popov, *Mechanics of Materials, 2/e*
Schneider and Dickey, *Reinforced Masonry Design, 3/e*
Wang and Salmon, *Introductory Structural Analysis*
Wolf, *Dynamic Soil-Structure Interaction*
Young et al., *The Science and Technology of Civil Engineering Materials*

Construction Cost Analysis and Estimating

Phillip F. Ostwald

University of Colorado at Boulder

Prentice Hall
Upper Saddle River, New Jersey 07458

Library of Congress Cataloging-in-Publication Data

Ostwald, Phillip F.
    Construction Cost Analysis and Estimating
Phillip F. Ostwald
        p. cm.
    Includes bibliographical references and index.
    ISBN: 0-13-083207-2
    1. Cost Estimating. I. Title.
    CIP DATA AVAILABLE

Vice president and editorial director of ECS: **MARCIA HORTON**
Acquisitions editor: **LAURA CURLESS**
Production editor: **IRWIN ZUCKER**
Executive managing editor: **VINCE O'BRIEN**
Managing editor: **DAVID A. GEORGE**
Manufacturing manager: **TRUDY PISCIOTTI**
Manufacturing buyer: **PAT BROWN**
Marketing manager: **DANNY HOYT**
Copy editor: **ROBERT LENTZ**
Vice president and director of production and manufacturing, ESM: **DAVID W. RICCARDI**
Director of creative services: **PAUL BELFANTI**
Cover director: **JAYNE CONTE**

© 2001 by Prentice Hall
Prentice-Hall, Inc.
Upper Saddle River, New Jersey 07458

The author and publisher of this book have used their best efforts in preparing this book. These efforts include the development, research, and testing of the theories and programs to determine their effectiveness. The author and publisher make no warranty of any kind, expressed or implied, with regard to these programs or the documentation contained in this book. The author and publisher shall not be liable in any event for incidental or consequential damages in connection with, or arising out of, the furnishing, performance, or use of these programs.

Printed in the United States of America

10 9 8 7 6 5 4 3 2 1

ISBN   0-13-083207-3

Prentice-Hall International (UK) Limited, London
Prentice-Hall of Australia Pty. Limited, Sydney
Prentice-Hall Canada Inc., Toronto
Prentice-Hall Hispanoamericana, S.A., Mexico
Prentice-Hall of India Private Limited, New Delhi
Prentice-Hall of Japan, Inc., Tokyo
Pearson Education Asia Pte. Ltd.
Editora Prentice-Hall do Brasil, Ltda., Rio de Janeiro

This book is dedicated to Mathew Phillip, Bethany Grace, Benjamin David, Luke Phillip, Clayton James, Andrea Grace, and Patrick Jonathan, my grandchildren.

το μυστηριον εφανερωθη τοιζ αγιοιζ αυτου.

Contents

Picture Lessons

Appendixes

Photo Credits

Image	Credit
The Historic Hoover Dam, an American Icon	Bureau of Reclamation
High Pressure Cracking Tower	Mike McPheeters, Director of Photography/Creative Resources, Black & Veatch, Kansas City, MO
Iron Bridge	Science Museum, London/Science & Society Picture Library, London
Islanbard Kingdom Brunel	National Portrait Galley, London
Concrete Arch Viaduct	National Railway Museum, York/Science & Society Picture Library, London
Saltash Bridge	Science Museum, London/Science & Society Picture Library, London
Thames Tunnel	Science Museum, London/Science & Society Picture Library
Pile Driving	Suzanne L. Burris, The Burlington Northern and Santa Fe Railway Company
Wooden Trestle	Photo by Jones Photo Company, Aberdeen, WA
Building the Railroad	Suzanne L. Burris, The Burlington Northern and Santa Fe Railway Company
Damn Big Dam	Bureau of Reclamation
Golden Gate Bridge	Library of Congress
Civilian Conservation Corps	Photo by Jones Photo Company, Aberdeen, WA
Penstock	Bureau of Reclamation
Interstate Highway System	Carl T. Sorrentino, Department of Transportation, State of Colorado
Worker Safety	Mike McPheeters, Director of Photography/Creative Resources, Black & Veatch, Kansas City, MO
Steel	Mike McPheeters, Director of Photography/Creative Resources, Black & Veatch, Kansas City, MO
Proud Moment	U.S. Army Corps of Engineers, Portland District
Power House	U.S. Army Corps of Engineers, Portland District
Airport Terminal	Denver International Airport
Turbine rotor	Mike McPheeters, Director of Photography/Creative Resources, Black & Veatch, Kansas City, MO
Concrete	U.S. Army Corps of Engineers, Portland District
Dome of U.S. Capitol Under Construction	Library of Congress (Photo by Corbis)

Preface

This first edition of *Construction Cost Analysis and Estimating* provides the latest principles and techniques for the evaluation of construction design. It is not a book about estimating only. Analysis and estimating must abide together, and the one must precede the other. It is this emphasis that makes this text different.

The book begins with four chapters devoted to analysis of labor, material, accounting, and forecasting. Then estimating is developed, and methods, work, and project chapters are given. An owner or contractor is concerned with bid assurance, analysis, and contracts and ethics, and these chapters are provided. This book organization develops the principles in a systematic way.

With the increasing importance of design over rote skills in contemporary construction courses, this text can be used for a variety of teaching situations: for lecture only, for lecture with a laboratory menu, or with professional mentoring with business, and with developed field trips. Courses that couple to on-line live or delayed video instruction can use this text, as the author has personal experience with these delivery modes. Further, lifelong learning programs for the professional in either formal or informal settings can use the text.

Academic requirements for this book/course may vary, and we believe that the text is suitable for a number of teaching approaches. It has been written to appeal to engineering/technology/construction management settings. The student needs a mathematical maturity of algebra and introductory calculus.

The instructor will notice internet requirements that search for information and apply it in practical contexts. We provide internet addresses for numerous assignments. (Regrettably, these addresses may change from time to time.) In the interactive environment of teaching, this book is a part of modern courseware. Word-processing and spreadsheet skills are assumed, and some CAD ability is always helpful. It would seem that the student must have access to a computer, and system requirements would be typical of more advanced personal or Pentium computers.

Various academic levels and backgrounds are appropriate, and the instructor will find this text suitable to a variety of teaching styles. The author attempts to involve the instructor in the leadership of many exercises, calling on you, the instructor, to localize the assignments to your construction needs.

The book has more material than can be covered in one semester or quarter, and thus chapters and sections can be selected to meet the objectives of each class. Chapter order can be adjusted. If the students already have an understanding of statistics, then Chapter 5 material can be excluded, for example. Other sections can be dropped, depending on student preparation and course objectives. Some sections are identified as "Optional," allowing instructor selection. The instructor will find that the text is versatile.

This text has a range of difficulty for Questions, Problems, More Difficult Problems, Practical Applications, and Case Studies. Throughout the text, the author has attempted to give the instructor opportunity for outcomes evaluation of student work with these many exercises.

There are 124 Questions in the eleven chapters. They are qualitative and require back-reading and a response of a few sentences for a thoughtful reply.

We stress construction as a design activity; therefore, the 237 Problems and 48 More Difficult Problems request computations or sketches. Whenever the student is asked to prepare open-ended designs, much learning occurs. The Problems have levels of difficulty.

We want the Problems and the More Difficult Problems to be tractable, either with calculator or spreadsheet, where the emphasis is on teaching concepts. It is not our desire to cause excessive computation, which is so prevalent in construction problems. Thus, this text ignores software encyclopedias that are found in construction for estimating designs. Those software applications restrict the learning of principles. Nor do we give much attention to the minutia of extensive take-off practices, as those temporal trade details can be learned on the job, if necessary.

There is an end-of-chapter addition, which we call Practical Application. The purpose of the Practical Application is to uncouple the student from books, libraries, and the classroom. As will be seen throughout the book, Practical Applications introduce the student to experiences in the real world. For example, they encourage field trips and communication through the internet to engineers, technologists, and other construction professionals. The instructor will appreciate this experiential approach, allowing him or her to use Practical Applications in exciting ways.

The end-of-chapter Case Studies are open ended, perhaps having several solutions. Students are often disturbed by this peculiarity, but instructors recognize that construction courses are unlike calculus courses with their singularity of answer.

The book contains Picture Lessons—they describe important historical contributions of civil engineering and construction. It is important that students have an appreciation of the grand heritage of our profession.

For the instructor, a comprehensive Solutions Manual is available. This manual can be requested from the Prentice Hall college representative or from Prentice Hall directly.

The author is grateful to many people. Their advice and information have made this a much better text. In writing a book of this magnitude, the author is aware that friends and colleagues are hidden but very important advisers. I am indebted to the following: Don Boyle, Boulder, CO; James E. Diekmann, Anthony Songer, Klaus Timmerhaus, and Paul Zoller of the University of Colorado, Boulder, CO; Rodney Ehlers, Boulder, CO; Marty Geist, Boulder, CO; John Heitkamp, Richardson Engineering Services, Inc., Mesa, AZ; Laurence D. Jacobs, Craftsman Book Company, Carlsbad, CA; John Ferguson, RS Means Corporation, Kingston, MA; Anthony Mason, California Polytechnic State University, San Luis Obispo, CA; Liang Y. Lui, University of Illinois, Urbana, IL; Barry McMillan and Christian Heller of the American Association of Cost Engineers, Morgantown, WV; Ruby Ostwald, Bowling Green, KY; Ted Plank, Boulder County Road Department, Longmont, CO; Wayne Shelton, International Brotherhood of Electrical Workers, Broomfield, CO; Neil Wagner, Thermopolis, WY; and Natalie Soulier Webster, American Society of Civil Engineers, Washington, D.C.

Finally, it needs to be mentioned that I am sincerely grateful to Irwin Zucker of Prentice Hall, who has attentively improved this book in numerous ways.

The names used in the examples and Case Studies are of real people, and they are mentioned because of my sincere regard for their contribution and friendship.

Phillip F. Ostwald
Boulder, Colorado

C H A P T E R 1

Importance

In this first chapter we consider several strategies to design, bid, and contract for a project. The student needs to understand the economic consequence of design in general, and in particular the dimensional and monetary differences between domestic and international construction.

Design is the major function for engineering and technology, and when considered in its broadest context, it causes a long chain reaction. From concept to the final stage, design, cost, and profit are meaningful to the enterprise. Whoever brings a grasp of these matters to his or her job is valued in construction.

1.1 CONSTRUCTION—WHAT IT IS, WHAT IT IS NOT

Construction is a process that sets up a portable plant, brings material to the site, and on completion of the work moves the plant away, leaving its output standing fast. The output of construction includes all immobile structures, such as airports, buildings, canals, dams, factories, municipal treatment plants, pipelines, power plants, roads, and tunnels. Those just mentioned are examples of large construction projects.

Construction plants vary in size and complexity, ranging upward from a worker, wheelbarrow, pick, and shovel to the plant employed to build the Hoover Dam. Construction plants are themselves the output of manufacturing plants, and they construct the manufacturing plants. Construction and manufacturing are inexorably linked—two important pieces of society's wealth engine.

Manufacturing plants are defined as fixed in location. The materials come to the plant, they are processed, and products move off the plant site. Such plants include a great variety of factories and job shops. Mistakenly, manufacturing plants are thought to engage only in mass production. This is true of some, such as automobile plants and beverage-container plants. In the latter, for example, production of the popular thin-wall aluminum 12-ounce soda and beer container exceeds many million cans per day worldwide. But manufacturing plants are better termed "lot or batch" plants, as quantity ranges from many units to a few to even just one. The average product-lot quantity is less than 100 units.

The agricultural plants of farms and ranches are similar to manufacturing plants except that the equipment moves on and off the plant site. The plant site is fixed.

Table 1.1 compares the three types of plants we have been discussing.

The construction plant focuses on the production of a single and unique end product. The product is stylized in design, varied in method of erection, and different in location.

PICTURE LESSON
The Historic Hoover Dam, an American Icon

As far back as 1902, government hydrographers and geologists had identified potential sites on the Lower Colorado with a dream of a Colorado River dam. In 1921 full-scale testing began to find suitable sites. Black Canyon was chosen for superior engineering features of minimal excavation, tunneling, narrow gorge of high-quality bedrock confirmed by extensive core samples, less concrete for the arch-gravity dam, larger reservoir area, and good gravel beds for concrete.

With tentative plans and cost estimates, which were developed by the Reclamation Service, debate started in Congress and finally concluded in the signing of the Act in December 1928 by President Calvin Coolidge. With six-state ratification of the Colorado River Compact, the mining engineer President Herbert Hoover authorized $165 million to build the dam and the All-American Canal. In January 1931, the Bureau of Reclamation released the plans and specifications to interested bidders at five dollars a copy. Each bid was to be accompanied by a $2 million bid bond and the winner was required to post a $5 million performance bond—enormous amounts in light of the depressed U.S. economy and banking system.

Listed in the specifications were 119 bid items: 3.7 million cubic yards of rock for excavation, 4.4 million cubic yards of poured concrete, 45 million pounds of pipe and structural steel, and so on. The government would provide all materials entering into the completed work, such as cement and steel, but it was left up to the contractor to furnish machinery, tools, vehicles, and supplies to carry out construction. Scores of firms were interested, but their enthusiasm dropped as they realize the magnitude of the job. Even the most aggressive of contractors were brought up short by the cost of the bid and performance bonds. It was simply too much money for any one company to handle alone.

Companies banded together—Utah, Morrison-Knudsen, J.F. Shea, Pacific Bridge, MacDonald & Kahn, and Bechtel-Kaiser-Warren Brothers—and incorporated under the name Six Companies. W.A. Bechtel was named vice president. Shortly, the key men met with the superintendent of construction, engineer Frank Crowe, in the smoke-filled hospital room of the dying president William Wattis, and combed over the details of the bid items. Accepting the engineer's cost estimate, the executives added 25% for profit for this extremely risky venture. The bid opening at the Bureau of Reclamation's office in Denver was tense, and movement in the room was difficult because of the crowding. The first bid read "$80,000 less than the lowest bid you get" and was disqualified. Another bid was disqualified because it was not accompanied by the required $2 million bid bond. The last opened bid, $48,890,955 with the required bid bond, was declared acceptable. It was only $24,000 more than the cost calculated by the Bureau of Reclamation engineers. Pandemonium broke out, camera flashes popped, and the crowd cheered. Six Companies had won the right to build the dam. The Hoover Dam—the Great Pyramid of the American West—was the supreme engineering feat of its day. New generations of bigger and more sophisticated dams have risen, yet Hoover Dam remains the benchmark and inspires awe. It remains the highest concrete dam in the Western Hemisphere. At this time the Boulder Canyon Project would be called the most ambitious government-sponsored civil engineering task ever undertaken. The view of Hoover Dam is just below Nevada (left) and Arizona (right) Powerhouses. Water is released from Lake Mead in a regulated flow to farms, homes, and factories downstream. The water passing through Hoover's turbines generates hydroelectric energy for markets in Nevada, Arizona, and California.

TABLE 1.1 Hypothetical comparison of three types of production plants

Plant Type	Plant and Equipment	Output	Location of Plant Site	Relative Value of Output
Construction	Mobile	Remains fixed and unique	Variable	Expensive
Manufacturing	Fixed	Moves off the plant site and is numerous	Fixed	Cheap to expensive
Agriculture	Fixed and mobile	Moves off the plant site and is numerous	Fixed	Inexpensive

While some homes, for example, are basically similar, the units are site adapted. It is difficult to neatly categorize the great spectrum of product that emanates from the construction industry, except to try the following oversimplified definition of the word "project." A *project* is the scientific planning for one end unit of constructed output.

Construction intersects all fields of human endeavor, and this diversity manifests itself in its projects and economic impact. In the United States, total construction employs about 5% of the national work force, or 8 million workers. It is a $1.2 trillion[1] industry.

Construction projects are broken down into four groups:

- Industrial
- Heavy engineering and infrastructure
- Commercial building
- Residential

In large measure these categories parallel the general specialties of designers and constructors.

Industrial.   Industrial construction has some of the largest projects and is dominated by very large engineering and construction firms. Industrial projects include automobile plants; the decommissioning of nuclear plants; fossil-fuel plants; heavy-duty manufacturing plants; mine development; oxygen-fuel plants; petroleum refineries and petrochemical plants; steel mills; synthetic-fuel plants, and other plants essential to our basic industries. Those projects are identified by the process of the plant. Industrial projects in the United States are privately funded. They are considered the most technical of the four categories, and comparatively few design firms and constructors are qualified to undertake them.

The second picture lesson shows the erection of a high-temperature, high-pressure tower for a petrochemical plant.

Heavy Engineering and Infrastructure.   This category includes many well-known structures: airports, bridges, flood control, dams and tunnels, highways, hydropower, irrigation, storm-water collection, water treatment and distribution, and urban rapid-transit systems.

Design and construction activities in this category are primarily the domain of civil engineers, but almost all of the engineering disciplines have roles. The equipment phase is equipment intensive and characterized by fleets of large earth movers, heavy trucks, etc., working with massive quantities of basic materials—earth, rock, concrete, steel, pipe, and so forth. Many of those projects are publicly funded, and this sector may spend $175 billion in a typical year. Projects tend to be long in duration. Construction contractors are well informed on geology and engineering, as compared to those involved in building and residential construction.

Commercial Building.   Examples in this category include churches, government buildings, grade schools, hospitals, small retail stores, shopping malls, urban development complexes, and warehouses. These structures interact closely with people. Economically, this category accounts for $200 billion of the total annual construction business. Like projects in the residential market, these are labor and material intensive, but their scope is more complex.

---

[1] Annual projection based on U.S. Census Bureau source: Sector values for subgroups of construction vary dramatically year to year.

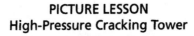

PICTURE LESSON
High-Pressure Cracking Tower

The picture shows the lifting of a high-pressure fractionator column prior to the vertical erection on pedestals. This unit is for high-temperature high-pressure cracking of oil crude for a petrochemical refinery project. There are several strategies to design and construct a project. In America the industrial and power sector has used the strategy of design-build for decades. In the European community and Japan, for example, design-build is the system of choice for nonresidential projects for more than 50% and 70%, respectively. The benefit for design-build is a single point of responsibility for quality, cost, and schedule. The design-builder is motivated to deliver a successful project by fulfilling multiple and parallel objectives, including aesthetic and functional quality, and budget, while the owner focuses on scope/needs definition and decision making, rather than on the coordination between the designer and builder. There are various advantages for the several contractual approaches to design and construction. In design-build, the owner's requirements are documented in performance terms. The design-builder warrants to the owner that it will produce design documents that are complete and free from error. By contrast, with traditional design-bid-build, the owner warrants to the contractor that the drawings and specifications are complete and free from error. The traditional approach relies on restrictive contract language, audit and inspection, and occasionally the legal system to ensure final project quality.

Design and construction professionals working and communicating as a team evaluate alternate materials and methods. Value engineering and constructability are enabled during the entire design process. Arguably, guaranteed construction costs are known earlier in design-build than in other project delivery systems.

With some exceptions, the private economy finances these structures. Design is coordinated by architects, who work with engineering specialists, such as structural, mechanical, or electrical.

Residential.   This category includes single-family homes, apartments, condominiums, and town houses. In a typical year residential construction totals about $300 billion. This industry is largely financed by private investment. It is characterized by a large number of contractors and subcontractors. Low capital and labor-intensive simple technology are commonplace. In leaner years, contractors may be underfinanced. If demand falls, there is a high rate of business failure.

Design is done by architects, drafting people, builders, or the home owner. There are PC software CDs that allow the home owner to design a house. They provide working drawings, isometrics, and a bill of material that is coupled with specifications for the contractor and the subcontractors.

The construction business is notable for its *free enterprise*. The *enterprise* involves business players, who include owners, designers, and constructors. Contractors are *free* to build or not, and how to build is their free choice. Further, the owner is *free* to conduct business with the contractor, or not. The owner is not restricted in choosing when or how to conduct business or with whom. For each enterpriser, the earning of profit is a primary goal for long-term survival in our free-enterprise system.

Our attention is naturally directed to the matters of technology and design. This requires application of science, business, and mathematics. We deal with technology and design as factors to handle increasing competition. The motivation for professional cost analysis and estimating is the necessity for profit in an environment of free enterprise and competition.

1.2   DESIGN AND ECONOMIC EVALUATION

At this point we shall define several terms. *Cost analysis and estimating* is concerned with evaluation of design. The term could well be "profit engineering," because for many players the assurance of profit is a top priority.

An *estimate* (noun) is the result of professional work. It implies an evaluation of a design, expressed as a cost or other economic metric and dimensioned in dollars. To *estimate* (verb) means to appraise or to determine.

We give the word "design" the broadest possible definition. It is not synonymous with "computer-aided design." Design requires creative engineers and is defined as follows: *Every design is a new combination of preexisting knowledge that satisfies an economic want.*

In this definition the phrase "is a new combination" emphasizes novelty and suggests the unique characteristics of the designer or design team and the circumstances. "Preexisting knowledge" relates to a design's (not designer's) intellectual background and to the knowledge base of the industry for which it has been created. When it "satisfies an economic want," the activity ultimately provides economic satisfaction.

The design procedure is illustrated in Fig. 1.1. It does not represent a precise sequential process. The design procedure is usually a hodgepodge of simultaneous continuous actions.

Problem.   The initial description of a new problem may be a vague statement addressing a want. It is necessary to transform this vague statement into an enlarged picture.

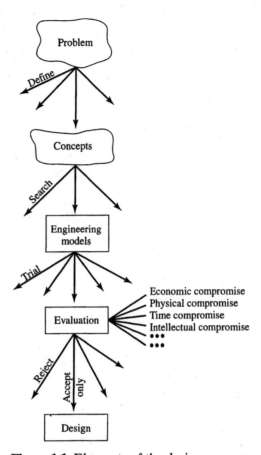

Figure 1.1 Elements of the design process.

Information (technical or nontechnical), costs, and other data are superficially gathered to give form to the problem. Suppliers, customers, competitors, standardization groups, safety and patent releases, and laboratories are sources for ideas. Questions may be asked such as, "Does this approach fit the company's needs, interests, and abilities? Are the people connected with the problem capable of carrying it to completion, or can suitable people be hired?" The aim is to arrive at a thoughtful and reasoned statement specifying the problem, which may lead to a more efficient result.

Concept.   Defining a problem sets the stage for the concept search. The quest may start with idealized notions such as the perfect gas laws, frictionless rolling, or perpetual motion. It is concerned with searching, learning, and recognizing, not with application, which comes later. A timely and fortunate search may uncover unapplied principles. It cannot, however, be exhaustive. When we consider that about a million scientific and technical articles are published annually, it should be clear that attempting simply to list (much less to study) all available information, even within a narrow field, is a hopeless cause. Nonetheless, there is a chance of finding the basic idea for a new development in patent disclosures, new texts, or journals in the field. A sense of where to start and when to stop in the search for concepts comes with experience.

Model.   The model involves application of creditable concepts selected from among those that have been uncovered. Models may range from casual back-of-envelope sketches to complicated physical shapes. The formation of a model is an engineering trait, manifesting a disciplined pattern of thought. Models, whether experimental or rational, can be manipulated for purposes of theoretical understanding or practical testing. An engineer selects the cheapest available way to formalize and understand a model.

Engineers use small-scale models to analyze larger problems. For example, an analyst may be unable to comprehend an actual system, and a simplified model may be a satisfactory substitute. Modeling allows discovery of pertinent variables, rejection or confirmation of prototypes, and comparison to a standard.

Laboratory testing provides numerical answers by using physical mockups. Data are obtained, results noted and conclusions stated.

Often it is either impossible or uneconomical to manipulate reality. For example, it is impossible to manipulate the market in a free economy, and uneconomical to manipulate a nuclear reactor for electrical generation.

Models allow us to predict reality through mathematical abstraction. Mathematical models use numbers and symbols to imitate relationships. Although such models are more difficult to comprehend, they are the most general. In an approximate way, they explain the real situation. It is customary to manipulate mathematical models according to the conventional rules of mathematics.

Mathematical models are desirable in engineering because they are easy to manipulate. An example of a model found in construction estimating is the unit-price formula. Whenever an engineer uses a recapitulation sheet, where labor, material, and overhead are summed, a mathematical model is used in a procedural sense. The discounted cash-flow model used to calculate a rate of return of an asset is one way of dealing with project estimates.

Models should be flexible to permit repeated applications. Once functional relationships and accuracy are assured, mathematical manipulation should be simple. Arithmetic is preferred over algebra. Algebra is preferred over calculus. Calculus is preferred over vectors. Computerized spreadsheets may be sufficient.

Evaluation.   Ultimately, the model reaches the evaluation point. Evaluation is influenced by economics, physical laws, social mores, ignorance, and the human fault of stupidity. Even moral questions may be debated. Although we cannot foresee all the future effects of a design, much can be clarified by asking the right questions. The owner and architect may contribute at this point by stating their concerns.

The cost analysis proceeds through stages similar to those that we have discussed. The problem is wisely stated (formulated in the design model), estimating ideas are considered, a cost model is selected, evaluation trials are started, and finally the cost estimate is completed.

Questions arise when performing economic evaluation. What is the total cost of developing this design up to the construction? Does the construction coincide with the abilities and experience of the contractor? What is the owner's profit from the total investment during the first few years of production and sales? How long will it take for the initial investment to pay back? A number of factors are noted when considering these questions. Experience, study, and a questioning attitude are vital.

For complex designs, later and subsidiary estimates are made as the design matures. If the task is at the subcontractor level, and the task is small, there may be no formal cost analysis, and the estimate may be merely a quotation sheet with essential numbers and prices.

Design.   Design is the execution of the plan into being and shape. Design involves, for example, computing, drafting, checking, and specifying to answer the question "How shall it be built?" rather than "How will it work?" Although we do not intend to overinflate its importance, designing occupies the greater part of the designer's time.

By using various methods, rules of thumb, simple calculations, and software, the analyst provides a quick and relatively inexpensive estimate. Its accuracy depends, of course, on the amount and quality of information and the time available to prepare the estimate. Solid facts and precision are lacking at this point.

This first cost evaluation is an important guide for the owner, engineer, architect, and contractor. An estimate has serious financial implications for the construction business. It has long been the case that anyone "...intending to build a tower, sitteth...down first, and counteth the cost, whether he have sufficient to finish it."[2] The first estimate screens designs and aids in the formulation of a budget. It is used to commit or stall additional engineering/architectural design effort, to decide on appropriations for capital equipment, or to cull out uneconomic designs at an early point. Although decisions based on the first estimate may not lead to a bid, or legal obligations, or authorization for capital spending, they can entail costly mistakes by unwittingly eliminating profitable designs.

The task is to represent the economic want of the design. A *want* is a value exchanged between competing and self-interested parties. It is the price a buyer is willing to pay for a residence, a contractor-owner agreement on the bid value of a building project, or the fiscal-year budget value for a rapid-transit system that the public authority proposes and elected representatives of public voters approve. As a new combination of preexisting knowledge, design satisfies an economic want. Cost analysis and estimating provide the metric for this compliance.

The dollar magnitude of an estimate varies with the design. A $0.10/unit price for a product and a 10^8 bid for a project are not unknown. We assert that the intellectual requirements for either estimate are the same. The 10^9 multiple from a minimum to maximum underscores the variability of the cost-analysis field.

1.3 STRATEGIES FOR THE ENTERPRISE

There are several strategies to design and construct a project. From the owner's point of view, the aim is to select which designer and contractor to hire, when to hire them, and under what type of contract. The owner's objective is proper and timely execution of construction work according to the contractual requirements. Further, this choice has a large bearing on cost analysis and estimating, which are required for owners, designers, contractors, and subcontractors.

Detailed engineering and design are large undertakings involving many professionals, such as those in the engineering disciplines (including chemical, civil, construction, electrical, mechanical, and other types as needed), architects, interior designers, and landscape architects. To some extent, engineer/architect consultants, design-constructors, professional construction managers, or program managers become involved in these stages. Along the trail from concept to the final stage, cost, profit, and economics are continuing concerns.

As the project broadens in scope, other organizations are involved. General contractors and subcontractors join in. Design and cost estimates continue to be important.

[2] Luke 14:28, The Holy Bible, King James Version.

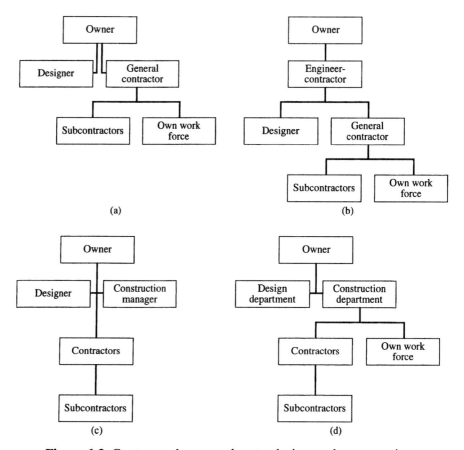

Figure 1.2 Contractual approaches to design and construction: (a) traditional, (b) design-build, (c) construction management, and (d) owner-builder.

There are four basic strategies for the enterprise: traditional, design-build, construction management, and owner-builder. Figure 1.2 illustrates these with a descriptive flow chart. Other approaches may be adopted, but for the sake of brevity we consider only these four. Some texts that provide further information are listed in the references at the end of this book.

Traditional. In this arrangement the owner hires a separate design firm, which prepares a set of contract documents. There is a negotiated professional fee for design services. Using the design package, the owner negotiates a price with a selected general contractor, or the owner solicits bids for the work. From the bids, a single general contractor is ultimately chosen.

The general contractor is totally responsible for delivering the completed project as defined in the work package. The general contractor may sublet work to subcontractors, where each subcontractor reports to the general contractor. The contractor and the subcontractors have the objective of performing the contract satisfactorily and making as much profit as possible.

Depending on the skills and needs of the owner, the design firm may be involved in overseeing the construction in the field. Typically, there is no direct formal relationship between the general contractor and designer. Communication is through the office of the owner.

The traditional method is popular, and most owners, designers, and contractors are familiar with the scheme. An advantage to the owner is the firm fixed price for the project before any work commences. Price competition between contractors is possible in an open market, which is an advantage to the owner.

Of course, the strategy has its disadvantages. Contractors and subcontractors have little opportunity to suggest improvements until after the award is announced. It may be difficult to phase or fast-track the project. There is less opportunity for interaction between the significant parties. Misinterpretations of the drawings and specifications can be difficult to eliminate, and unforeseen conditions may be more likely to emerge.

Design-Build. In the design-build strategy, the designer and the construction company are from the same firm or are connected through a joint venture to form a single company for the duration of the project. A joint venture is a legal contract between two parties for a specific purpose. The terms *turnkey* and *design-manage* are generally equivalent to design-build.

A major advantage for the design-build arrangement is the seamless communication that may occur within the single firm encompassing both design and construction. The contractor is able to improve constructability. Scheduling is more effective, and fast-tracking is likely. Design changes are simpler. The owner is less involved, perhaps.

The owner may not have a firm fixed price in hand early, though budgetary estimates may be an option. With the owner less involved, questions arise about knowledge and awareness of the project and some of the crucial decision making.

In the traditional method there are checks and balances between the designer and the contractor. The contractor is measured against the set of contracts, working drawings, and so on. The design-build method is more vulnerable from this viewpoint.

Construction Management. In this method the owner hires both a design firm and a construction-project management firm early in the conceptual stage, assigning responsibilities to each. A major advantage is effective communication among the three parties. The contractor can review the design of the project, and similarly the designer has a part in contractor selection and in reviewing work in the field. The method allows simultaneously phasing various tasks of the project in a coordinated effort. Further, the owner is able to benefit from competitive bidding by contractors and subcontractors.

If any of the parties are uncommunicative, the advantages quickly become disadvantages. The owner must have advanced, professional expertise and abilities.

Owner-Builder. The owner is responsible for design and construction. The functional organizational chart may have two departments, design and construction. There will be contractors and subcontractors in addition to the owner's own work force.

1.4 INFORMATION

The engineer needs information of various kinds in order to estimate the economic want of a design. Two extremes exist. Visualize the case where virtually no information is available. It is unlikely, then, that an estimate can be made. At the other extreme, assume that all data are available, which implies that the money has been spent for the design and an estimate is unnecessary. The after-the-fact cost analysis is likely to be accounting and not engineering cost analysis. The engineer works between these extremes in a cost-data and design regime where only partial information is available and unknown information must be estimated.

We separate the accounting from the engineering cost-analysis field on the principle that the accountant deals with cost quantities that have been spent and consistently recorded. The engineer reckons in cost quantities that have not been spent for a design. Cost accounting and engineering cost analysis are specializations that work together and have much in common.

Some cost data are *historical*. Accounting reports are historical because they emphasize the transactions recorded through cost-controlling accounts that may be kept in a computerized ledger system. Money is expended, and materials, labor, services, and expenses are received. Specific accounting procedures must be provided for recording the acquisition and disposition of materials, the use of labor, and the assignment of these transactions to the accounts. Cost accountants are primarily responsible for historical information. Accounting is a major source of information, but other departments external and internal to the construction business also provide historical information.

Construction crafts are the producing organizations and are concerned with doing. The foreman knows the operating details at that moment. Frequently, he or she is a direct source for information. The foreman may often assist in obtaining data on special forms that report extraordinary costs of process equipment, manning, efficiency, scrap, repairs, or down time. Sometimes the foreman is the oracle for a "guesstimate" on operations with which he or she is familiar. Specialized subcontractors are sources of significant information, such as provided by courtesy quotations.

The personnel department, charged with the handling of employees, interprets the union contract (where unions exist), conducts labor-contract negotiations, and keeps personnel records regarding wages and fringe costs.

The purchasing department in many organizations is responsible for spending money for bought materials. Some companies believe that the purchasing department is responsible for the outside subcontractor. The purchasing department knows about purchasing and shipping regulations and can be a frequent source of information.

Some information data are *measured*. Engineers may find that work-measurement methods give information that is amenable to estimation, either in time or dollar dimension. Data sets are popular books of information to estimate construction work. Readily obtainable, they are prominent in construction-work estimating and are helpful for reckoning straightforward construction. These publications go back a long time, perhaps to the early 1900s, when the first attempts were made to provide estimating data. They describe the design, job, site conditions, and labor, material, equipment, and performance.

Material quantities, which are calculated from design drawings and specifications, are a form of measured data.

Finally, some data are *policy* and have the property of being fixed for engineering purposes. They are accepted as factual and often are unchallengeable by the engineer. The origins of policy data are varied. Union-management wage settlements or union-hall hiring of construction labor, where predetermined policies dictate the wages and types of labor on equipment to be operated, are examples of policy data. Budgets, legislative restrictions, and laws ranging from the municipal to the national level dictate codes of conduct and cost. The federal government requires a social security tax from the employer for the purpose of providing old-age benefits. An unemployment compensation tax is collected by the states to provide funds to compensate workers during periods of unemployment.

A variety of basic economic facts and trends are reported by the U.S. government. The Bureau of Labor Statistics (BLS) reports trends in the prices of materials and labor.

Data are available from manufacturers' agents and jobbers, who, although they promote special interests, are willing to release information given to them by their clients. Trade associations, subsidized by groups of businesses sharing a common need, are typical organizations that publish data for cost analysis. The references at the end of this book provide an extensive listing of these informative helps.

The student of engineering must be informed about worldwide systems of units and money. Understanding of international trade and dimensional and monetary conversions is required. U.S. customary and metric units and the dollar, pound, euro, peso, and so on are used in discussions throughout this book to provide this important familiarity. With the exception of the United States, almost all of the world uses the international system of units (SI), and the student needs to be exposed to SI in many contexts. We now consider this important topic.

1.5 INTERNATIONAL SYSTEM OF UNITS

The International System of Units, known officially worldwide as SI,[3] is a modernized metric system and incorporates many advanced dimensional concepts. With SI it is possible to have a simplified, coherent, decimalized, and absolute system of measuring units. While it may be decades before the United States deals exclusively with SI units in engineering, the student needs to appreciate that construction worldwide transacts in terms of SI, not the English system.

The SI system has seven base units (meter, kilogram, second, ampere, degree Kelvin, candela, and mole), two supplementary units (radian, steradian), and many derived units. Basic to SI are the following definitions: 1 newton is the force required to accelerate a mass of 1 kilogram at the rate of 1 meter per second squared; 1 joule is the energy involved when a force of 1 newton moves a distance 1 meter along its line of action; and 1 watt is the power that in 1 second gives rise to the energy of 1 joule.

The SI units for force, energy, and power are the same regardless of whether the design is construction, civil, mechanical, electrical, or chemical. Confusion sometimes arises in the U.S. customary use of both pounds-force and pounds-mass, but that is avoided in SI. The SI system uses a series of approved prefixes and symbols for decimal multiples, which are shown in the book's end papers.

There are many aspects of SI practices to become familiar with. One is capitalization of SI symbols to avoid confusion: for example, K is Kelvin but k is kilo. Further, to separate digits into groups of three, SI uses a space (e.g., 1 000 000) instead of a comma, as shown in the book's end papers. Also, the European practice is to use commas where we in the United States use decimal points. In this text, in order to avoid confusion, we do not adopt the European practice.

The use of mixed units is to be avoided—for example, kilograms per gallon.

---

[3] SI is short for *Le Système International d'Unités*.

PICTURE LESSON
Simple Tool

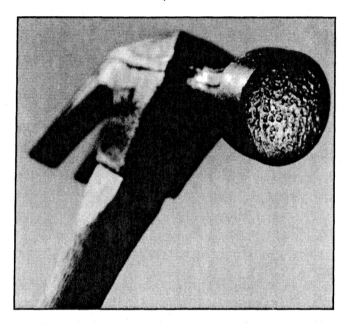

Known for millenia, the hammer is the simplest tool for construction. But humility need not dishonor this tool. In the hands of a competent tradesman, it remains an effective implement. The hammer has changed over the ages. Beginning with the stone mallet, there are other head materials, such as plastic, rubber, wood, iron, brass, lead, and several alloys of hardened steel that blunt the force impact.

A short handle allows for a sharp stroke by the carpenter to drive nails, or in the case of a 16-pound sledgehammer, the long handle is wielded with a powerful swing. The form of the head may be small and rounded for a peening hammer, or it may be wedge shaped for cracking and separating. If the head is sharpened to a fine edge, it becomes a scraper. The claw is back-curved for the carpenter for pulling out nails, but it is flatter and has a keen edge for the mason, allowing the breaking of brick.

The basic hammer has cousins: rip framing, wall board, straight claw, California framing, mustang claw, milled faced (the one shown in the picture), tack, brick, maul, shingling, and maul are popular varieties. Can you name other types?

SI makes use of prefixes (such as *kilo-*) before unit names to show orders of magnitude, thus eliminating insignificant decimals and providing a convenient method for writing powers of 10. For example, 12.3×10^3 meters becomes 12.3 kilometers.

Nominal dimensions name the item, and SI equivalents are hardly necessary. A 2×4 stud is such in name only; the "2×4" refers to the approximate inch dimensions of a rough-sawn, green piece of timber, whereas the dimensions when the timber is seasoned and finished are less.

Data for engineering units and money units come from a variety of sources and have been recorded with varying degrees of refinement. Specific rules are observed when engineering or cost data are added, subtracted, multiplied, or divided.

Consider the example of adding three numbers for engineering analysis, where the first represents data reported in millions, the second in thousands, and the third in units, as shown below in example (a):

(a)	163,000,000	(b)	163,000,000
	217,885,000		217,900,000
	96,432,768		96,400,000
	477,317,768		477,300,000

If those numbers are pure engineering data, then they should first be rounded to one significant digit beyond the rightmost one of the least accurate number, summed as in example (b) (to 477,300,000), and then rounded (to 477,000,000).

If the numbers are pure cost data, then the choice between approaches (a) and (b) depends on the intended use of the information. This author encourages example (a) as preferred, but he would accept example (b) if only an approximation were required. If the intended use of cost data is to present a bid to a potential customer, then the preferred practice in this book is (a).

The rule for multiplication and division of numbers representing engineering data is as follows: the product or quotient must contain no more significant digits than does the number with the fewest significant digits. The rule for addition and subtraction, on the other hand, requires rounding digits that lie to the right of the last significant digit in the least accurate number.

Multiplication:	$113.2 \times 1.43 = 161.876$, which rounds to 162
Division:	$113.2 \div 1.43 = 79.16$, which rounds to 79.2
Addition:	$113.2 + 1.43 = 114.63$, which rounds to 114.6
Subtraction:	$113.2 - 1.43 = 111.77$, which rounds to 111.8

In the examples above, the product and quotient are rounded to three significant digits, because 1.43 has only three. The sum and difference are rounded to four significant digits.

Numbers that are exact counts are treated as though they consisted of an infinite number of significant digits. When a count is used in computation with a measurement, the result is rounded to the number of significant digits in the measurement. If a count of 113 is multiplied by a value of 1.43, then the product is 161.59. However, if 1.43 were a rough value accurate only to the nearest 10 and, hence, had only two significant digits, then the product would be 160.

Rules for cost analysis are similar, although there are exceptions for dealing with the exaggerated precision sometimes required for construction contracts that involve large-quantity considerations.

Rules for rounding are the same as for estimating and engineering practice. If 3.46325 were rounded to four digits, then it would be 3.463; if to three digits, then 3.46. If 8.37652 were rounded to four digits, then it would be 8.377; if to three digits, then 8.38. If the digit discarded is exactly 5, then the last digit retained should be rounded upward if it is odd or not if it is even. For example, 4.365 when rounded to three digits becomes 4.36, and 4.355 becomes 4.36.

This book follows the practice of avoiding cents as a dimension in order to avoid confusion. For instance, a unit of a construction is expressed as $1.43 and never as 143¢.

Conversions from U.S. customary units to SI, and vice versa, are made by using the tables given in the end papers. The tables are approximate. Professionally preferred seven-figure-accuracy conversions can be found from information provided by the American Society of Testing and Materials.[4] When converting, remember to round in accordance with the accuracy of the original measurement. Consider the following examples: convert 12.52 ft to meters, 17.2 ft^3 to cubic meters, 5.15 lbm to kilograms, 2.005 in. to millimeters, and 2.4637 mile to kilometers.

Here are the answers:

$12.52 \times 3.048 \times 10^{-1} = 3.8161$ m,   which, on using the rule of precision of the original measurements, becomes 3.82 m;

$17.2 \times 2.831 \times 10^{-2} = 0.48693$ m^3 becomes 0.487 m^3;

$5.15 \times 4.53 \times 10^{-1} = 2.3330$ becomes 2.33 kg;

$2.005 \times 2.54 \times 10^{1} = 50.927$ becomes 50.93 mm, because a two-place decimal for the millimeters has the same degree of accuracy as contained in the original inch measurement;

$2.4637 \times 1.609 \times 10^{3} = 3964.093300$,   which, when rounded to three decimal places, becomes 3964.093 km because of similar degrees of accuracy implied by a measurement.

Now that we have summarized the conversion of measurements to dimensional units for SI, the next step is the consideration of currencies other than the dollar, either U.S. or Canadian.

1.6 INTERNATIONAL CONSTRUCTION

Construction business is international, even though the student sees mostly the domestic scene. International construction business thinks in terms of many national currencies. The importance of worldwide economics requires that we study what is happening elsewhere on this planet. Newspapers discuss international business events. For instance, the European Union and the euro currency, tariff agreements and politics, technical standards and associations, professional licensure requirements for the engineer, intellectual property, and the U.S. dollar exchange rate are a few topics that may stimulate the student into additional reading. As one illustration, the 15 EU Member States have a 40% greater population and an 18% greater gross domestic product than the United States.

Business opportunity exists in international markets for larger contractors. But cross-frontier business inevitably causes foreign-exchange exposure for one or both of the contracting businesses. Consider the example of a contractor who wants to purchase equipment

---

[4] American Society for Testing and Materials, 1916 Race St., Philadelphia, PA 19103. See also the web site, www.ASTM.org.

from Germany. We assume that the purchaser is in the United States and has the choice of currency for buying and paying for the machine. Three options are available: The company may agree to buy in euros, U.S. dollars, or some third-country currency, such as Canada dollars. If the purchase is in U.S. dollars, then the buyer has avoided a foreign-exchange problem, but a problem remains for the seller, who will receive dollars and sell them for euros. On the other hand, if the equipment is invoiced in euros, then the buyer will buy euros before buying the machine. A foreign-exchange transaction is involved either way. If the goods are invoiced in a third-country currency—for instance, Canada dollars—then the U.S. buyer arranges payment in Canada dollars to the German supplier, who in turn sells the dollars and converts the money into euros. In this event we have two foreign-exchange transactions. It is axiomatic that movement of goods or services across a frontier causes a foreign transaction.

The prevailing foreign-exchange rates between any two countries are the prices at which bills of exchange of one of those countries will sell in the currency of the other. The bills of exchange are expressed in terms of the price or rate at which the currency of one country is exchanged for that of another. In theory, a currency's value mirrors the fundamental strength of its underlying economy, relative to other economies. Table 1.2 illustrates foreign-exchange rates. For instance, it takes \$1.0884 U.S. to exchange for 1 euro.

Free and uncontrolled foreign-exchange rates fluctuate daily. Low exchange (meaning a lower value of domestic versus foreign currencies) normally indicates a strong demand for foreign currency or a heavy offering of U.S. dollars. A distinction is also seen in Table 1.2 between spot and future exchange, perhaps 1 year from the date of the spot value. When an importer purchases spot exchange, delivery is actually taken at the time of purchase for a definite amount of foreign exchange, for which the contractor pays the rate then quoted for the company's particular bill of exchange. When the importer purchases a future exchange contract, the importer agrees to purchase a given amount of exchange on a fixed date in the future or within a fixed period, paying for it at the rate specified in the future contract. This future rate may be higher or lower than the spot rate.

Consider the following examples. A material is priced currently at \$3.65/lbm U.S. Find the equivalent spot market in British pounds sterling and SI.

$$\left(3.65\,\frac{\$}{\text{lbm}}\right)\left(\frac{1}{1.6593}\,\frac{\pounds}{\$}\right)\left(\frac{1}{0.4535}\,\frac{\text{lbm}}{\text{kg}}\right) = \pounds4.851/\text{kg}$$

German 27.5-MPa concrete material is currently priced at 48.5 euro/m³. Determine equivalent values 1 year hence in U.S.-dollar and cubic-yard dimensional units.

$$\left(48.5\,\frac{\text{euro}}{\text{m}^3}\right)\left(1.1139\,\frac{\$}{\text{euro}}\right)\left(1.3080\,\frac{\text{m}^3}{\text{yd}^3}\right) = \$70.67/\text{yd}^3$$

TABLE 1.2 Illustrative exchange rates, U.S. dollar per unit of country's currency, i.e., \$1.6593 U.S. to exchange for 1 £ British

	British Pound Sterling	Canada Dollar	China Renminbi	European Union Euro	Hong Kong Dollar	Japan Yen	Mexico Peso	Russia Ruble	Saudi Arabia Riyal
Spot	1.6593	0.6812	0.1208	1.0884	0.1291	0.007095	0.1116	0.1604	0.2666
One Year	1.6149	0.7034	0.1108	1.1139	0.1106	0.009921	0.1093	0.1503	0.2718

Observe that the trailing decimals on the right side of the equation are increased over the euro amount because of the exact nature of currency conversion.

1.7 A LOOK AT THE BOOK

This book presents the kinds of thinking that are found in cost work. The circular riddle—do problems provide the stimulus in finding solution methods, or do techniques discover and solve problems?—is really never answered. An engineer would not seriously consider redesign of the wheel or feel any guilt in exploiting its theory and practice.

An effort has been made to assimilate theories and practices that are broadly attractive to students, whether they are (or are to be) employed in research, development, design, construction, sales, or management. Though cost practices vary among the several fields of technology, the principles do not. This becomes clearer when we look at the organization of the book.

Labor comprises one of the most important items of a construction job. And to forecast the productivity for work, we study procedures that identify, describe, and measure—this is an intermediate goal to analysis and estimating of labor costs. Chapter 2, for the most part, deals with the measurement of construction labor cost.

The panoply of materials with its engineering, construction, and cost effects requires alertness by owners, contractors, and subcontractors. Successful bidding and earning of profit are coupled closely to the effective use of material resources. The student at one time or another is likely to be engaged in these activities and, accordingly, will want to pay attention to the principles of Chapter 3.

Accounting, which is covered in Chapter 4, is the means of analyzing the money transactions of business. Accountants prepare periodic balance sheets, statements of income, and other information to aid the control of cost and the finding of overhead. Although cost-accounting data may have been wisely and carefully collected and arranged to suit the primary purposes of accounting, these raw data are incompatible with estimating and construction pricing.

Despite all statements to the contrary, "emotional estimating" or "guesstimating" has not disappeared from the construction estimating scene, nor has estimation by formulas and mathematical models been universally nominated as a replacement. Somewhere between those extremes is a preferred course of action. In Chapter 5 we consider ways to enhance construction cost analysis by graphical and statistical techniques.

Early chapters develop fundamental ideas about construction business. Even though construction designs differ greatly, estimating methods are remarkably similar. This is seen in Chapter 6, which discusses general estimating methods and their advantages and shortcomings. Some methods call mainly for experience and judgment, others emphasize mathematics.

Chapter 7 presents the principles for the work estimate. Owners, contractors, and subcontractors, while differing in many important ways, need a framework for planning this estimate. Students need to know this step. Labor, material, tools, and equipment costs are the grist for the detailed work estimate, and these elements must be determined efficiently. The work estimate is the starting point for the final project estimate, which gives the very important bid.

Techniques for estimating projects are strikingly similar, despite large differences in the design and in the sums involved. The principles in Chapter 8 are germane for either an owner or a contractor.

An estimate's accuracy, reliability, and quality are important. If the bid leads to winning a sale, then there is an opportunity to verify its general goodness. Chapter 9 describes the analysis of estimates when compared to a so-called actual value. There are lessons to be learned—for instance, in improving the capture rate, understanding the importance of human behavior, and employing other techniques.

The preparation of the analysis is unique for each design. Nevertheless, certain underlying principles aid this work. The student will want to learn the principles given in Chapter 10. Cost analysis is performed during construction "tradeoffs." These tradeoffs happen early in important construction projects—perhaps 60% to 80% of the consequential cost/trades are made before and during design.

If the construction bid is favored by the customer, then the owner and contractor proceed to contract discussions. Legal requirements are important, and that is why the engineer needs an understanding of contracts, as described briefly in Chapter 11. The last chapter emphasizes goals above rigid rules.

SUMMARY

With a liberal interpretation, cost may be defined as any of several economic measures of want. Many believe that a building, computer, or rocket design has economic value as the first and last requirement. The thought is this: physical and real-world economic restrictions are a constant companion to the design. Engineering, construction, marketing, sales, and finance conform to the design, because the drawings and specifications are the authority for construction and operation. Design, understood in its broadest context, initiates a long chain reaction. The student who brings to his or her job an understanding of the economic consequence of design is valued in business.

In this first chapter we couple design to its business environment. In this text we assume that design neither leads nor lags its economic shadow. Vital to the design is the cost information on which decisions must be based. With the design and cost data at hand, the engineer builds a corresponding cost-design structure. The next chapter begins the development of the cost-design structure.

QUESTIONS

1. Define the following terms:

Construction	Want
Manufacturing	Policy data
Heavy engineering and infrastructure	Euro
Design-build	Spot rate
Turnkey	SI
Profit engineering	Engineering models

2. Prepare a list of career opportunities in construction engineering from the classified want ads of a newspaper or a professional trade magazine.

3. How do competition and failure of the enterprise interact with principles of cost analysis and estimating? Discuss.

4. How would you define profit? Discuss fully. What are the consequences of negative profit (loss)? How can profit be rendered appealing to the individual?

5. List positive and negative results of failure in business. Should government prevent business failure? Does company size and political power affect your answer?

6. How do economic laws and physical laws differ? Are the well-ordered cause-and-effect relationships separable in business and engineering fields?

7. Describe how computers have improved "back-of-envelope" techniques in cost analysis.

8. Relate the role of cost analysis and estimating to design engineering. Contrast these roles.

9. What do exchange rates between countries reflect? Why do they fluctuate?

10. How do extractive plants, such as mining or forestry, compare to the description of construction, manufacturing and agriculture plants?

11. In what ways does the mission of ASTM coinside with construction? (Check out the web site for answer.)

PROBLEMS

In Problems 1.1 through 1.11, convert from U.S. customary units to SI units, or from SI to U.S. units. Show the correct abbreviations.

1.1 (a) 17 ft^2, 2.4 ft^2, 450 in.^2, 5000 ft^2 to meters^2 (m^2)

   (b) 0.15 in.^2, 0.035 in.^2, 20.61 in.^2 to millimeters^2 (mm^2)

1.2 (a) 280,000 British thermal units (Btu) to joules (J)

   (b) 7,500,000 kilowatt-hours (kWh) to J

1.3 (a) 18 pounds force (lbf) to newtons (N)

   (b) 180,000 lbf, 1.8×10^6 lbf to newtons with appropriate prefix

1.4 (a) 18 ft, 2.0 ft to meters (m)

   (b) 10 in., 0.01 in., 100 in., 0.00015 in. to millimeters (mm)

1.5 (a) 15 ounces mass (ozm) to kilograms (kg)

   (b) 25 tons to kilograms (kg) (*Hint*: Use a short ton, or 2000 lbm.)

   (c) 1400 kg to pounds (lb)

   (d) 83 short tons to tonne (*Hint*: U.S. practice discourages the use of the metric tonne and instead adopts the Megagram or 1 Mg.)

1.6 (a) 15 pounds-mass/foot^3 (*Hint*: lbm/ft^3 = mass/volume = density) to kilogram/meter^3 (kg/m^3)

   (b) 180 kg/m^3 to lbm/ft^3

1.7 (a) 2500 pounds-force/foot^2 (lbf/ft^2) to megapascals (MPa)

   (b) 180 pounds per in.^{2} (psi), 1750 psi to MPa

   (c) 17 MPa to psi

1.8 (a) 200 °F, 1000 °F to C (*Hint*: Do not use the degree symbol with Celsius.)

   (b) 1500 Celsius (C), 200 C, 1000 C to Fahrenheit (°F)

1.9 (a) 180 feet/minute (fpm), 500 fpm to meters/second (m/s)

   (b) 180 inches/second (in./s), 1855 in./s to meters/min (m/min)

1.10 (a) 0.37 ft^3, 125 ft^3, 700 ft^3 to meters^3 (m^3)

   (b) 0.01 in.^3, 12 in.^3, 150 in.^3 to millimeters^3 (mm^3)

   (c) 1000 yard^3 to m^3

          (d) 250 mm^3, 1500 mm^3 to in.^{3}

          (e) 183 bft, 17,500 bft to m^3

          (f) 51 gal (*Hint*: U.S. liquid) to m^3

1.11 **(a)** 800 ft^3/min, 65 ft^3/sec to m^3/s

          (b) 1000 m^3/s to ft^3/min

          (c) 65 yard^3/min to m^3/s

1.12 An architectural garnish costs $1750 U.S. per unit. What is the spot value of the garnish in euros? In Mexican pesos? In Canadian dollars?

1.13 If a catalyst is worth $850 per gallon in the United States, then what is an equivalent international cost for the China renminbi and the Japan yen in one-year values? (*Hint*: The U.S. measure is liquid.)

MORE DIFFICULT PROBLEMS

1.14 Steel for low-rise buildings can be purchased on the international market. For instance, Chinese companies offer structural grade A-36 steel in many sizes and lengths on west coast U.S. or Vancouver, Canada, docks. The business transports these commodities on container ships from ports in China, selling the steel at competitive local prices. There are no special orders, there is no distinction of grade and size for purchase, and only the cafeteria sale of the product that is loaded on the container ships is available. The steel is offloaded from the ship to train gondolas or flatbed trucks hired by the buyer. The buyer pays the transport cost to the construction site. The Chinese company will only consider orders in excess of 25 metric tonnes from listed supplies.

     A Canadian contractor wishes to buy 185 tons of steel for a Calgary, Alberta, project. Also, a Calgary steel warehouse supplier quotes a cost of Canada $2985 per ton delivered to the site, and there is no extra FOB cost. The buyer determines that rail freight costs are $17.31 per ton from dockside to the construction site. Determine the not-to-exceed bid offer for the Chinese company's steel. Express the offer in terms of China currency. (*Hints*: This is an immediate offer, and spot exchange rates are used. FOB means free on board where the material is loaded to the customers flat bed, for example, and it is not an additional charge to the buyer.)

1.15 A business woman travels from New York to four countries. She will start with U.S. currency and exchange her dollars in each country by using that country's prevailing exchange rate. In each country she will buy the next air fare and will incur business expenses. The travel and expense budget expressed in the currency of the country is as follows:

U.S. to Canada	Canada to England	England to France	France to Germany	Germany to New York
$875 (U.S. dollars)	$2200 (Canadian dollars)	£1600 (United Kingdom pounds sterling)	12,000 E (euro)	12,700 E (euro)

What minimum amount of U.S. cash does she need in New York?

1.16 An oxygenated high-pressure ultra high-temperature reactor-vessel assembly can be fabricated in Germany, England, or Canada and shipped to Saudi Arabia for integration in the construction of a refinery. Quotations are received from the fabricators in those countries and ratioed to the U.S. estimate in terms of U.S. currency by using Table 1.2 and spot exchange rates. According to the contract, payment is electronically transferred from the design-constructor to the fabricator on the day the reactor reaches the port of entry of Saudi Arabia. The exchange rate at the time of the scheduled arrival is, however, expected to be different and is shown as a relative change.

	Germany	England	Canada
Ratio of subcontract bid to U.S. estimate	0.9962	1.0065	1.0062
Exchange rate on arrival compared to Table 1.2	+0.40%	−1.1%	−0.05%

The U.S. estimate for the reactor assembly is $197.5 million. In which country should the vessel be built, and what is the cost in U.S. currency? What is the dollar penalty if the next-lower-estimate country is selected? Where does nonnumerical judgment enter the decision process for the selection of the country in which to fabricate the reactor?

PRACTICAL APPLICATION

The internet's educational value for construction-cost analysis and estimating prompts this practical application. Cyber skills can enhance your learning in the course you are taking. This first internet application is one of several that progress to increasingly difficult assignments.

You are to examine several internet search engines and hunt for WWW addresses that associate closely with the course objectives that your instructor will identify. Once you have located useful sites and information, transmit this information to another student (or to the instructor or to your college department, or to your company, if you work for a business that might be interested). Consider the following examples: conference proceedings of the American Society of Civil Engineers; news bulletins from The Association of General Contractors of America; contracts from the American Institute of Architects; and information from other professional organizations and construction-cost consultants. There are other internet home pages for you to find.

The transmission of the information can be to the intranet email system, or to the mailto function on the college's or the instructor's or your companion's home page on the internet. Your instructor will discuss the desired information, a plan for your work, and rules on grading.

CASE STUDY: PROFESSOR JAIRO MUÑOZ

"Good morning, Professor Muñoz. I'm Rusty Green, and I'm in your eight o'clock construction class."

"Uh huh," replies the Professor without looking up from the desk. "What can I do for you, Rusty?"

"Well, it's like this. I'm not sure that I belong in your class." Rusty smiles and continues as the Professor looks up. "Oh, it's not you, Professor, but I really want to be a (*make your own selection*) and it's unclear what this course will do for me later on."

"Yes, go on," says Professor Muñoz, leaning forward in his chair. "What are you wanting?"

"I do want a career-oriented program, but is construction going in that direction?" Rusty looks at the Professor expectantly.

"That's an important question, Rusty. There are many, many considerations."
Help the professor answer Rusty's question. Consider the following:

- Call and talk to a construction practitioner.
- Check the newspaper want ads, trade journals, technical magazines, and your college career office for employment opportunities.
- Determine the educational requirements and the experiences needed for these careers.
- Consider tangential opportunities that can develop from the construction stem.
- List the job titles and their relationship to construction. Do these situations change after 1 year, 5 years, or 10 years?
- Which courses—basic and engineering science, mathematics, humanistic and social science, technical elective, or design—should you emphasize to reach your goal?
- What do you enjoy doing?

Once the investigative part of the case study is concluded, word-process a prospectus for an internship that you may submit to a company. Provide a copy of the prospectus to your instructor.

CHAPTER 2

Labor Productivity and Analysis

Labor is one of the most important items of a construction job. And to forecast the productivity for work, we study procedures that identify, describe, and measure—this is an intermediate step toward analysis and estimating of labor costs. This chapter, for the most part, deals with the measurement of construction labor time and cost. In later chapters that information is modeled for cost analysis and estimating.

2.1 LABOR

Labor has received intensive study, and many recording, measuring, and controlling schemes have been developed in an effort to manage the cost of labor. Labor is classified in a number of ways—for instance, direct-indirect, recurring-nonrecurring, designated-nondesignated, exempt-nonexempt, wage-salary, management-blue collar, and union-nonunion. Social, political, and educational factors and type of work are other ways of classifying labor. Further, payment for wages may be based on attendance or performance. For this text we select the direct-indirect classification for cost analysis.

There is the simple qualitative formula

$$\text{Labor cost} = \text{time} \times \text{wage} \qquad (2.1)$$

Once time values are known for construction work, they are multiplied by the wage or the wage and fringe costs. The "time" of Eq. (2.1) may be for individuals or for crew work, and it matches the requirements of the job.

Time is expressed relative to a unit of measure such as stud, piece, bundle, container, 100 units, 1000 board feet, and so on. The unit of time may be minute, hour, day, month, or year. Thus, we may use 14 man hours per frame for building construction, 15.025 hours per 100 units in manufacturing, or 1.5 man years per reactor tank for chemical-plant construction. Neither the application nor the magnitude of the time affects the generality of Eq. (2.1).

In early estimates, a building design may be known only roughly by the total square feet and type, and large quantities of time are used. In some situations the estimate of time is a guesstimate and is unrelated to any measured, referenced, or analyzed data. A guesstimate is based on observational or rough experiences. Circumstances exist where personal-judgment values are unavoidable. As the design becomes more detailed, so does the refinement and precision of time.

The second part of Eq. (2.1), wage, is defined in the context of the job. The design may be for one worker and his tool box, or a crew with one equipment, or a crew with several units of equipment. In the simplest case, one on one, the worker title (say, electrical wire-man) and job description are known.

Dimensional units for the wage are compatible to the time estimate. If the time estimate is in hours per unit, then the wage is expressed in dollars per hour. The wage may be the amount that the worker sees in the paycheck, or it may include all or part of the fringe costs. This choice is coordinated with accounting analysis.

Labor productivity and analysis are concerned with direct and not indirect labor. Direct-labor time for work means that the worker is working on the design. Typical titles of direct labor are carpenter, mason, electrical wireman, or roofer. In some situations, engineers are classified as direct labor for project designs. Titles of indirect labor are time-keeper or superintendent in construction projects. Stated differently, direct labor "touches" the design, and indirect labor is supportive of that effort. Indirect labor and its cost are covered by overhead, a topic discussed in Chapter 4.

2.2 THE MYTHICAL MAN HOUR

There is an almost unquestioned dependence on a historical value of time in construction. The ever-present and the most popular metric is the "man hour." This basic unit is defined as one worker working for one hour. Examples of man-hour units are given as:

> Number of welder man hours per inch of field welding
>
> Number of carpenter man hours to erect 100 ft^2 of wooden framework
>
> Number of road crew and equipment hours per mile of 28-ft-wide asphalt road construction

Literally, thousands of varieties of man-hour units are found in construction analysis.

Though the number of working days varies in a month or year owing to holidays, vacation, sick leave, leap year, and so on, for cost analysis the *man year* is defined as 52 weeks of work with the popular 40 hours in each week, or 2080 hours. A *man month* is 173.3 hours per month, or $(40 \times 52)/12 = 173.3$. Man minutes or man weeks are uncommon units and are discouraged in cost-analysis applications. Man hours are regularly used for estimating. Regrettably, this use is not as effective as it should be. For the most part, there has been little measurement of the man hour. In theory, man hours can be effective if they are relatively constant, preknown, and unaffected by changes in wage rates, overtime, connectivity between the trades, and so forth.

When applying the man hour, its credentials need to be openly understood, such as

- Guess from experience
- Time spent as actual clock hours
- Adjustment of actual clock hours to allowed time
- Measurements and analysis of the worker's time with respect to constructive and nonproductive effort, and idleness
- Data book

Many estimates go awry because of several misconceptions. We confuse effort with progress, hiding the assumption that men and hours are linearly interchangeable. For example, if twenty man hours are proposed for a task, or one worker will consume 20 hours,

PICTURE LESSON
Iron Bridge

Very early materials of construction were wood, stone, natural cement and lime for lime mortar, and later iron. The development of cast iron preceded wrought iron and steel. It was not until late in the eighteenth century that iron was used for structures as distinct from machines. By the turn of the century Great Britain led the world's iron industry of the world. Coalbrookdale, one of England's earliest iron-making areas, provided the pieces for the first iron bridge, as shown by the picture. The bridge weighs 378 tons and took only three months to assemble in 1779 over the river Severn. It is an iron arch of 100-foot spans carried by a series of semicircular ribs, each of which was cast in two pieces. The bridge still carries pedestrians and cyclists today.

Without rivets, bolts, and welding technology at that early time, the bridge was assembled using the joinery of carpentry. Its joints and fastenings were the metal equivalent of the slots, dovetail joints, and mortise-and-tenon used by furniture makers.

Cast iron is a general term applied to iron-carbon-silicon alloys in combination with small percentages of manganese, phosphorus, and sulfur. The cast iron for the Iron Bridge contains so much carbon, or its equivalent, that the material is not considered malleable.

Early iron making used iron ores from bogs or mines, charcoal or coal, clam or oyster shells or limestone as a flux, and air to combust the mixture. The resultant molten iron was poured either into pigs and ingots or into sand molds which were shaped into the bridge's components.

During this period the first rolling mill was started, but that process is unable to roll cast iron. Like some important innovations, the use of steel was slow to become established. The Bessemer process of steel conversion was announced in 1866, but it waited until 1877 for its first civil engineering structure of any magnitude, the Forth Bridge. The Bessemer process forced air through molten iron contained in a pear-shaped brick-lined vessel. As the air moved upward through the molten iron, it burned out the impurities, resulting in a ferrous steel alloy.

we assume that a crew of two will consume 10 hours. This divisibility is an accepted but an unproved hypothesis. While cost does vary as the product of the number of men or women and man hours, progress may not.

Further, there is a linkage between consecutive tasks in construction. The quality of carpentry work and the estimated hours influences the drywall installer and her actual man hours. The man hours for the many tasks are assumed to be independent, which is fallacious reasoning, as there are interface dependencies. There is a philosophy of estimating that pretends that the sum of the individual man-hour estimates will equal a total for the job without any slippage of the man-hour constants.

The construction industry gives allegiance to the man hour for other mythical reasons. Much of the estimating information is based on no quantitative measurement, supported by little data, and certified chiefly by the hunches and experience of the people doing the estimating. The incompleteness and inconsistencies become apparent only during the implementation stage of the construction. There is entropy in the system as tight estimates go steadily up—these being the consequence of no measurement and no backup data to substantiate the measurement. Overestimates do not change.

Finally, the man hour is mythical for another reason. While we might like to substitute the term labor hour, person hour, people hour, humanity hour or even humankind hour for the terminology *man hour*, our view of language is to accept the designation man hour as gender neutral.

2.3 PRODUCTIVITY MEASUREMENT

There is a school of thought that asserts that "time is the measure of productivity." Although the slogan is debatable, there is an element of truth when applied to the needs of construction cost analysis and estimating. For there is an almost unquestioned desire for an *objective* measure of time for productivity. Historical records provide costs about labor, supervision, methods, and a host of endeavors but are subject to significant error.

The two categories of work are direct and indirect. Construction jobs can be divided into one or the other, inasmuch as the direct-indirect category is determined to some extent by practice and definition. Discussed in this section are techniques of measuring labor. The measurement of time is limited mostly to direct labor. Though time is measured, it is cost that is ultimately required.

Although engineers may not be involved personally with the measurement of construction labor, they depend on the results from productivity measurement. Engineers are satisfied if such labor measurements are objective, as far as possible, and are willing to use the information, provided that professional techniques were used in the finding of time for work.

Four methods for the measurement of construction time are developed in this textbook:

- Job-ticket reports
- Nonrepetitive one-cycle time study
- Multiple-cycle time study
- Work sampling

Although we may argue in favor of a particular method, each is suitable and necessary for different occasions.

There are several ways in which actual time data are collected, analyzed, compiled, and used. Definitions to guide the study of the methods are given in Table 2.1.

TABLE 2.1   Definitions for productivity analysis

Actual time	The time reported for work, which includes delays, idle time, and inefficiency, as well as efficient effort.
Allowance	An adjustment to the normal time providing for personal needs, fatigue, and justifiable delays inherent in the work.
Allowed time	Reported man hours adjusted for work skill, weather conditions, and the like, based on analysis and judgment by engineering, management, etc.
Avoidable delay	Time-consuming interruption in the work that the worker could have avoided or minimized by using better skill or judgment.
Constant element	An element whose normal time is constant with respect to various independent effects on the element.
Continuous timing	A method of time study where the total elapsed time from the start of the study is recorded at the end of each element.
Cycle	The total time of elements from start to finish.
Delay allowance	One part of the allowance included in the standard time for interruptions or delays beyond the worker's ability to prevent.
Element	A subpart of an operation or task separated for timing and analysis; beginning and ending points are described, and the element is the smallest part of an operation observed by time study. The length of the element time can vary from minutes to hours to days, etc.
Elemental breakdown	The description of the elements of an operation in a measurable sequence.
Equipment time	Time required by equipment. See also *machine time*.
Fatigue allowance	An allowance based on physiological reduction in ability to do work, sometimes included in the standard time.
Foreign element	An element unrelated to the task and removed from the time study.
Frequency of occurrence	The number of times an element occurs per operation or cycle.
Idle time	An interval in which the worker, equipment, or both are not performing useful work.
Machine interference	Idle machine time occurring as a result of worker attention to another machine in multiple-machine work.
Machine time	Time required by a machine. See also *equipment time*.
Man hour	A unit of measure representing one person working for one hour. See also *person hour*.
Normal time	An element or operation time found by multiplying the average time observed for one or multiple cycles by a rating factor.
Observed time	The time observed on the stopwatch/electronic clock or other media and recorded on the time study sheet or media tape during the measurement process.
Operation	Designated and described work subject to work measurement, estimating, and reporting. See also task.
Personal allowance	An allowance included in the standard time for personal needs that occur throughout the working day.
Person hour	A unit of measure representing one person working for one hour. See also *man hour*.
Productivity	The amount of work performed in a given period. Usually measured in units of work per man hour, or man month, or man year.
Productivity rate	A unit rate of production; the total amount produced in a given period divided by the number of hours, or months, or years.
Rating factor	A means of comparing the performance of the worker under observation by using experience or other benchmarks; additionally, a numerical factor is noted for the elements or cycle; 100% is normal, and rating factors less than or greater than normal indicate slower or faster performance.
Regular element	An element that occurs once in every cycle.
Select time	An elemental time chosen from a time study. It may be a single observation or an average observed value.
Snap back	A method of timing that records the elemental time at the end point of the element.
Snap observation	A virtually instantaneous observation as to the state of the operation (i.e., idle, working, or the nature of the element).
Task	Designated and described work subject to work measurement, estimating, and reporting. See also *operation*.
Standard time	Sums of rated elements that have been increased for allowances.
Variable element	An element whose normal time varies or depends on one or more dependent effects.

2.3.1 Job-Ticket Reports and Man-Hour Analysis

Information for man-hour analysis is sometimes obtained from the job ticket or the foreman report. Workers may complete their own job ticket, which may be verified by a time clock with computer-terminal entry or simply collected by the owner, superintendent or foreman. The job ticket is just a time card with additional information. The foreman's report, on the other hand, has similar information and content but is compiled by first-level management or a job foreman. While the collection of the information can be computerized or manual, the fundamental approach is not altered. The job ticket or foreman's report is used initially to find the number of man hours for a job. The same reporting document is later used for cost and time control. Man hour, man month, and man year are the measures of time eventually desired from those reports.

Consider an example in a welding operation by a subcontractor where job tickets are used. Observe from Table 2.2 that the items of interest are operation, elapsed time, and units completed. These are simple descriptions of the operation, and are not difficult to identify for a worker or foreman familiar with the work and skill. Note that this job ticket has no foreman's approval, timekeeper's mark, or any external verification. The operation calls for the welding of three steel gussets to columns on the second floor of a building. The work involves overhead welding, a more difficult skill.

Several of these job tickets can be collected for a variety of welding jobs. From the job tickets, and with the aid of other instruction sheets, construction codes, or engineering drawings, the engineer will examine the data for consistency, completeness, and accuracy. The observer may check back with the welder or foreman on questions that arise. In a field welding site there is little repetition of similar work, so the technologist may choose to purge or alter information. This reworking of field data is risky, but in analyzing historical work data, the first step is to clean up the information as rationally as possible. If raw data were used as received, then other problems might arise later.

Job tickets are collected for similar work, where that may be possible, and a spreadsheet is devised for the data. Notice Table 2.3, where arc-welding operations for a building are collected. A spreadsheet brings together the facts and prepares the data for subsequent mathematical analysis. (No single best way exists to devise a spreadsheet.) At

TABLE 2.2 Illustration of foreman's report of an arc-welding operation of gussets to building columns

Name *Neil Wagner* Employee No. *505 30 9710*
Date *June 11* Order No. *101*
Job No. *—* Location *2nd floor of building*
Part No. *6682* Part name *Gusset, A 36*
Time started *9:15* Time stopped *11:43* Quantity *3*
Operation *Arc weld gusset, 36 x 18 x 1/2 to columns. 54 in. of double-vee welding. Overhead welding.*

TABLE 2.3   Spreadsheet example for the analysis of arc-welding operations for a building

Part No.	Material	Thickness × Length	Method	Job Ticket Actual Man Hours	Remarks	Allowed Man Hours
6682	Steel	3/4 × 24 in.	Arc weld	2.47	Fillet 2 sides	1.24
7216	Steel	1/2 × 8.4 in.	Arc weld	1.06	No change	1.06
8313	Steel	3/16 1/4 1/2 × 34.4 in.	Arc weld	1.18	Increase by 25% for various thicknesses	1.48

this point judgment is necessary. From Table 2.3 we see that the welding operations are varied and that the engineer chooses to analyze the general operation welding rather than divide the process into one of the many welding methods. Observe that several thicknesses, welding methods, and three part numbers are posted on the spreadsheet. The mixing of the welding processes and handling time into one lump time value is not an immaterial detail. But the engineer may reduce, leave as is, or increase the three job-ticket times to give allowed time. The allowed time is then used for man-hour estimating data. We leave it to the student to find the allowed man hours for welding from the data, as a problem is provided.

Job-ticket reports and man-hour analysis will vary with the different construction trades. Observe Table 2.4, in which the foreman reports the daily work associated with a residence, elapsed time, and weather conditions. There are two workers, carpenter and apprentice, and the foreman. The foreman is nonworking and spends only 15% of his time on the job. The foreman reports that the weather is cold during the framing erection of a wooden partition for a residence, and he gives the size of the wall with the rough opening sizes for two windows.

At this point the observer gathers other foreman's reports. After evolving a spreadsheet, the work is matched against code 06111.10, "rough framing of partitions." Not all details are given by the job-ticket reports, and the engineer will recapitulate the essence of the work to satisfy the requirements of the residence architectural drawings. He or she may visit the job site from time to time and assess the skills of the crew who are working

TABLE 2.4   Foreman's report of rough framing of a residence

Foreman's Daily Report

Foreman: *Jim Diekmann*   Date: *Dec 5*   Address: *6682 whaley*

Workers: *Walter Meyer, carpenter*
           Tony Songer,   Apprentice

Time Start: Description:
  0800   *Construct wall, wooden partition on wooden d*
                         8 x 26 ft.   92 1/4 in. studs, 16 in. on
                         centers. Frame rough opening for window,
                         2 - 3x4 ft.

  16:30   *4 extra corner studs.*

Weather: *Stormy, cold 40-50°F.*

Other unusual condition: *None*

on the job. In the remarks column the engineer will indicate the efficiency of the labor and the percentage of the work devoted to the major tasks or elements of the job to construct a partition wall.

Observe that in Table 2.5 the elapsed time of 8 hours × 2 crew = 16 total hours is reduced by 80% for cold weather, which results in an allowed total of 12.8 hours. This is subdivided into work percentages required for an 8 × 26 ft stud partition, 2 rough window openings of 3 × 4 ft, and extra corner studs. The man hours are matched to important units of output.

If man-hour spreadsheets are to be of value as a permanent measured record, then their backup data must include information to permit the engineer to weigh deviations from the observations. For instance, vital factors that need to be known in construction work include the weather conditions, skill of the crews (experienced or green, native or imported), equipment, hazards, location of work, material condition, and type of construction. Remarks on these factors appear on the foreman's report, and the engineer makes office adjustments for those effects. Those adjustments lead to *allowed time*, which is an intermediate value from observed hours. The adjustments may be derived from engineering instructions, other information, or, as a last resort, the engineer's experience. Sometimes the foreman's report lumps together elements of work unsuitable for this analysis.

The job tickets are reconstituted into spreadsheets and then the engineer determines the productivity rate. The rate is then used for future cost analysis and estimating of jobs that are similar. Further, the productivity rates may be entered into the company's esti-

TABLE 2.5 Spreadsheet example for the analysis of the rough framing of a residence

Cost Code	Description	Elapsed Time	Crew Size	Remarks	Allowed Man Hours
06 111.10	Frame west side of 2 story house. Standard 8 x 26 ft. partition.	8.0	2	Cold, 80%	
				60%	7.68
	2 window openings, 3 x 4 ft.			35%	4.48
	4 extra corner studs			5%	0.64
06 111.10	Rough carpentry wall, first floor	5.55	2	Normal, 100%	
	Standard 8 x 31 ft. partition			40%	4.44
	1 window, 8 x 4 ft opening			30%	3.33
	1 double door, 6 ft. wide rough opening			14%	1.55
	8 extra studs			16%	1.78
06 111.10	Rough frame side wall of house	9.0	3	Green labor, 60%	
	Standard 8 x 32 ft partition			50%	7.9
	1 window 3 x 4 ft opening			12%	2.0
	Glass door 6 ft wide opening			10%	1.6
	1 window 6 x 4 ft. opening			15%	2.5
	10 extra studs			13%	2.2

mating system, and others are instructed in its use and the appropriate cases where it is considered accurate.

Small contractors and subcontractors provide the best opportunity for job tickets. In these cases there is a greater connection between work reporting and the intimate details that are required to make job tickets effective in measuring productivity.

In summary, man-hour reports are derivative time values adjusted by the engineer or owner from job tickets or foreman reports. Perhaps the least accurate methods are those which depend on the worker or foreman to collect the raw information. That the data are nonhomogeneous are evident, and judgment and skill are indispensable to transform the work to allowed man hours. But job-ticket reports and man-hour analysis are improved by other approaches to work measurement, to which we now turn.

2.3.2 Nonrepetitive One-Cycle Time Study and Man-Hour Analysis

Sometimes called all-day time study, nonrepetitive time study is useful for construction, direct-labor long-cycle types of work, and indirect labor. This class of work is called nonrepetitive, because it is difficult to preplan in any fine detail. Often we are unable to study more than one cycle of a construction job, and in some cases the study may be only a fraction of the task's effort.

Nonrepetitive time study provides information for cost analysis and estimating. Indeed, it is possible to glean information that has utility to the contractor in excess of its cost of acquisition. There are numerous situations to which nonrepetitive one-cycle time study is well suited. It is applied to all construction trades. Tight statistical reliability of the results is unlikely; however, the method gives information superior to guesstimating and job-ticket analysis, which is why students need to understand it.

The study entails continuous timing with an electronic stopwatch or video camera. A time recording is essential. The equipment for taking a time study is simple: a clipboard and an electronic timer or decimal minute or hour/day stopwatch or the wrist watch. A camcorder with video playback is acceptable, and video tape can be rerun to determine acceptable methods and the time required for work. In rare cases even a monthly calendar can be adopted, as a time study is simply "describing the job and indicating the time." The basic procedure for a calendar study includes checking off the days as the job proceeds, along with providing a description of the work and workers.

Field and office procedures follows these steps:

- Inform the labor that is to be time studied in order to analyze and, if possible, improve construction methods.
- Record job, equipment, worker(s), environmental conditions, and other circumstances before and during the timing of the work.
- Time the work and write down the job elements as the job proceeds.
- As the key steps of the job change, record the new elements along with the elapsed time.
- Record the time consumed by each element as it occurs.
- Rate the pace or tempo at which various elements of work are performed.
- Calculate the normal time.
- Adopt preapproved allowances, calculate the allowance multiplier, and extend normal time to standard time.
- Express the standard in common units of productivity.

The time study begins by informing those who are involved in the time-study: superintendent, foreman, worker(s), etc. Being honest and forthright with those who time and those who are timed is the best policy. Laying out the purpose, and informing the workers how the study is being done and that they will have an opportunity to see the results, may help the process. It is not unusual for workers to resent being timed. Clearly, an honest purpose to assert for them is to help productivity, lower costs, and and improve cost analysis and estimating. Time study is best done in an open-handed way—it should not use sly or sneaky methods.

Time study does not pretend to engineer the job to be optimal; it only evaluates the time associated with a particular method. Of course, among several candidates that have been evaluated by the time-study procedure, a "best" method may be selected.

It is not usually possible to alter the method of construction before or during the time study, so the ongoing method is timed under the actual situation. Skilled or unskilled workers? Bad or engineered methods? Motivated or unwilling workers? Poor or able management? These are the controlling factors for the time-study technician.

The second step involves the recording of the general information for the job. Equipment, worker(s), weather, and the many details that influence the work are identified.

The third step records the information on the time-study form. After the title-block information is written down, the elements of work are identified and recorded in sequence. When breaking down the job, the technician needs to make the elements as short as possible, but long enough for timing accuracy. Wherever possible, manual worker time is separated from equipment time. During the equipment time, if the back hoe is moving dirt, for example, it is helpful if the volumetric size of the dipper is indicated. Wherever possible, elements that are constant (or nearly so) are separated from variable elements. Elemental start and stop time should be easy to identify.

Crews are usually the object of timing in nonrepetitive one-cycle time study. A two-worker crew may be side-by-side and doing the same elements. Larger crews may be doing different elements, even though they are engaged in collateral work. Describing the elements under these circumstances can be a daunting task. The observer may have a camcorder and preview the tape on the monitor, or she may voice-record the time-study readings and describe the ongoing work are techniques for making the time study. Later she summarizes in the office by recording the significant events.

In a nonrepetitive one-cycle time study, the observer may not know for certain what will happen after an element is completed, since the observer is not giving work instructions for the crew. Further, a time-study element, whether it takes a few minutes or a few hours to complete, will not have its starting point and ending point identified beforehand. Repeated occurrences of the work in one cycle are not anticipated.

The rating factor is a judgment made by the observer about the worker's speed and skill. If the worker is skilled and motivated, then the observer will rate him as above normal, using a rating factor above 1.00. If the worker is unskilled and slow, then the observer will assign a value below 1.00. If the worker is considered normal, then the rating is 1.00. Rating factors in construction typically range from 50%–60% to 120%. This rating activity is a lightning rod for criticism; it is important that it be learned and handled with professionalism and be consistent, fair, and objective.

Once the time study is complete, the observer determines the nature of the element and chooses or calculates a "select" time for it. This office calculation is the net time less any reductions for foreign elements.

Normal time is found by multiplying a selected time for the element or cycle by the rating factor using

$$T_n = T_o \times RF \qquad (2.2)$$

where $\quad T_n$ = normal time, hours, days, etc.

$\qquad T_o$ = observed time, hours, days, etc.

$\qquad RF$ = rating factor, arbitrarily set, number

If the worker is judged to be fast, then the rating factor is greater than 1.0, say for an example as 110%. If the selected cycle or elemental time is 1.2 hr, then the normal time is 1.32 hr (= 1.2 × 1.10). If the worker is rated as 85%, then the normal time is 1.02 hr. The rating factor recognizes the observed worker as a "sample" observation. It allows adjusting the observed value for nomal workers to arrive at a typical and fair value. Technicians can be trained in these matters, and often they have practical experience in the activity that they observe.

Normal time does not include provision for legitimate interruptions. Allowances for such interruptions are divided into three components: personal, fatigue, and delay (PF&D).

In the process of timing the cycle, idle time is considered a foreign element and expunged from the approved work. Of course, workers need breaks for restroom, body, and functional necessities. The purpose of the personal portion of the allowance is to add back 4% or 5%, which is the customary amount in the United States.

The fatigue portion of the allowance provides for the human weariness that work engenders. Fatigue does not mean simple "tiredness," but rather the physiological weariness that hard work can bring. The fatigue allowance can vary anywhere from 0 to 50%, according to practice. Heat or cold, noise, percent of time under load, and ergonomic factors would increase the number over that for typical work conditions. A typical allowance for fatigue is 5%. Fatigue is not subject to precise measurement. Fatigue allowance percentages are determined by practice or union negotiation or common sense.

Delays are another component of the allowance. Approved delays are those caused by external influences on the job and not by poor worker skill. Such delays may be due to conditions such as misdirection by the foreman, shortage of materials, weather, and so on. Delays can vary from 0 to 25%.

PF&D allowances are necessary for work to continue throughout the work day, as the normal time is insufficient without them. Once they are included, we have what is called a *fair day* work standard—fair both to the worker and to the contractor.

The allowance is allocated in cost-analysis procedure as a percentage of 480 minutes. Because productive time in the work day is inversely proportional to the amount of PF&D allowance, the allowance is expressed as a percentage of the total work day.

PF&D allowances are generally in the range of 10%–20%. An allowance multiplier is found using

$$F_a = \frac{100\%}{100\% - \text{PF\&D}\%} \qquad (2.3)$$

where $\qquad F_a$ = allowance multiplier for PF&D, number

$\qquad$ PF&D = personal, fatigue, and delay allowance, percentage

The final step is to find the standard productivity. This is the time required by a trained and motivated worker or workers to perform construction tasks while working at a normal tempo.

$$H_s = T_n \times F_a \tag{2.4}$$

where H_s = standard time for a construction job per unit of effort, hour, day, etc.

In engineering, the term "standard" implies an *almost* immutable and desired practice; in productivity measurement it is understood more broadly. In the job-ticket approach, as discussed in Sec 2.3.1, the output parameter is identified as "allowed man hours." But in the nonrepetitive one-cycle approach we understand that the measurement process is stricter, and we speak of "standard man hours."

Let us look now at an example. Refer to Table 2.6, and notice the general layout of the time study. "Paper" forms have many styles, and this one provides the barest of details. Simply, a time study describes what the operation is showing, and then provides the time to do that work. Many styles of time-study forms can aid that objective.

The operation is the delivery of 55 yd^3 of 4,000-psi concrete to wooden forms, which have reinforcing steel installed. The concrete is pumped to the forms by a concrete pumper truck. Supporting the operations are a crew of six who work with the concrete at the forms and a number of concrete trucks that deliver concrete to the hopper of the pumper truck, but the study focuses on the productivity of the pumper truck and its operator.

The title block gives various information: location, worker, skill level, and the specifications of the pumper truck, which is a 300-hp diesel powered truck with a maximum delivery rate of 195 yd^3. There is one equipment and one worker, which is the simplest combination of machine and worker.

The entry columns are clock reading, time, element description, rating factor, element type, and remarks. A wrist watch is the timing device. Timing is hours and minutes; seconds are ignored. Of course, other electronic or mechanical clocks or timing video recorders for live TV can be used. The initial reading is 0, which indicates the start of the operation.

The work is described in the description column, and the observer writes down there what he sees. Only the barest of details need to be shown. The skill of selecting these details is learned with on-the-job practice.

While the work is going on, the observer is rating the elements. Here the ratings are 100%, because the job output is mostly equipment controlled, rather than worker controlled.

The observer also judges the type of element that is being performed. For this example there are four choices for the elements: setup and teardown, foreign, constant, and variable. The element setup and teardown is the time taken to start and then, after the work is completed, to tear down the equipment and make the pumper truck ready to move on to the next job.

During this operation there are foreign elements, which are purged from the time study and do not affect the pumper truck's productivity. The foreign elements here are the delays caused by the poor scheduling of the concrete trucks to the job site by the concrete supplier and by a breakdown of the pumper truck's plumbing. The job requires 3 hours and 7 minutes to pump 55 yd^3.

Once the observation in the field has been completed, the office analysis to find the man hours for the productivity begins. The summary is given by Table 2.7. Foreign elements, as mentioned above, amount to 1 hr 19 min out of 3 hr 7 min, or 42%. The principal causes for the foreign elements are poor scheduling by the concrete supplier and equipment breakdown. Normal time is found by multiplying the net time after the foreign elements are purged by the rating factor, 100%. The allowance for PF&D is 15%, which becomes an allowance multiplier of 1.176.

Multiplication of normal time by the allowance factor gives standard time. Now we have productivity expressed as:

TABLE 2.6   Illustration of nonrepetitive one-cycle time study for concrete pumper truck

Construction Time Study - One Cycle

Time by : *Tom*   Date : *7/13*   Location : *Sun Mirosystem Interlocken*

Description of operation : *Pump 55 yd³ of 4000 psi concrete to forms with 6-person crew*

Operator or operators: *Norm Eifler*   Skill level: *Average*

Equipment: *1200-32 Truck - 300 HP Mack, diesel, 120 ft. boom Maximum delivery rate = 195 yd³*

Units for time study: *Minutes*

Clock Reading	Time	Element Description	Rating Factor	Element Type	Remarks
0	0	Start			
0:30	30	Position and jockey truck. Position boom to forms.	1.0	Setup	
0:45	15	Concrete truck arrive	0	Foreign	Wait for concrete
0:50	5	Concrete truck into position	1.0	Constant	
0:60	10	Load 12 yd³	1.0	Variable	
1:07	7	Wait for new truck to position	1.0	Constant	
1:18	11	Pump 12 yd³ and truck leaves	1.0	Variable	
1:34	16	Trucks of concrete - wait	0	Foreign	Wait for concrete
1:44	10	Pump concrete 11 yd³	1.0	Variable	
1:56	12	Wait for truck	0	Foreign	Wait
1:60	4	Truck arrive, pumping	1.0	Constant	
2:36	36	Pumper truck breakdown Hydaluic cooler not working.	0	Foreign	Breakdown of equipment
2:43	7	Pumping continues 12 yd³	1.0	Variable	
2:49	6	Concrete truck exchange	1.0	Constant	
2:59	10	Pour, fill voids, 8 yd³ Pouring finished	1.0	Variable	
3:07	8	Job completed. Boom retracted. Pads up. Drive off.	1.0	Teardown	

TABLE 2.7   Summary of nonrepetitive one-cycle time study for concrete pumper truck and worker

	Time Study Time	Hours
Total cycle time	3 hr 7 min	3.117
Foreign elements	1 hr 10 min	1.317
Net cycle time	1 hr 48 min	1.800
Rating factor: 1.00		
Total normal time	1 hr 48 min	1.800
Setup and teardown	38 min	0.633
Net operation cycle time	1 hr 10 min	1.167
Constant operation time	22 min	0.367
Variable operation time	48 min	0.800
Units produced during variable operation time: 55 yd^3		
Normal time:		
Setup and teardown	38 min	0.633
Constant	22 min	0.367
Variable	48 min	0.800
Rate of productivity, hr/yd^3	48/55 = 0.87 min/yd^3	0.800/55 = 0.015
Allowance, 15%		
Allowance multiplier: 100/(100 − 15) = 1.176		
Standard time:		
Setup and teardown		0.745
Constant		0.431
Variable		0.942
Rate of productivity, hr/yd^3		0.942/55 = 0.017
Utilization of equipment		(0.745 + 0.431 + 0.942)/3.117 = 68%

$$\text{Time per job} = \text{Equipment setup} + \text{Supply setup}$$
$$+ \text{Concrete truck pumping rate} \times \text{yd}^3 \qquad (2.5)$$

where
$$\text{Equipment setup} = 0.745 \text{ hr}$$
$$\text{Supply setup} = 0.431 \text{ hr}$$
$$\text{Concrete truck pumping rate} = 0.017 \text{ hr per yd}^3$$
$$\text{Time per job} = 1.182 + 0.017 \times \text{yd}^3$$

Using the equation, we are able to find the productivity to pump concrete with pumper equipment: Time per job = 1.182 + 0.017 $\left(\text{yd}^3\right)$. For a job requiring 55 yd^3, the standard time is given as 1.182 + 0.017 × 55 = 2.117 hr, which, when comparing standard to actual, gives an equipment utilization of 68%. The student will learn more about estimating by formula methods in Chapters 5, 6, and 7.

The arithmetic of finding the productivity rate can be simpler, though not as accurate. For example, we can sum 0.745 + 0.431 + 0.941 = 2.117 hr, and then divide by 55 yd^3 to get 0.038 hr/yd^3. This is the approach to data development typically given by the estimating manuals, a topic covered in Chapter 7. Probably, if job tickets were used to find productivity for this equipment, then the productivity rate would be the total cycle time divided by the concrete quantity supplied, or 0.057 hr/yd^3 (=3.117/55). These approaches are not as accurate as the linear equation example.

The rated maximum concrete delivery for the pumping equipment is 195 yd^3 per hour. But when the equipment is supplying concrete, as observed by this time study, the equipment is working at a rate of 59 yd^3 per hour (=1/0.017). Operation performance is being set by factors other than equipment.

On occasion several cycles are possible. Examples are short-cycle tasks such as masons laying blocks or carpenters installing 4 × 8 ft panels. More cycles are involved in the time study.

2.3.3 Multiple-Cycle Time Study and Man-hour Analysis

Multiple-cycle time study is observing several cycles, perhaps on the same occasion or on different ones. Improved definition of the work content is an advantage. There is a wish to improve the statistical reliability over that which we calculate using nonrepetitive one-cycle time study or job tickets, but we leave to other textbooks the statistical equations of the confidence limits, reliability, and the number of observations. The observations, timing, rating, and finding of the standard man-hour productivity are similar to what is described above. Wrist watch, timing clock, clipboard, and voice recorder or TV camcorder are the paraphernalia employed.

In multiple-cycle time study the observations can be "snap back." This means starting and stopping the watch for each element as it occurs. In nonrepetitive time study, the observations are continuous, i.e., we let the stop watch run without stopping or reinitializing it. If the definitions of the elements are known in advance, along with the starting and ending description—as they are more likely to be in multiple-cycle time study—the observer knows when to snap back the crown on the watch.

In some cases the units of time in multiple-cycle time study, H_s, are in standard minutes, and then

$$\text{units per hour} = \frac{60}{H_s} \tag{2.6}$$

Now that we have discussed its background, we show an example of a repetitive multiple-cycle time study. Look at Table 2.8. and examine the general layout of the time study. These time studies have many styles, and this one is a summary of actual practice.

The work deals with an electrical utility and its distribution system. Equipment is identified for the crew of three. The job description specifies the experience and nature of work that is required of a crew. In this example, the crew and equipment are drilling holes in a rocky strata for wooden power poles, setting the poles, and tamping the crumbles that have been removed. After completing work on a pole, they move on to the next location and repeat the process.

The observer rides along in the truck cab with the crew. At each location he times the crew's activities, using a decimal hour stop watch. The timing is "snap back," meaning that at the end of each element, the watch is reset to zero, and the timing is begun again for the next element. The timing is suspended if the workers are idle, or on rest break, or involved with some other element, either productive or foreign.

Three elements are described, and there are four cycles, or repetitions, over a two-day period. Computations entered on the time-study form are total time, number of readings, average time, rating factor, and normal time. Elements 1, 2, and 3 give 0.27, 2.61, and 0.33 normal hours, and for the crew of three the normal man hours are 9.63 hr (=3.21 × 3).

TABLE 2.8   Illustration of multiple-cycle time study for electrical distribution construction

Construction Time Study - Multiple Cycle

Timed by: *Mark*   Date: *6/17-18*   Location: *Aberdeen, WA.*

Description of operation:
    Wood poles in rock   holes, set pole, 30 ft. high.
             Truck dig and set.

Crew: *Luke, Andrea, Amy*   Skill level: *New, below*

Equipment: *10 ton, Ford V-10 line   average*
        truck, 40 ft. boom.

Units for time study:
             Decimal hour stop watch. Snap back.

Element:	1	2	3	4	5	6
Cycle						
1	0.38	2.60	0.29			
2	0.52	3.15	0.41			
3	0.17	2.17	0.28			
4	0.13	3.68	0.48			
5						
Total Time	1.20	11.60	1.46			
No. of readings	4	4	4			
Average time	0.30	2.90	0.37			
Rating factor	90%	90%	90%			
Normal time	0.27	2.61	0.33		3.21 hr	x 3 =	9.63

Description of Element

Element 1: *Move truck, level and pads down.*

Element 2: *Truck drill hole to depth and clean out hole.*

Element 3: *Set pole, pack down — Hold pole straight.*

Element 4:   —

Element 5:   —

The PF&D allowances for this work are 17%, and $F_a = 1.205$. Subsequently, the standard productivity for a composite element description of "truck move and position to a new location, drill hole, set pole, and tamp" is 11.6 hr (=9.63 × 1.205) for the crew of three.

The work of electrical distribution construction can be described and planned in advance. A series of elements can be determined, as in Table 2.8, and a campaign conducted to determine standard productivity for this work. The time for setting wood poles, 30–50 ft high, in rock is calculated as 11.6 standard hours, the value obtained by repetitive multiple-

TABLE 2.9 Illustrative man-hour productivity for electrical-line distribution construction. Equipment: Line truck. Crew: 3. Standard man hours per occurrence. (Does not include travel to job site)

Task Element	Dirt	Sand	Rock
1. Setting wood pole, 30–50 ft high			
Truck dig and set	1.5	5.5	11.6
Hand dig and set	10.2	16.0	15.3
Back hoe and set	3.0	3.0	3.4
Back hoe dig and truck set	2.1	3.4	8.0
2. Setting anchor			
Truck set	1.6	5.5	—
Hand set	7.1	10.2	—
Back hoe set	2.9	3.4	2.9
	—	All	—
3. Framing			
a. Single-phase armless tangent			
Deenergized line on ground		1.6	
Energized line in air		2.0	
Dead-ending line and hot tap		2.0	
b. Three-phase wood cross-arm tangent			
De-energized line on ground		3.0	
Energized line		5.2	
4. Miscellaneous work			
Transformer mount		4.4	
Install single aerial conductor per single span		1.7	
Street light		1.3	
Lightning arrester		1.8	

cycle time study in Table 2.8. Notice Table 2.9, which is the productivity for electrical line distribution.

Once the standard productivity has been developed from the time studies, the data are printed or electronically published and job costs can be consistently found throughout the network by others who find the costs for the customers. For example, a new home is to be connected to the main electrical system. Two poles will be set in dirt, and two are in rocky terrain. A total of two soil anchors will be sufficient. Framing is single-phase armless tangent when deenergized on ground. One transformer and a lightning arrester are required. Cost for the crew and truck is $95 per hour. The time and cost of the labor and the truck at the site are shown by Table 2.10.

So far, we have discussed ways to find productivity using time study and job tickets. There is one more method available, which makes sampling observations of a large work force over a period of days, weeks, or months. It also provides productivity information. It is called work sampling.

2.3.4 Work Sampling and Man-hour Analysis

Work sampling is a technique for gathering information about large segments of a work-force population. Its objectives are to measure labor efficiency and equipment utilization

TABLE 2.10 Finding task man hours and cost for residence application using standard productivity information.

Task Element	Task Man Hour	Quantity	Total Man Hour
Set pole in soil	1.5	2	3.0
Set pole in rock	11.6	2	23.2
Anchor	1.6	2	3.2
Framing	1.6	3	6.4
Transformer	1.8	1	1.8
Run wire	1.7	3	6.8
Total crew and equipment hours			44.4
Direct labor and equipment cost			$4218

and their cost, and to make adjustments for improved cost performance. Advantages of work sampling include the following:

- Productivity measurement and improvement
- Statistical definition
- Relatively low cost

It is a counting method for quantitative analysis in terms of time or percent of the worker or machine activities, or any observable state of a job. It is useful in analysis of construction-work activities. It is relatively inexpensive to obtain and convenient to perform and can be conducted without the need for a stopwatch or historical reports.

In time-study and full-time observations of tasks, the entire span of work is observed. But in work sampling, only discrete observations are made, such as in samples photographs of the work. Can "still" photographs of a work task taken at several or more instants in time be indicative of the total work? The answer is yes, if the number of the photographs are descriptive of the work, and the voids of observations are insignificant.

A work-sampling study consists of a number of field observations that pertain to the specific activities of the person(s) and equipment at random intervals. Those observations are classified into predefined categories directly related to the work situation. The technician instantaneously makes tally marks, such as "working," "idle," or "absent," by walking around the job site and noting the state of the worker/equipment during the course of the work-sampling study. A variety of technical methods are available to provide the observations. For instance, the sampling observations can be made by a remote TV unit. Later, the video tape is analyzed as if a technician were making the direct observations by walking the job site.

The key to accuracy is the number of observations, for which the requirements may vary. One survey may require very broad areas to be investigated, in which case relatively few observations are required for meaningful results. On the other hand, many thousands of observations may be needed to establish construction standards. To determine the number of observations necessary, the engineer or owner defines the accuracy of his or her results. Four thousand observations will provide more reliable results than 400. However, if accuracy is unimportant, 400 observations may be ample.

Because work sampling is a statistical technique, the laws of probability must be followed to obtain an accurate estimate. In this approach, an event such as "equipment working" or "idle" is instantly tallied. For this "snap" or instantaneous observation mathematicians define a binomial expression, where the mean of the binomial distribution is equal to Np_i, with N equaling the number of observations, and p_i the probability or relative frequency of event i occurring. The variance of this binomial distribution is equal to $Np_i(1 - p_i)$.

As N becomes large, the binomial distribution approaches the normal distribution. Because work-sampling studies involve large sample sizes, the normal distribution is an adequate approximation to the binomial distribution. The purpose for this approximation is that the normal distribution is well understood and easier to work with than the binomial distribution.

In work sampling we take a sample of size N observations in an attempt to estimate p_i:

$$p_i' = \frac{N_i}{N} \tag{2.7}$$

where p_i' = observed proportion of occurrence of event i, decimal
        N_i = instantaneous observations of event i, number
        N = total number of random observations

Equation (2.7) is related to the binomial distribution. This random variable is called a proportion. As shown in textbooks on probability, the standard error of a sample proportion for a binomial distribution is given by

$$\sigma_{p'} = \left[\frac{p'(1 - p')}{N}\right]^{1/2} \tag{2.8}$$

where $\sigma_{p'}$ = standard deviation of the proportion of the binomial sampling distribution, number

Bias and errors occur in any sampling procedure. This results in a deviation between the estimated p' and p, or the true value. A tolerable maximum sampling error in terms of a confidence interval I commensurate with the nature and importance of the study is preestablished. For instance, if $p' = 62\%$, and if a maximum interval of 4% is desired, then $I = 4\%$ for 60 to 64%, or $62\% \pm 2\%$ where $I/2 = 2\%$. The relative accuracy is $0.02/0.62 \times 100 = 3.2\%$.

This confidence and interval may be viewed by examining Fig. 2.1. The factor 1.645 is obtained from a table of values of the standard normal distribution function (see Appendix 1) and is the normal distribution for a confidence of 90%, which is usual for work-sampling studies.

The total area under a normal curve is 100%, and the opportunity for a sampling value p' to fall within the tails is given in Appendix 1 and is equal to

$$\pm Z \propto 2 \times \text{(probability from the table for a given } Z) \tag{2.9}$$

For instance, if Z ranges from -1.645 to $+1.645$, then the probability that the true value p will be between the limits is $2 \times 0.45 = 0.90$, while the probability that the true value will be outside the limits is $1 - 0.90$ or 10%. The value 0.4500 is determined from Appendix 1. Some values of Z corresponding to confidence areas follow:

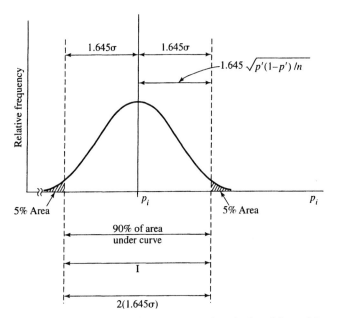

Figure 2.1 Analogous curve from "normal" relationship to binomial work-sampling practices.

Area Between Limits (%)	− Z to + Z	Area Outside Limits (%)
68	±1.000	32
90	±1.645	10
95	±1.960	5
99	±2.576	1

The sampling interval is given by

$$I = 2Z\left[\frac{p_i'(1 - p_i')}{N}\right]^{1/2} \tag{2.10}$$

where I = interval, decimal

Z = value of the standard normal distribution function for a chosen confidence, number

We expect that the true value of p falls within the range $p' \pm 1.645\sigma_{p'}$ approximately 90% of the time. In other words, if p is the true percentage of the work estimate, then the estimate will fall outside $p' \pm 1.645\sigma_{p'}$ only about 10 times in 100 owing to chance or to sampling errors alone.

Equation (2.10) may be solved for the sample size when the other factors are either assumed or known, as

$$N_i = \frac{4Z^2 p_i'(1 - p_i')}{I^2} \tag{2.11}$$

The value of N_i that is maximum from the events i is chosen as N for the overall work-sampling study. Relative accuracy is found as $I/2p'$.

As an example of work sampling in construction, assume a job where carpenter crews are forming layup walls on a dam. Those concrete retaining walls are numerous and standard. In addition, the carpenters are doing other work, but their principal activity relates to the layup walls. Once the work content is understood, we define the five job elements as

1. Form layup walls
2. Inspection waits
3. Set up for form work
4. Crane waits
5. Miscellaneous

A preliminary work-sampling study, such as Table 2.11, helps to uncover problems before the major study is started. An observer walks around the job site and makes the observations.

The study sells the idea, and, importantly, a percentage of observations falling into each activity gives a useful, albeit rough, estimate of the universe percentages. Note in Table 2.11 that we might decide that inspection waits are not an important enough element to consider separately, especially if we have little control over that task, so we decide to add inspection waits to the "miscellaneous" category. But we find that material waits in the miscellaneous category deserve a separate identification, and we may have some management control over those waits. Rearranging the data, we would have (1) form layup walls, 62%; (2) set up for form work, 16%; (3) crane waits, 12%; (4) material waits, 4%; and (5) miscellaneous, 6%.

With the job elements now defined, the next step deals with calculating the number of observations for the full work-sampling study. This number depends on the percentage of observations in each element and the size of the desired confidence interval. For example, we might estimate that an element will occur somewhere close to 60% or 64%, and a 90% confidence interval on a plus-minus tolerance of 2% or less is desirable. Our first es-

TABLE 2.11 "Stick chart" of a preliminary study of construction work

Element	Observations	p'
1. Form layup walls	THL THL THL THL THL THL I	62 %
2. Inspections	III	6 %
3. Setup for form work	THL III	16 %
4. Crane waits	THL I	12 %
5. Miscellaneous	II	4 %
		100 %

TABLE 2.12   Finding sample size for construction study by using 90% confidence interval

Job Element	Rough Proportion, p'	Desired interval I	Observations for Job Element, N_i	Relative Accuracy $\dfrac{I}{2p'} \times 100(\%)$	90% Confidence Interval
Form layup walls	0.62	0.04	1594	±3.2	0.60-0.64
Set up	0.16	0.03	1616	9.4	0.145-0.175
Crane waits	0.12	0.02	2858	8.3	0.11-0.13
Material waits	0.04	0.02	1042	25	0.03-0.05
Miscellaneous	0.06	0.02	1527	17	0.05-0.07

timate of sample size in our example is based on preliminary percentages, and Table 2.12 indicates the appropriate tolerances. The largest sample size will control the study, and in this case we need about 2858 observations. Note that it is element 3, which is not the largest proportion.

Our next step is to spread the observations equally among the days and then randomly within the working day, excluding breaks and lunch periods. The snap observations will then commence.

In retrospect it becomes possible to determine the magnitude of the sampling error after the study is underway or concluded. Assume that for element 2, set up, and partially through the study when $N = 875$, a total of $N_2 = 184$ tally marks are indicated, and $p' = 0.21$. This would give an interval of

$$I = 2(1.645)\left[\frac{0.21 \times 0.79}{875}\right]^{1/2} = 0.0453$$

We can say that the true value lies within 0.21 ± 0.023 with a probability of 90%. Observe that the lower value 0.187 ($=0.21 - 0.023$) does not lie in the interval of element 2 $(0.145, 0.175)$ as given in Table 2.12. The student may want to explain this observation in terms of sampling errors, interval, and probability.

The final result of work sampling can be the percentages for each of the categories. Management would then find ways to improve the work performance. But there is an additional feature of the work-sampling process, which is important to analysis of construction.

With work-sampling information about the event or job and a percentage fact for each of the job elements, it is possible to compute labor cost. A model using those data is

$$H_s = \frac{(N_i/N)HR(1 + \text{PF\&D})}{N_p} \tag{2.12}$$

where H_s = standard man hours per job element i
N_i = event i observations, number
H = total man-hours worked during work sampling study
R = rating factor, decimal
PF&D = personal, fatigue and delay allowance, decimal
N_p = construction units accomplished during period of observing this event, number

Assume a construction job where carpenters are making layup walls for the concrete for the dam. Carpenter work is expected to last for several months, and the work-sampling study occurs at the start of the overall work. The study runs for 3 weeks for a crew of 14 carpenters.

Because of the close proximity of the carpenters, it is possible to have a roving observer making the observations. Ratings for the work speed, which are similar to time-study practices, are made during the course of random work sampling, and $R = 0.96$ is determined. A total of 17 frames are finished, and work sampling indicates that this event $N_i/N = 62\%$ of the total effort of the gang. For an allowance of 20% the time per frame is found as

$$H_s = \frac{0.62(14 \times 3 \times 40) \times 0.96 \times 1.20}{17} = 70.6 \text{ man hours per frame}$$

Work sampling is a tool that provides broad opportunities for construction. It allows the observer to get the facts about some large construction activity in an easy and fast way. In summary, the following considerations should be kept in mind for work-sampling analysis:

- Explain and sell the work-sampling method before putting it to use.
- Isolate individual studies to similar groups of equipment, jobs, or activities.
- Use as large a sample size as is practical, economical, and timely.
- Observe the data at random times.
- Take the observations over reasonably long periods, (e.g., 2 weeks or more), although rigid rules must bend with the situation and design of the study.

Methods to determine productivity are now complete. We now study wage and fringe rates, which must be rendered in dimensional units compatible with the productivity rates, and after the product is taken, we have the direct cost for the task.

2.4   WAGE AND FRINGE RATES

Labor costs, which are dollars paid for wages or salaries for work performed, are a major ingredient of an estimate. In heavy construction, for example, labor costs constitute 40% or more of the bid on a building.

A wage is paid or received for work by the hour, day, or week, and, once denominated by a period of time, it becomes a wage rate (e.g., $32.75 per hour). A salaried employee is paid for a period of time, say a week, month, or year.

Payrolls cover two classes of workers. First we have management and general administrative employees, who may or may not be on a salary basis. General administrative employees may be management, engineers, superintendents, inspectors, and computer programmers. Second, hourly labor may be those workers who "touch" the job's direct material, i.e., the real and tangible material that appears in the job. Hourly labor includes many trades—for example, plumbers, tile layers, painters, and bricklayers' helpers.

Hourly labor is sometimes classified as direct or indirect. "Direct" refers to employees who can be associated with a job directly, such as a millwright or glazier; that is, those

PICTURE LESSON
Islanbard Kingdom Brunel

This inventive British civil engineer, Islambard Kingdom Brunel (1806–1859), a dreamer and risk taker, applied his genius to building tunnels, bridges, railway stations, ships, docks, and water towers—and many are still in use. The picture shows I. K. Brunel in front of forged and heavy chains. He designed the Royal Albert Bridge of Saltash and London's Paddington Station. He introduced broad-gauge railways and initiated the then modern era of transatlantic steamships. He persuaded his employer to build the first steamship, the *Great Western*, to make regular voyages across the Atlantic. One of his ships laid the first transatlantic cable. Brunel died during his term as President Elect of the Institution of Civil Engineers.

doing any work that can be preplanned or designated. Foreman are usually designated as direct, as are carpenters, electricians, and roofers. There are many job descriptions that can qualify as direct labor. Direct labor is estimated, and it is a part of the formal estimate.

"Indirect" labor refers to workers who are generally performing undesignated work, such as that of computer programmer, time clerk, or superintendent. In a cost-allocation sense, their work is usually for a variety of tasks performed somewhat simultaneously during the pay period, making it difficult to designate precisely what portion of their work contributes to a particular job. In the main, the indirect worker is not clearly identifiable to a particular task. Indirect labor is not estimated, and it becomes a part of the overhead, a topic covered in Chapter 4.

Fringe benefits are one of the costs of labor. In the past, we concluded that wages or salaries constituted the total sum of labor costs. This is not so today. Fringe benefits, which are related to wages and salaries, constitute as much as 30% or more of the actual cost incurred for labor.

Variety exists in the calculation of wages and fringes. If a deduction occurs from the employee wage alone, then it is not explicitly identified as an addition to the job cost, because it is included in the employee wage rate. Being a reduction in the wages paid to the workers, it is not a contractor's cost of doing business. However, a variety of additional costs to the wage must be estimated and budgeted for any job. There are labor costs not included in the wage rate that are known loosely as fringes, and it is necessary to calculate their content and value.

Once the job is known, the work is described, and job descriptions of employees may be needed. Those job descriptions indicate the skill, knowledge, and responsibility required of the worker. Contractors, unions, employer agents, or the Bureau of Labor Statistics have occupational descriptions that are available. Table 2.13 lists the essential duties and skills of journeyman wiremen. The occupations are graded in terms of skill, and a contractor will have a pay scale that increases from the simple occupation to the most difficult. In some cases, unions or labor-management contracts may be the source of information for wages in construction.

Wages and fringe effects are calculated by one of two methods:

- Wage only
- Wage and fringe combined, or gross hourly wage

In the *wage only* method, the fringe effects are collected in overhead; these methods are described in Chapter 4. By either method, fringe costs must not be ignored.

TABLE 2.13   Illustrative job description of a journeyman wireman[1]

Planning and initiating job; establishing temporary power during construction; establishing grounding system; installing service to buildings and other structures; establishing power distribution within a job; planning and installing raceways systems; installing new wiring and repairing old wiring; providing power and controls to motors, HVAC and other equipment; installing receptacles, lighting systems, and fixtures; troubleshooting and repairing electrical systems; installing and repairing traffic signals, outdoor lighting, and outdoor power feeders; installing fire alarm systems; supervising journeymen and apprentices; establishing OSHA and customer safety requirements; installing instrumentation and process-control systems; erecting and assembling power-generation equipment; installing security systems; installing, maintaining, and repairing lightning protection systems; and installing and repairing telephone and data systems.

[1] Job description courtesy of the International Brotherhood of Electrical Workers.

Wage payment can be classified further into two general groupings: those that pay for attendance and those that pay on performance. In time-attendance wage plans, gross wages are figured easily. The time in attendance is multiplied by the rate. An engineer who earns $54,000 per year earns $4500 per month. If he or she starts or leaves within the month, the pay is prorated to the number of calendar or working days, depending on company policy.

The qualitative formula given by Eq. (2.1) is formally expressed as

$$C_{dl} = H_a \times R_h \tag{2.13}$$

where C_{dl} = time-rate cost for direct-labor work, dollars

H_a = actual hours

R_h = wage rate, dollars per hour

Naturally, R_h can be found from annual, monthly, or weekly pay scales. For a $4500 monthly scale, the weekly scale is $4500 \times 12/52 = \$1038.46$. There are $173\frac{1}{3}$ hours per month, and the hourly scale is $25.96. Those calculations use the popular 40-hour week. In attendance-based plans (sometimes called *day work*), the worker is paid on the amount of time spent on the job.

Of course, the worker is interested in as large a wage as possible. The employer, on the other hand, is interested in the labor performance. If the employer is able to encourage increased output from the worker, the employer might be willing to pay higher wages. In these performance wage plans, the worker's earnings are related to productive output and are called incentive, piece-rate, bonus, or subsidy plans. Incentive plans that increase wages for performance above standards are uncommon in construction. Unfortunately, direct labor generally fails to achieve even average construction performance. Alas, this poor situation stems from a decline of worker productivity, lack of productivity standards, and inadequate management.

There are numerous labor laws that are important. For example, the Davis-Bacon Act requires payment of the prevailing wage rate and fringe benefits on federally financed or assisted construction. The Walsh-Healey Public Contracts Act requires payment of minimum wage rates and overtime pay on contracts to provide goods to the federal government. Federal and state laws regulate wages and salaries paid by employers. Some groups are exempted from the wage-hour law, such as management and engineers. The Contract Work Hours and Safety Standards Act sets overtime standards for service and construction contracts. The Family and Medical Leave Act, which entitles eligible employees of covered employers to take up to 12 weeks of unpaid job-protected leave each year, with maintenance of group health insurance, for the birth and care of a child, for the care of a child, spouse, or parent with a serious health condition, or for the employee's serious health condition. There are equal-pay provisions which prohibit sex-based wage differentials between men and women employed in the same establishment, who perform jobs that require equal skill, effort, and responsibility and are performed under similar working conditions.

The wage-hour law specifies a minimum per hour as the lowest paid wage. For example, the wage minimums as established by Congress constitute a floor for employed labor in most categories. Many other labor laws exist, of course, relating to matters such as discrimination, limiting work hours for children, safety and sanitation laws, and unemployment insurance, taxes, and social security taxes. We refer the students to texts that describe these laws; several are listed in the references.

Contractual agreements may specify additional requirements, such as number of holidays, time off, sick leave, and uniforms. In addition to wages earned by employees, the employer must pay the appropriate government agency or insurance carrier an additional amount.

A partial listing of fringe costs may include (1) legally required payments, such as social security, unemployment insurance, and workers' compensation; (2) voluntary or required payments, such as group insurance and pension plans; (3) wash-up time, paid rest periods, and travel time; (4) payment for time not worked, holidays, vacations, and sick pay; and (5) profit-sharing payments, service awards, and payment to union stewards. Fringe costs depend on local situations and must be determined individually for each case.

Other deductions from employees pay—income tax, for example—are mandated by federal and state law and are significant. This text does not discuss these topics, except for the administrative cost of handling these income taxes, such as the withholding amounts and their electronic forwarding to the Internal Revenue Service.

Social Security and Medicare. The Social Security Act requires that businesses and the employee pay a tax for retirement, medical, and some death benefits to survivors , called social security tax, or FICA (Federal Income Contribution Act). This amounts to a percentage of a fixed sum of gross earnings of an employee. The employee is required to contribute an amount equal to that paid by the employer.

This rate and the base on which it is levied are subject to change by Congress and have marched steadily upward since the first law in 1935. Initially, the employee paid 1% of the first $3000 earned and the employer paid an equal amount. Recently, though, the employee deduction was calculated as 6.20% of eligible income to a ceiling of $68,400 for FICA[2] and 1.45% for Meditax. There is no ceiling for Meditax. These deductions change whenever Congress finds it necessary. There is no free lunch for social security—the people pay, and Congress sets the rates and uses any surpluses to its own advantage. Inflation and financial recklessness by the government hurt the future recipients of the retirement payments—you.

Social security is not entirely an investment plan for people's retirement. Social security is social insurance: Americans are required to set aside some of their wages so that elderly persons—as well as the disabled and spouses and children of deceased workers—get a modest stipend.

The employer's percentage amount is added to the cost of business. The employee's share of the FICA and Medicare tax is not a business cost; it is a reduction in real wages that is borne by the employee. For special self-employed workers the rate is 15.3%.

Discussion of social security and Medicare and all of the implications is left to other books and the newspapers. Its financing, the solvency of the trust funds, and the politics of benefits and social consciousness are interesting topics. What student doesn't know his or her social security number? The student will want to keep informed about these matters.

Workers' Compensation. Workers' compensation deals with the health, safety, and rights of workers. If an employee is hurt on the job, he/she may need medical attention. *Workers' compensation* is the structure of fees and benefits for the contractor and employee. Each state enforces the Worker's Compensation Act and typically will have an office charged with this responsibility.

Almost all private and public employers must provide workers' compensation coverage, if one or more full- or part-time persons are employed. Workers' compensation is a levy against employers for continuation of income to employees in periods when they cannot work because of accidents occurring on the job.

---

[2] The amount was $68,400 for the year 1998 for FICA. But Congress may choose to raise the rate in the years ahead.

Employees are grouped into 20–40 or so work types, and insurance rates based on experience of risk incidence are established for each type. Contractor payments to cover this risk may be made to the state compensation insurance fund or a state-approved insurance company. Those employers that meet strict financial and loss-control standards may self-insure this risk.

In the event of injury or death, the insurance carrier provides financial assistance to the injured person or to the survivors. Employee medical costs caused by injury and illness events that are job related are covered through a medical provider.

Unemployment Insurance.   Another federal-state tax on employer-paid wages is for unemployment insurance, as established by the Social Security Act of 1935. Setting up the employment services was delegated to the states, which collect the major portion of the tax to operate their own unemployment offices. State legislation has to conform to the requirements of the federal Act, as the federal government sets up minimum standards for the states.

The unemployment insurance program provides temporary and partial wage replacement to workers who have become unemployed through no fault of their own. The intent of the program is to aid economic stability within a community by safeguarding the income and purchasing power of the unemployed worker. The program is funded by employer-paid taxes and provides benefits to those workers who meet the eligibility requirements of the various states.

Built into the system is a merit-rating procedure to reward "good" employers (i.e., employers who operate their labor pool to prevent repetitive hirings and firings and who terminate employees only for sound economic reasons). Typically, when a business is established or enters the system, it pays the maximum state rate. If its employment record becomes stable, then the rate drops until some minimum rate is reached. The employer is rewarded for good employment practices. All employers who come under the provision of the state unemployment insurance laws must pay taxes on their payrolls up to a prescribed amount per year for each employee. The tax rate is adjusted up and down based on the employer's experience and the amount of funds available in the national and state trust funds.

Some employees, such as farm workers, casual labor, federal or municipal workers, or domestic help, are not covered. If an employee is "laid off," then he or she is entitled to unemployment "pay" for a limited period. The unemployment insurance pays for this involuntary employee cost. A formula is used to calculate the amount of the dollar benefits based on the wages paid to the claimant during a prescribed number of periods. To be eligible to receive unemployment benefits, the worker must be unemployed through no fault of his/her own and must be able to work, available for work, and willing to seek and accept suitable work. The amount of unemployment varies from state to state, and the benefits are limited as to both weekly amount and the number of weeks they are payable. In some states, unemployment benefits are paid to workers who are voluntarily on strike.

Fringe Benefits.   Fringe benefits include expenditures, other than payroll taxes, workers' compensation insurance, and unemployment insurance, that benefit employees individually or as a group. Must an employer provide vacation, sick leave, holiday pay, retirement, health, or life insurance benefits? Typically, state law neither prohibits nor requires the granting of such benefits. Because those benefits are not legally required, they vary from employer to employer and are based on labor-market competitiveness, industry practice, management's social consciousness and attitude toward employees, or management-labor contract.

One such benefit is the medical and dental insurance paid by the employer. Supplemental *medical insurance* may cover the employee and family for all or part of ordinary illness, or may apply only to catastrophic illness. Usually, a schedule of payments by the employee and the employer is established, a percentage of present employees are required to initiate the plan, and all new personnel are included automatically. The company's portion of the insurance premium needs to be considered for estimating the cost of this benefit. If the employee contributes to this insurance, the contribution is not considered a business cost.

Another supplemental benefit is *life insurance*. A schedule is established showing the employee's and employer's portion of the premium. In initiating this plan the personal histories of present employees are secured, and a stated percentage of the work force must participate. All new personnel may be automatically included. The company's portion of the premium needs to be considered as a business cost.

Vacation pay and holiday pay are also considered a fringe benefit. Vacation policy varies as to length. When the employee becomes eligible to participate, the amount of vacation earned but yet unpaid is a company business cost. There are exceptions. In management-union agreements often found in construction, deductions are made against the earnings of the employee, not the employer. The contractor then forwards the money to a secured depository, where the funds are available for the employee for withdrawal at such a time he or she desires.

Sick pay is similar to vacation pay. It may not be formally set up on the accounting books as such but is charged to overhead when taken. It can, however, be estimated on experience.

Another fringe benefit is an employee supplemental *pension plan*, which provides retirement benefits in addition to social security. The plan may be self-funded, in which case the company agrees to invest contributions, or the plan may be funded by an outside agency, such as an insurance company or a mutual fund company. In the latter case, the company makes the payments required under the plan. In a third arrangement, the plan may be fully funded by the company or partially by employees. In management-union agreements, there are deductions against the employer. These deductions will vary depending on the union local, area, and other negotiation agreements.

Another fringe benefit is the *stock-purchase* plan, in which the company encourages employees to become shareholders in the company by allowing them to purchase stock at less than market value. The immediate benefit the employee receives is the difference between market price and the price paid. Typically, these employer benefits are limited in application, usually to owners and long-term employees.

Other miscellaneous fringe-benefit expenses include Christmas bonuses and maintaining incidental employee conveniences. Labor costs for foreign projects may include family living allowances, extra midday rest periods, severance pay, subsistence for work away from home, social taxes, and perfect-attendance-record bonuses.

Labor costs can be wage only, or wage plus required and voluntary fringe costs. A contractor needs to know the *full cost of labor*. In this text we define the full cost as the *gross hourly wage rate*. Consider an example of finding the effective gross hourly wage for a journeyman wireman. A typical job description for a journeyman wireman is given by Table 2.13. If we know the job title, and if a contractors association and union collective bargaining agreement exists, it is possible to figure out the gross hourly wages for the required and fringe costs, as described above.

A journeyman wireman receives $24.50 per hour, but this *bare wage* is not full cost until other costs of employment are added, such as those given by Table 2.14. After ad-

TABLE 2.14   Illustrative wage-fringe schedule for journeyman wireman under union contract

	Element	Rate
1	Wage	$24.50/hr
2	FICA	6.2% of annual wages up to $68,400
3	Meditax Act	1.45%
4	Workers' compensation (cost based on contractor's gross annual estimated payroll)	$29.18/$100 of straight-time pay[a]
5	Unemployment insurance	5% of straight-time pay[b]
6	Union employees benefit agreement	3% of gross payroll[c]
7	Union health and welfare	$2.00/hr
8	Union pension	$1.25/hr
9	Industry service charge fund	0.4% of total wage payroll[d]
10	Apprentice training	0.6% of monthly payroll[e]
11	Promotional and cooperation fund	$0.06/hr

[a] The rate is usually applied against total payroll and is influenced by the history of the employer and the statistical patterns of the type of work. The money is usually payable to a state-approved insurance carrier.
[b] Often the rate is applied for only a portion or base amount, which is usually less than the amount of the employee earnings.
[c] Gross monthly payroll of those enrolled in the bargaining unit.
[d] Includes overtime.
[e] For gross monthly payroll for journeyman and apprentice.

justment the bare wage becomes the gross hourly wage, which is considered the real cost of employing the tradesman.

Some of the cost factors are on a basis other than the worker's base wage, but for simplicity the base rate is used (Table 2.15). Additional complications arise if work is on overtime. Our example proceeds with journeyman wireman working 40 hours per week, which is called *straight time*.

Workers' compensation is the most costly of the nonwage listed items. These costs vary among the states and various construction trades. Rates vary from approximately $5 to $40 per $100 of payroll. For instance, steel erection is one of the more costly occupations according to a survey.[3]

TABLE 2.15   Procedure to find the gross hourly wage for journeyman wireman

	Element	$/Hour
1	Wage	$24.50
2	Social security tax	1.52
3	Medicare tax	0.36
4	Workers' compensation: $24.50 \times 29.18/100$	7.15
5	Unemployment insurance	1.23
6	Union employees benefit agreement	0.74
7	Union health and welfare	2.00
8	Union pension	1.25
9	Industry service charge fund	0.10
10	Apprentice training	0.15
11	Promotional and cooperation fund	0.06
	Gross hourly wage rate	$39.06
	Increase over wage	59%

[3] ENR, Annual Quarterly Roundup, September issue.

Another calculation method to find the gross hourly wage is more useful for management-type employees or for workers not involved within union contractual requirements. For example, we are interested in finding the gross hourly wage for a land surveyor. The forthcoming budget year consists of 52 40-hour weeks, and overtime at a wage scale of "time-and-half" is seasonal for 4 weeks, consisting of Saturday work of 8 hours. The contractor pays for 4 holidays and 10 days of vacation at regular wage. Five days of sick leave are paid. A 10% subsidy for performance, although preplanned, may not be paid. Nonhourly costs include FICA and Meditax taxes, unemployment insurance of 2% based on $15,000, workers' compensation at 2% of regular wages up to the base of $15,000, additional accident insurance of $500, medical plan of $2400 annually, term life policy cost of $400, and Christmas gift of $100. There is a retirement mutual fund cost for a 401 k plan of $3800. The hourly wage is $24.50. Find the effective gross hourly wage. What is the percentage increase due to "fringe benefits" over the wage rate? Consult Table 2.16 for the calculation to find the gross hourly wage for the land surveyor.

It is also possible to figure gross hourly wages for a class of common employees, using averages for entitled vacation, overtime, holidays, and all of the items listed in Table 2.16.

TABLE 2.16 Procedure to find the gross hourly wage using annual hours

	Item	Calculation	Annual Hours	Annual Cost, $	Annual Excess Cost, $
1	Regular paid clock hours		2080	50,960	
2	Planned overtime hours	4 × 8	48	1,176	
3	Overtime cost	50% of 2			
				588	588
4	Subtotal		2128	52,724	
5	Holidays		32	784	784
6	Entitled vacation		80	1,960	1,960
7	Paid sick leave		40	980	980
8	Subtotal	(5 + 6 + 7)	152	3,724	3,724
9	Total	(4 + 8)	2,280	56,448	
10	Expected performance subsidy	10% × 4		5,272	5,325
11	Total FICA base	(9 + 10)		61,720	
12	Chargeable hours	(4 − 8)	1,976		
13	Nonhourly cost				
	FICA	6.2% × 61,720		3,827	3,827
	Meditax	1.45% × 61,720		864	864
	Workers' compensation	2% × 15,000		300	300
	Unemployment insurance	3% × 15,000		450	450
	Supplemental risk insurance			500	500
	Supplemental life insurance			400	400
	Supplemental health insurance			2,400	2,400
	Bonus, gifts			100	100
	Retirement plan			3,800	3,800
	Profit sharing				
	Subtotal	of 13			12,641
14	Excess costs	(3 + 8 + 10 + 13)			22,226
15	Excess hourly rate	(14/12)			11.24
16	Wage			$24.50	
17	Effective gross hourly wage			$35.75	
19	Increase over base wage			46%	

SUMMARY

We have studied methods to find productivity for construction. There are four methods—job tickets, nonrepetitive one-cycle time study, repetitive multiple-cycle time study, and work sampling. Those techniques can be applied by small to large contractors and subcontractors. Then there are two procedures to find wages and fringes—wage only and the gross hourly wage. These two effects, labor productivity and the gross hourly wage, are the grist for future cost analysis and estimating. If these steps are incompetently done, then the contractor is unlikely to able to achieve optimum profit results in the construction business.

QUESTIONS

1. Give an explanation of the following terms:

   Man hour   Snap-back timing
   Direct labor   Job tickets
   Repetitive multiple-cycle time study   Day work
   PF&D   Work sampling
   Productivity   Apprentice
   Normal distribution   Workers' compensation
   Unemployment insurance   Job description
   Gross hourly wage   Fringe benefits
   Davis Bacon Act

2. Why do we separate labor into direct and indirect categories?
3. What is the purpose of rating? A job is time studied and is rated greater than 100% and the productivity rate is used for a second worker. What does the original rating of over 100% mean to the second worker?
4. Why do we use job-ticket reports? Name some types of work that are appropriate.
5. Determine the current FICA rate and the base for the employee and employer. What are the political ramifications of social security as it pertains to employer costs? List mandatory and voluntary types of fringe costs.
6. Discuss the success of measurement of productivity in the construction business.
7. Describe the purpose of the allowances for productivity.
8. Work sampling has pros and cons as a technique. Develop a listing, and then word-process the contrasting differences in these advantages and disadvantages.
9. The man hour is much criticized, yet no better replacement is in sight. Discuss the dilemma of the man hour.
10. Check out the ENR website for details on workers' compensation.

PROBLEMS

2.1 (a) A construction cycle is 29 minutes. Find units per hour at 100% and 90% efficiency.
   (b) If the hourly productivity is 11 units, find the hours per 1 and 100 units.
   (c) The construction rate is 9.5 units per hour. Find hours per 1, 100, and 10,000 units.
   (d) Find pieces per hour and minutes per unit for 15.325 hours per 100 units.

2.2 (a) During a work week, carpenters expended 400 man hours and constructed 5000 ft^2 of contact area of forms. Find the productivity expressed in units of 100.
   (b) Five iron workers install 10 tons of structural steel in one day. Find the productivity in terms of ton units of structural steel.

(c) A bricklaying crew is composed of five masons and three helpers. Historical records show an average of 550 bricks per day for the masons. Find the person hours per brick. Discuss the nature of the crew hour per brick as a contrast to person hours per brick.

2.3 (a) Find the direct cost of labor for structural steel placement if 35 tons of steel are placed for ironworkers who have a gross hourly wage of $41.72 per hour. Productivity is 4 man hours per ton.

(b) Find the direct labor cost for laying 3000 bricks with a productivity rate of 0.0233 man hours per brick. The labor rate for masons is $25 per hour.

(c) Find the total number of crew hours if productivity is 4 worker hours per ton. There are 35 tons to be placed, and 5 ironworkers in the crew.

(d) The productivity for laying brick is 0.0233 worker hour per brick. There are 3000 bricks necessary for the job. The crew is made up of 8 workers. Find the number of crew hours.

(e) Find the unit price of labor if productivity is 4 man hours per ton and the gross hourly wage rate is $41.72 per hour.

(f) Find the unit price of labor if the productivity for this grade of work is 0.0233 man hours per brick and the labor rate is $25 per hour.

(g) A new job will require 20 tons of structural steel to be erected. If the unit price of the work is $166.88 per ton, what is the quick estimate for labor?

(h) A new job having 5000 bricks is to be estimated for labor cost. The unit price is $0.58 per brick. Find the cost of labor.

2.4 (a) What is the duration in days to install 6000 ft^2 of 5/8-in. drywall with a crew of two carpenters? The productivity as given in a manual of estimates is a daily output of 2000 ft^2 with man hours of 0.008 per ft^2.

(b) Find the duration in days for the information given in (a) above, and assume a crew of three carpenters using the following equation:

$$\text{Duration (days)} = \frac{N \times H_s}{(8 \text{ hr per day})(\text{workers in crew})}$$

where N = quantity, units

H_s = man hours per unit

(c) Find the duration in days for a masonry job that has 24,000 bricks with a crew of five masons and three laborers. The work day is 10 hours. The productivity of the crew is 0.103 man hour per unit.

(d) Steel erection takes 0.95 man hour per ton. A crew is to install 50 tons during an eight-hour working day. How many iron workers are necessary for the crew?

2.5 A two-person carpenter crew is installing $\frac{1}{2}$ in. 8 × 4 ft plywood as roof sheathing during an eight-hour working period. A cycle is defined as the installation of one panel. The crew installs one panel in 12.1 average minutes. If the *unit of measure* is ft^2, find the *productivity factor* expressed as min/ft^2 and hr/ft^2 for the crew. What is the daily production expressed in ft^2? How many days are necessary for 8,000 ft^2?

2.6 A mason requires 12 mason-hours per 100 ft^2 for the bottom 4 ft of wall, and a hod carrier requires 14.5 helper hours to mix the mortar and carry mortar and brick to the masons. The masons lay the brick, square and tool the joints, and clean up at the end of the day. Wages for the mason and helper are $22.50 and $13.50 per hour. Find the cost per square foot.

2.7 Three thousand five hundred and seventy-eight 8-in. concrete blocks are required in a running bond. The crew is 5 masons ($22.50/hr) and 3 helpers ($13.50/hr). The mason productivity is 140 units/day. Find the hours per unit. How many hours are necessary? What is the direct labor cost for the job?

2.8 (a) A farm is to be wired from the main-line electrical distribution system. The electrical distribution planning department indicates that 10 poles truck dug and set in dirt, four anchors which are truck set, seven three-phase wood cross-arms, one transformer, a lightning arrester, and a farm light are included in the bid. If the crew and equipment costs $90 per hour, what is the job cost for direct labor and equipment? Use Table 2.9.

(b) Let the crew size be two instead of three and unit man hours be increased 60% because of unbalanced crew work. If crew labor and the line truck costs are $40 and $25 per hour, what is the estimated direct labor and equipment cost for (a)?

(c) A storm has ravaged an area, and the unit man hours for a task are acceptable if they are increased by 25%. A light commercial building needs four poles in rock, three-phase cross-arms, four anchors, one transformer, and two parking-lot lights. Costs for the crew and equipment are $75 per hour. Find the direct labor and equipment costs for this repair, which is charged against *overhead contingency* and not billed to the customer.

2.9 Concrete is delivered by powered buggies at the rate of 5 mph. Average one-way distance from the elevator chute for the high-rise commercial building is 100 ft. Loading and dumping of the concrete requires 0.5 min. Find the travel cycle time expressed in minutes.

2.10 Concrete is delivered by powered buggies at the rate of 4 mph over an average distance of 300 ft. Each buggy is operated by one laborer. Loading and dumping requires 0.5 min. A one-cycle haul delivers 9 ft^3. How much direct-labor time is required for 100 yd^3 if efficiency is 90%?

2.11 A 36-in.-wide trench is to be excavated for a 21-in.-dia. sewer pipe. The depth of the trench is 12 ft in hard tough clay. The ground material is stable and water seepage is not anticipated, nor is shoring required. These conditions will allow a ladder-type trenching machine. Examination of the path reveals obstructions that will reduce the digging speed to about 60% of the theoretical optimum 1 fpm. A crew consists of a worker ($20/hr), three common laborers ($15/hr), and a foreman for 50% ($30/hr). The hourly cost of equipment $97.50. Find the cost of the crew labor and equipment per linear foot.

2.12 The placing of form work on concrete columns using 24 × 24 in. plywood, which will have one use, has a daily output of 190 ft^2 of contact area. The crew consists of 3 carpenters at $25/hr and 1 laborer at $20/hr. A cost element of power tools is $35 daily. Find the productivity for the *bare* direct-labor cost. (*Hint*: The term "bare" cost is jargon in the construction trade, meaning that the direct labor is unloaded, or without overhead and profit. Use a standard day of 8 hours.)

2.13 Engineering aides prepare *job tickets* of their daily work, and after a sufficient period of data collection, classification of the data gives the following information:

Description of Work	Minutes per Occurrence	Frequency per Drawing
Post drawing numbers	7.0	1/1
Duplicate drawings	19.0	1/1
Correct computer entries	38.0	2/1
Phone calls and email	6.0	1/4
Update CAD changes	27.0	1/2

A personal allowance of 10% is used for this category of work. Fatigue and delay are not considered likely.

(a) Determine the *standard person hour* per engineering drawing.

(b) A new building design is anticipated and a staff will be hired for this work. This design will result in about 2,500 prints over a 1-year period. How many aides are required, and at $25 per hour, what amount do you estimate for this activity in a project overhead budget?

2.14 A foreman collects mason productivity via the *job ticket* for the placing of concrete blocks. Job-ticket information is summarized as follows:

Blocks per Mason Hour	Floor & Suspended Scaffold	No. of Blockouts	Blockout ft^2
31	1	3	155
24	2	8	445
36	1	0	0
14	3	2	195
26	1	2	405
19	1	6	740

A job is to have 1,000 blocks each on the first and second floors, where the first floor has 3 blockouts and 200 ft^2 of openings, while the second floor has 5 blockouts and 500 ft^2 of opening. Find the mason hours to perform the work. (*Hints*: Use the nearest example for guidance. The usual assumption is that more labor time is required to place block on a second floor because the mason works from a suspended scaffold, and blockouts interrupt the mason's productivity when placing blocks in a straight wall.)

2.15 Determine the allowed time for arc welding of A36 carbon steel gussets with butt welds, where the data are given by Table 2.2. Discuss those items of the job tickets that are common and those pieces of data that are dissimilar to analysis. How would you improve the collection process if there was a next time? (*Hints*: Set up a spreadsheet and be sure to give dimensions for the productivity rate. Assume that any productivity rate will be constant rather than variable work.)

2.16 Determine the allowed time for the rough framing of a residence by a carpenter and apprentice and a green crew, where the data are given by Table 2.5. Discuss those items of the job tickets that are common and those pieces of data that are dissimilar to analysis. How would you improve the collection process if there was a next time? (*Hints*: Find the productivity rate for partitions, windows, extra studs, and doors. Assume that any productivity rate will be constant rather than variable work.)

2.17 The fire department is concerned about the speed of the crews assigned to the all-purpose trucks. The chief wants to know the time required after receiving the alarm before starting to fight the blaze. To answer the chief's questions, a time study of the activities of one crew on seven different alarms is conducted and the following is the summary:

Crew Task	Alarm						
	1	2	3	4	5	6	7
Start timing	.00	1.51	2.33	3.24	4.12	5.04	5.91
Get dressed	.12	1.58	2.43	3.36	4.25	5.15	6.01
Board truck	.29	1.82	2.60	3.55	4.41	5.32	6.19
Start engine	.44	1.94	2.76	3.70	4.55	5.48	6.32
Drive to fire*	—	—	—	—	—	—	—
Unload hoses	.64	2.00	3.02	3.89	4.78	5.63	6.50
Connect hoses	.89	2.33	3.24	4.12	5.04	5.91	6.75
Unload ladders	1.08						6.95
Position ladders	1.51						7.31
End timing for this alarm	1.51	2.33	3.24	4.12	5.04	5.91	7.31

(*Hints*: On the observation sheet, continuous (no reset) electronic watch readings in hundredths of minutes are recorded, indicating full minutes only when it changed. *Watch stopped because of variable nature of distances. Time to fight fire is ignored because of the variability of this task.)

(a) Determine the average time for the tasks.

(b) If the crew is rated at 110% for all elements, find normal time in minutes per occurrence.

(c) Determine each task time standard for a 20% allowance.

(d) Find the standard cycle time.

(e) Discuss the similarity and differences of this work as compared with construction.

2.18 A piling contractor conducts an informal nonrepetitive one-cycle time study of his operation. Because of a fluctuating crew size (the average crew number is 7), a large job site of a river crossing area, a roving time-study technician, and inaccuracies inherent in timing, work rating of the crew is impractical. This study will last a few months, which is a daunting task for the observer and a significant commitment by the contractor.

Deductions for foreign elements are considered more crucial than work rating. Nonetheless, management will use the information for later job bidding and continuous improvement.

The tapered and hollow A-36 carbon steel piles are driven by steam-hammer energy using a mandrel, and later the piles are filled with concrete. The essential elements of the job include material handling of the piles, driving the piles to either required resistance or penetration depth, cutting the piles at cutoff elevation, and filling with concrete. The time study excludes moving the equipment and materials on and off the job site.

Find the efficiency of net cycle direct-labor time to total labor time, and the productivity of crew man hours to length of piles driven. A summary of the time-study data is as follows:

	Measured Time
Total time, hour	5848
Foreign elements, hour	2580
8-in. pile by 40-ft depth. Qty = 87; total hour	2436
10-in. pile by 60-ft depth. Qty = 27; total hour	832

2.19 Find the productivity in terms of standard man hours per linear foot using two nonrepetitive one-cycle time studies of the following assembly: Walls with 2×4 in. studs, $\frac{1}{2}$-in. drywall interior ready for painting, which is over $3\text{-}\frac{1}{2}$ in. R-11 insulation with a 5/8-in.-thick rough-sawn 4-ply, 4×8 ft panels exterior-grade plywood siding on the outside face. Wall studs are 16 in. center to center, double top plates, single bottom plates, fire blocking, and nails. The time for framing headers and posts for window and door openings are not a part of this productivity analysis, although the time study will see that effort. A crew of a carpenter and helper is employed. Use an allowance of 14% for this operation.

(*Hints*: Timing of the operation is via a robotically controlled TV camera, with the feed being saved for later analysis. From time to time a technician will adjust limit controls, start and stop, and the aiming point and view of the camera. Compare this viewing recorder to the TV cameras that are found in bank. Total cycle time = wall assembly time + header, post and opening time + foreign-element time.)

The TV time studies are summarized below:

TV Time Study	1	2
Total cycle time, hour	183.0	123.4
Foreign elements, hour	32.0	26.0
Header, post and opening time, hour	29.60	42.5
Wall length, ft	90.2	70.0
Rating factor, %	85	80

2.20 A repetitive time study is taken of the installation of $\frac{1}{2} \times 48 \times 96$ in. plywood on a 3/12 roof by one carpenter. The conditions for the study have the material stored on the roof at an average distance of 20 ft from the point of installation. One cycle includes picking up one sheet, moving it to the location on the roof, adjusting, air nailing, and returning empty handed to the pile ready for the next sheet. Eight cycles are observed and timed. The times are recorded as decimal minutes and are given as 10.70, 8.38, 12.70, 13.88, 8.47, 15.83, 9.72, and 13.57 min. The observer rates the work as 85%. If the allowance is 20%, find the average, normal, and standard cycle time expressed in minutes. What is the time expressed as units/hr and hr/100 units of work? Find the man hours expressed in ft^2.

2.21 A repetitive time study is performed for installing 4 × 8 ft wall forms for a straight wall. The cycle timing is continuous, five cycles are timed, and the time study is in units of decimal minutes. The crew consists of three carpenters. The observer breaks down the operation into the following elements:

1. Walk to get form
2. Pick up form
3. Take form to location on floor
4. Adjust form and nail 16 times into place

Element	1	2	3	4	5
Start timing	0.00	15.69	31.73	48.89	63.87
1	1.35	17.79	33.14	50.50	65.26
2	4.79	22.23	36.59	53.65	68.87
3	8.44	25.73	40.64	56.86	73.80
4	15.69	31.73	48.89	63.87	80.72

At the conclusion of the time study, the observer rates the work as 110%. PF&D allowances for this work are 18%. Find the average, normal, and standard element times in decimal minutes. How many panels are expected per hour? Find the productivity in terms of carpenter hours per square foot of contact area.

2.22 A work-sampling survey of a construction job, which is designated into 12 job categories, has the following observations:

Job Item	Count	Job Item	Count
1	92	7	24
2	99	8	33
3	37	9	3
4	11	10	22
5	25	11	8
6	14	12	32

If this sample covered a span of 25 days for 8 hours per day, what are the percents and expected hours per item of work?

2.23 The pediatrics department in a hospital is work sampled for period of 608 hours. In a preliminary study the following categories were found to be significant, and in a final study the observations are given as:

Job Element	Count	Job Element	Count
Routine nursing	496	Other	79
Idle or wait	263	Feeding	52
Unit servicing	183	Bathing	22
Report	129	Elimination	11
Personal time	128	Transporting	8
Intervention	102	Housekeeping	7
Unable to sample	91	Ambulation	7

Find the percent occurrence, percent cumulative occurrence, element hours, and cumulative hours for each work category. Compare and discuss the similarities and differences of nursing with construction in work sampling.

2.24 A work-sampling study is taken of a construction job site with the following information obtained: number of sampling days, 25; number of trips per day, 16; number of people observed per trip, 3; and number of items being sampled, 4. The four sample job tasks are broken down as A, 80; B, 320; C, 1600; and D, 2800.

(a) How many man days and observations were sampled?

(b) What are the percentages and equivalent hours for the activities?

(c) For a confidence level of 90%, what is the relative accuracy of each of the items?

2.25 Suppose that we want to determine the percentage of idle time of a large construction crew by work sampling. Assume that a confidence level of 95% and a relative accuracy of $\pm 5\%$ are desired, and a rough estimate of 25% is suspected for idle time. Previously, a preliminary study of the activities of "inactive because of foreman fault," "no materials," "missing," and "idle" were work sampled, and those categories are collected under the title "idle," which leads to the need for an accurate work-sampling study.

How many observations are necessary to satisfy this specification? Now let the relative accuracy be $\pm 2\frac{1}{2}\%$. How many observations are required? What happens to the number of observations as relative accuracy becomes less?

Suppose you are the construction manager, and the idle category as suspected is realized. Discuss corrective policies that you would implement.

2.26 **(a)** To get a 0.10 interval on work observed by work sampling that is estimated to require 70% of a construction crew's time, how many random observations are required at the 95% confidence level? Repeat for 90%.

(b) If the average "handling and moving" activity during a 20-day study period is 85% and the number of daily observations is 45, then what is the interval allowed on the activity? Use a confidence level of 90%. Repeat for 99%.

(c) Work sampling is to measure the "not-working time" of a road construction crew. A preliminary study suggests that "not-working-time percentage" is around 35%. For a 90% confidence level and a desired relative accuracy of 5%, what is the number of observations required for this study? Compare to a 95% confidence level.

2.27 A five-person electrical crew is installing switch gear components for an electrical power utility yard. A work-sampling study is undertaken, and the following observations of work elements are recorded over a 15-day, 8-hour period:

Task	Count
Set up and teardown	312
Electrical wiring	264
Mechanical fastening of components	204
Move materials with portable crane	324
Idle	96

A rating factor of 90% for the entire period is found. The number of switch gear completed during this period is 26. This contractor uses an allowance value of 10% for work of this kind. Average journeymen wage is $30 per hour. Find the elemental costs. What is the standard labor cost per switch gear? What is the actual cost? (*Hint*: A standard unit cost would not include any "idle" effort, as this is contributed back into the cost as a part of the allowance.)

2.28 An eight-person CAD department, which drafts A, B, and C drawing sizes, is work-sampled over a 4-week period. The categories are summarized as follows:

Work Task	Count
Computer drafting	778
Calculating dimensions and takeoff quantities	458
Check prints	110
Training	125
Professional time off	172
Personal time, idle	270

Fifty-five drawings (A = 20, B = 25, and C = 10) were produced during this period with a total payroll of $52,800.

(a) If the policy is to accept professional and personal time as necessary to the drafting of prints, then what are the person-hour and cost factor for each drawing size? (*Hints*: For this question we use the measured count and do not include a PF&D allowance. The print relative sizes are A = 1, B = 2A, and C = 2B.)

(b) Now if personal and idle time are prejudged at 10%, then what is the per-print size factor? (*Hints*: Remove the personal and idle time, as if it did not exist. In its stead, substitute the 10% allowance factor.)

2.29 A concrete-steel building is under construction. The total elapsed time for turnkey construction is estimated to be 23 months. During the early months of construction, the contract manager desires to know the effectiveness of the ironworker trade, and he uses work sampling to glean information.

   The building is six stories with a basement and uses 935 tons of steel. Stories are 15 ft, bays are 40 ft, the live load is 200 lb/ft^2, and steel material is typically A36 grade. Average gross hourly wages for the ironworkers are $33.94. The iron workers are estimated to require a total 58,000 man hours for erection during construction. At the start of the erection a total of 6375 man hours have already been worked during the period of the study. The construction management firm establishes the following delay categories:

1. Design engineering, specification, or management
2. Weather
3. Equipment
4. Labor
5. Material

   (*Hints*: A trained observer notes whether the worker/equipment is working or not working, and for the not-working category, she tallies the reason as one of the five. This choice can also be made following the observation, i.e., confirmed in the office rather than entered in the field. As an example, the carpenters are working, while the iron workers are idle. If the iron workers are idle, then one of the five reasons is selected. Are they not working because of a weather problem, or it is labor laziness or one of the other causes?)
   Results of the work-sampling study are shown below.

Work Category	Primary Count	Secondary Count
Working	709	
Not working	431	
Design, specification or management		8
Weather		102
Equipment		18
Labor		260
Material		41

During this sampling period 84 tons of steel were erected.

What is the most significant reason for nonproductivity? Find the cost of steel per ton placed as found by the study. At this rate what is the expected actual cost for the iron-worker classification? What is the labor efficiency? Based on the study, what do you recommend for management action to improve iron-worker productivity? Describe the details of a program to implement your suggestions.

2.30 Determine the gross hourly wage of an iron worker. Additionally, find the percentage increase to the base rate. The iron worker is paid according to a management-labor agreement. The work area is in a subsistence zone and the iron worker works the second shift. The base rate is $19.20 per hour. He works six 10-hour days per week. The workers work 7 hours and are paid 8 hours under the shift pay agreement. Include the cost effects for FICA and Meditax, in addition to the craft and insurance rates for iron workers shown in the table below,

Wage Element	
Union pension benefits	$1.14/hr
Union health and welfare benefits	$1.30/hr
Vacation (employer requirement)	$1.00/hr
Apprentice training	$0.14/hr
Workers' compensation	$29.18/$100 of straight-time pay
Unemployment insurance	5% of straight-time pay
Public liability, $50,000 coverage	$4.47/$100 of straight-time pay
Property damage, $100,000 coverage	$2.44/$100 of straight-time pay
Subsistence	$20.00/day

Discuss the point that labor costs carry public costs that proceed not from any management-union negotiation but from public concerns.

(*Hint*: It is necessary to find the gross hourly wage per week, which includes the base rate + fringes + workers' compensation + unemployment insurance + public liability + property damage + FICA + Meditax + subsistence.)

2.31 A five-person crew is working 52 weeks, and the contractor has assembled selected labor costs. Find the cost per man hour for these benefits. (*Hint*: Use the typical 40 hour work week.) The budget is given below:

Wage Supplement	Annual Cost, $
Personal liability insurance	$4,812
Medical benefit insurance	21,650
Workers' compensation insurance	11,203
Vacation and apprenticeship allocations	4,750
Union pension fund	16,500

2.32 Find the cost per man hour for the following composite crew. If the job is estimated to take 83 hours, what is the crew labor cost?

Crew	Hourly Wage	Hourly Burden Wage	% Time on Job
Foreman	$28.65	5.20	25
Operating engineer	28.15	3.45	12
Iron worker	26.12	3.15	45
Oiler	18.47	2.85	5
Laborer	12.91	2.65	45

MORE DIFFICULT PROBLEMS

2.33 A large office with a 14-ft high ceiling by 250-ft concrete sides is to be initially painted, and redecoration is required every 4 years. Beforehand all holes are filled and sanded, and one prime coat and one finish coat are applied by a spray gun. Initial protection is nominal. The redecoration process calls for the walls to be washed and one finish coat to be applied with brush and roller.

A crew of three is employed (one foremen and two journeymen). The foreman has a job rate of $40 per hour, and each painter has a job rate of $25 per hour. Job elements are:

1. Scaffolding time: 0.90 man-hour per scaffold, erection; 0.60 man hour per scaffold, dismantle. A three-section scaffold is 21 ft long × 5 ft wide × 8 ft high. The scaffold is moved 12 times per wall, and it takes 10 minutes to move for a new unpainted wall and 12 minutes for redecorating.
2. Sealing and sanding time: 0.50 man hour per 100 ft^2.
3. Spray paint surfaces: 0.65 man hour per 100 ft^2.
4. Paint surface by roller: 0.75 man hour per 100 ft^2.
5. Washing walls time: 0.87 man hour per 100 ft^2.

Forecast total direct-labor expenses for both the initial and redecorating processes, and compare labor cost per square foot for the initial and redecorating processes. Which task element is the most costly?

2.34 A brick wall is shown in Fig. P 2.34. The two 8-ft-high × 40-ft-long brick sections of exterior double width are to be estimated for labor time and cost.

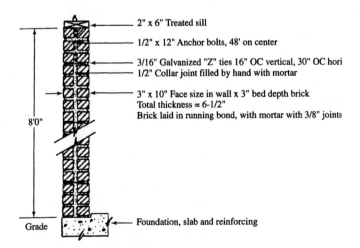

Figure P2.34

A quantity take-off is given as

Description	Quantity	Man hours/unit	Gross Hourly Wage Rate, $/hr
3×10 in. face brick, $640\,\text{ft}^2$	3072 ea	0.023	$24
3/16 in. Z ties	115 ea	0.006	$24
1/2 in. collar joint	14	0.200	$24
1.2 $\times$ 12 in. anchor bolts	10 ea	0.050	$24
2 $\times$ 6 in. treated sill	40 ft	0.08	$25

Find the total direct-labor cost and the productivity factors for lineal foot and square foot. A similar wall is 95 ft long. What is its cost?

2.35 Bulldozers move earth in making highway cuts or stripping top soil. You are to determine the output or *utilization* of a bulldozer for the following conditions: The material is sandy loam top-soil, weighing 2700 lb/yd^3 bank measure. It has a swell of 25%. The haul distance is 100 ft over level ground, and the bulldozer is operating in a slot, which improves its earth-moving capacity. The mold board is 9 ft 6 in. long, 3 ft high. The rated capacity for loose volume is 3.6 yd^3. Use a 50-min hour for working equipment.

In the pushing state, the dozer operates at 1.5 mph, while returning it moves at 3.5 mph. The time to load and shift gears is fixed, or 0.35 min. Determine the total minutes per cycle for 100 ft. Find the output per hr expressed in yd^3 bank measure.

(*Hints*: The rate of output for labor is referred to as productivity, but for equipment the efficiency is called *utilization*. For equipment of this nature, and rather than to give equipment a PF&D allowance, the 50-min hour is used. Net mold-board capacity is 3.6/1.25 = 2.9 yd^3 bank measure to allow for swell of the material.)

2.36 A subcontractor has two methods of hanging steel siding panels. Each panel is 2 feet wide, and the height varies from 14 to 18 feet. The method is the placement of two panels, one a liner, and then a face panel. A foreman collects job-ticket information from six different jobs, and a summary is given as:

	Job 1	Job 2	Job 3	Job 4	Job 5	Job 6
No. of crew members	3	4	3	4	3	4
Method of placement						
1. Both panels together		x	x			x
2. One panel at a time	x			x	x	
Type of scaffolding						
1. On wheels		x	x		x	
2. Hung	x			x		x
No. of panels placed per crew hour	18	25	22	20	19	22

Which method gives the highest productivity?

2.37 A *nonrepetitive gang time study* is taken of a subcontractor crew for a concrete-pumping operation, and what follows is a tabulated summary of the method and time.

(*Hints*: It is unrealistic to rate this kind of work, and PF&D allowances are seldom a part of the calculation. Crosses in the table indicate that the job element is necessary for the worker or the equipment. *Cost of concrete truck and driver are subsumed in the cost of the concrete material and are not included in the cost of subcontractor, except for the cost of the concrete.)

Crew and Equipment

Task	Min.	Foreman	Pump Operator	Hopper Man	3 Workers	2 Vibrating Crew	Concrete Truck & Driver	Truck Pumper Rental
Make ready	60	x			x			
Travel to job site	30	x			x			
Pump machine in position	45	x	x	x	x			x
Set up machine	20	x	x	x	x			x
Inspect	15	x	x	x	x			x
Adjustment	30	x	x	x	x			x
Pump concrete to forms	255	x	x	x	x	x		x
Normal delay for set of concrete	20	x	x	x	x	x		x
Dismantle pump	45		x					x
Put tools away	15	x	x	x	x			
Return pump	30		x					x
Clean up	50	x		x	x			
Cost per hour		30	20	16	15	17	*	75

(a) The work consists of placing 400 yd³ of 3000-psi concrete, using a pipeline truck. Determine the direct cost of the subcontract job and of placing a cubic yard of concrete. Concrete costs $65/yd³. (*Hints*: Assume that when not assigned to this job, as shown by the "x" in the table, the worker is doing something else profitable. Pay time and a half for overtime of any assigned work in this job. Rental of truck pump equipment is on a contract basis and does not involve overtime premium. The task "make ready" starts a new work day.)

(b) Assume that the union contract requires the entire crew (exclusive of concrete truck and driver and concrete pump truck) to be at the job site during the entire scheduled time. What is the actual cost per yd³ then? What is the cost of this nonproduction?

2.38 Three nonrepetitive one-cycle time studies are made of arc-welding operations of A36 steel gussets to building columns. A summary is given below:

Part No:	6682	7216	8313
Total cycle time, hour	2.47	1.06	1.18
Foreign elements, hour	0.15	0.40	0.05
Material thickness and length	$\frac{3}{4}$ × 24 in.	$\frac{1}{2}$ × 8.4 in.	$\frac{3}{16}$ × 6 in.[a] $\frac{1}{4}$ × 9 in.[b] $\frac{1}{2}$ × 19.4 in.[c]
Welding time, hour	2.12	0.49	[a]0.12 [b]0.26 [c]0.65
Non welding operation time, hour	0.20	0.17	0.10
Rating factor, %, of welder	95	90	110

Use an allowance of 18% for this operation. Find the standard-productivity equation for each material thickness. (*Hint*: Total cycle time = welding time + nonwelding operation time + foreign-element time.)

2.39 A repetitive time study is taken of a structural steel crew consisting of four iron workers. A cycle consists of the following work steps:

1. Load structural steel column or beam or girder from truck to ground
2. Lift member from ground to position of placement
3. Position
4. Make initial connection
5. Level and align member
6. Tighten connections

While each cycle involves differences in the type, weight per foot, and length of the structural member, whether a column, beam, or girder, the method of handling, cranes, tools, and other assists did not change between the members. A W 14 × 68 30-ft beam is selected as average for the six cycles. The time study, which is dimensioned in decimal minutes, is given below, and the work steps are shown for their element time:

	Cycle					
Work Step	*1*	*2*	*3*	*4*	*5*	*6*
1	2.83	3.10	3.65	3.46	3.73	3.13
2	4.50	3.89	5.43	4.78	5.03	5.22
3	9.10	10.78	8.14	8.37	8.13	8.73
4	11.42	10.40	10.89	9.93	9.77	10.33
5	20.73	18.09	18.63	18.47	16.43	18.53
6	21.62	15.86	16.43	16.43	18.53	15.30

The technician rated the iron workers as faster than normal by 10%. The contractor uses a 17% allowance factor to cover personal, fatigue, and delay. Find the standard cycle time for the crew as if it were operating as one iron worker, and then find the standard for the cycle for the crew. Determine the standard productivity per ton.

2.40 Repair of rural asphalt roads having low traffic volume involves the operations of patching, crack sealing, sand distribution, sweeping, and chip sealing. Each operation is mobilized with different equipment and crew size. For instance, the chip sealing of a 24-ft-wide road has a crew number that depends on the distance from the raw material depot of sand and tar-oil and the length of the road that is to be reconditioned.

A road resurfacing maintenance campaign of 10 8-hour days is planned for country roads that are an average of four miles distant from the supply center of materials. During this period 21 miles of double-lane road are "chip sealed." A work-sampling study is conducted of the efficiency of the crew during this period.

The crew is composed of the following:

Job Title	No. in Crew	Wage, $/hr
Truck driver for pea gravel	4	15
Chipper driver	3	15
Traffic control & flagger	2	10
Roller & packer	1	18
Sweeper	1	14
Foreman	1	20

The work-sampling elements and their "instantaneous" observations are given as:

Element	Count
1. Daily setup and teardown	410
2. Work—use tools and equipment	1763
3. Travel	817
4. Wait—for materials and equipment	902
5. Idle—breaks, not working	612

The elements are not rated for purposes of this study. An allowance of 15% is used. What is the percentage of "good" work? Find the standard crew man hours per lane mile. Find the standard cost per lane mile. What is the actual cost per lane mile? Determine the cost per square yard. Find the *labor efficiency*.

(*Hints*: The standard crew hours per lane mile are found without elements 4 and 5. A lane is 12 ft wide, or half of the road width. For this study a trained observer samples from a pick-up truck, and while random observations are a goal, the success of randomness is not assured, nor will the study make statistical claims, except that the information is suggestive for manpower planning and cost analysis. Some items of equipment, such as trucks, spreader box attachments, and sweepers, are linked to the crew, but equipment utilization is not the purpose of the study. The crew have their assigned duties, which differ, although similar job titles have similar assignments. The work-sampling elements are neutral to the job-title requirements, and they focus on the labor efficiency of the work.)

2.41 Find the gross hourly wage of a boilermaker. The craft rate is $20 per hour. The work is in a subsistence area. The boilermaker works the second shift on a two-shift project where a contract requires a "work 7 pay 8 hour" pay basis for straight time. He works 6 days, 10 hours a day. In addition to time and a half for overtime Monday through Friday, the contract requires double time for work on weekends. Other information is provided in Prob. 2.30.

2.42 We are interested in finding the estimated cost for the *job description* of superintendent. The contractor employs key people throughout the year, but hires and fires nonmanagement labor depending on business activity.

The preplanned year consists of 52 40-hour weeks, and overtime at a wage scale of "time-and-half" is seasonal for 12 weeks consisting of Saturday work of 8 hours. The contractor pays for 4 holidays and 10 days of vacation at regular wage. Five days of sick leave are paid. Expected nonchargeable hours are 5%. A subsidy for performance is 10%; it is preplanned but may not be paid. Nonhourly costs include FICA and Meditax taxes at the current federal rate, and unemployment insurance of 2% and workmen's compensation at 3% of regular wages up to $15,000, additional accident insurance of $500, medical plan of $2400 annually, and Christmas gift of $100. The hourly wage is $30. Find the effective gross hourly wage. What is the percentage increase due to "fringe benefits" over the wage rate?

PRACTICAL APPLICATION

This practical application is the second module in the development of internet skills for construction cost analysis and estimating. The student will find a logo for the college or company that he or she is associated with. That logo is incorporated using various graphic formats, GIF, TIF, PCT, JPEG, and so forth, and is used as a header for homework, project assignments, and reports. This requires the student development of Web skills. It should be noted that this development is useable for a number of courses, and it may be learned in other places.

Instructor evaluation of the practical application is the preparation and submission of a header page for a homework problem given in this chapter.

CASE STUDY: HIGHWAY CONSTRUCTION

Asphalt highway construction is equipment intensive, i.e., controlled by equipment performance. We measure the *utilization* of equipment using pseudo work-sampling methods to assess the performance of this equipment.

The specification of the road is a new two-lane high-traffic-volume 3-in.-lift asphalt road. The road is a 30-ft-wide mat. Grading and leveling of the road base are concluded before the lay-down of the asphalt. For a job that is within a ten-mile radius of the stockpile of asphalt, the preferred equipment is as follows:

Equipment	Specification	No. of Units	Cost per Equipment Hour
Lay-down machine	Blaw-Knox, 8 ft min., 16 ft max, adjustable width	1	$275
Roller compactor	Caterpillar pneumatic tire; Caterpillar 6 ft wide drum	2	$70
Semi trailer truck, 20 ton	Mack & Kenworth	6	$75
Distributor	Etnyre, 2000 gal capacity, 0.1 oil gal/yd^2 tack coat	1	$85

(*Hints*: While labor cost is significant at any time, for equipment utilization analysis we add the gross hour labor cost to equipment cost. For instance, the lay-down machine has a crew of 1 operator and 5 workers, and direct-labor cost is included into the equipment cost. For work sampling to obtain equipment utilization, labor rating is ignored. Work quality of the equipment is not considered pertinent to the objective of this study, as county road specifications set minimum quality levels.

Geographical separation of the equipment units requires advanced methods of observation. A walking observer is impractical. But a work-sampling survey of highway construction is possible using a pickup-truck observer with a cell phone hookup, helicopter surveillance and radio talk, self monitoring by the equipment operators on an alarm reminder basis, or rpm sensors on the engine power output, where on-board computers upload the telesignal via geoposition satellite to the central computer system, which performs the random selection for the digital work-sampling observations. Engine rpm sensors will read high rpm, interpreted as full-load working, while low or moderate rpm suggest waiting or unloaded travel. A no-rpm signal indicates down time. Each equipment sensor unit is engineered for an appropriate rpm response to working conditions.)

Translated equipment sampling observations are as follows:

Sensor/Equipment Count	Lay-down	Rollers	Trucks	Distributor
No rpm	398	888	5596	178
Low rpm	707	507	3069	534
Moderate rpm	133	634	3249	712
High rpm	972	1141	6137	801

The project is a period of two six-day weeks and ten-hour-day working time. The length of the road is 9.15 miles.

Which equipment is the highest or lowest in utilization in terms of downtime/uptime percentage? In terms of downtime/uptime cost? Find the overall equipment performance in terms of cost. Find the cost of waiting for the equipment. What are equipment and direct labor costs per mile? Discuss other measures of utilization that an analysis of this kind may supply.

(*Hints*: Utilization for this problem is percentage of time exclusive of downtime. Performance is measured in terms of uptime cost to total cost. For this analysis we ignore the costs of right-of-way purchase, materials, supervision, survey, engineering, etc., and other indirect costs.)

C H A P T E R 3

Material Resources and Analysis

Knowledge of materials with its engineering, construction, and cost effects requires alertness by owners, contractors, and subcontractors. Successful bidding and earning of profit are coupled closely to effective use of material resources. The student will confront these issues often in practice and will need to apply the principles presented in this chapter.

This chapter is concerned with understanding the types of material, finding their description and quantity, and selecting cost. Together with Chapter 2 it provides the labor and material underpinnings for later discussion of cost accounting, estimating, bidding, and cost analysis.

3.1 OVERVIEW

A broad review of materials is necessary before we consider their refinements. Materials are classified into commodity, engineering, semiengineering, and normative.

Commodities are traded on the commodities exchanges and the price is volatile. Examples are iron ores, foodstuffs, precious metals, and timber. Methods for forecasting the costs of commodities are seldom required for construction.

Engineering materials are commodities that have undergone substantial manufacturing processing (e.g., iron ore to pig iron to hot-rolled and cleaned I-beams). The engineering design may specify CA-610, C1020, M type HSS, or A36 material. The terminology is not the only aspect of materials that may be complicated. For instance, an I-beam has a variety of manufacturing costs—for the melted metal, molding into ingots, rolling, cleaning and pickling, heat treatment, and so forth. Further, to fully understand an I-beam, the engineer must be acquainted with the material and hot/cold-rolling specifications, the inspection requirements, and the dimensions. Knowledge of the physical properties, such as tensile strength, yield strength, and elongation and chemical composition, and so on, must be clarified. It is not, however, the I-beam process costs but the product cost that is needed for construction.

Semiengineering materials, insofar as their cost is concerned, behave as either engineering or commodity materials. Copper, for example, is a semiengineering material, because occasionally its price is steady, like that of an engineering material, and at other times the price flip-flops and follows the roller coaster of commodity prices.

Examples of *normative* materials are crude oil and certain metals, for which the price is normalized or fixed by various governments or cartels.

In this book we are concerned only with engineering materials, because they are used in construction and have reasonable price stability.

Information about engineering materials is so vast that we refer the inquisitive student to other texts. In this book we discuss classification, take-off of material, losses, and the cost of the material in a ready-to-use condition.

We define *materials* as the substances being altered. This may be the iron ore, coke, oxygen, and limestone to a blast-furnace industry for producing pig iron; ladles of molten iron to the basic oxygen furnace; steel ingots to a steel rolling mill for refining; and the rolling of the bloom into I beams. The scope of what constitutes materials depends on the situation. Materials have been purchased, not manufactured or constructed, by the business that adds value to them. Thus an A36 steel I-beam is a product from a rolling mill but is a material to a steel warehouse, which inventories the I-beam for sale to a steel contractor specializing in structural subcontracting.

Besides the drawings, engineering/architect design documents are the engineering bill of material and specifications. An engineers' bill of material or parts list accompanies the blueprints and consists of an itemized list of the materials for a design. The list may be prepared on a separate sheet or may be printed directly on the drawing. Frequently, it is downloadable as a file from the computer system. For example, a parts list contains the part numbers or symbols, a descriptive title of each item, quantity, material, and other information such as number, stock size of the materials, and weight or volume of materials.

There are several CAD practices in those design documents. Material details are listed in general order of size or importance or with the special-design parts first and the standard parts last. The list continues from the bottom upward on the drawings, so that new items may be added later.

In a project where the contract must be executed, the materials must be bought quickly to have them available when needed at the job site. In some cases the engineering department determines all the materials and job requirements. Inasmuch as the engineers are conversant with the drawings and specifications and are already familiar with the design and contract, they begin the job of calling out material requirements. Quantity take-off must be complete and include all descriptions necessary in order to estimate and obtain (if necessary) the materials. In more advanced systems, the materials are ordered through electronic-commerce methods on the internet.

Specifications are considered apart from the bill of materials and the engineering drawings. In construction, specifications are specific statements as to construction requirements. In manufacturing, specifications relate to performances, or materials, or special requirements. Basically, the specifications state the technical details about the item. Specifications are used for the proposed work so that bids may be compiled. Specifications are used as a guide or a book of rules during construction and manufacturing. Finally, specifications are legal documents to the construction industry.

Along with the specification of the material, it is important that correct quantities be known, such as the dimensional units, weight, volume, and so on. To have a correct cost, such as $2700 per ton of I-beam material, and then improperly specify the number of tons is a serious goof. Finding those correct quantities is what "taking off" is about, and we discuss that shortly. In construction, one type of a contract is called *unit price*. It is based on the placement of certain well-defined materials. Imagine a contract where the number of tons of steel in a building depends firstly on the cost per ton of steel placed. Then accuracy of the quantities of the take off is critical, as the costs of labor and the mobilization for the equipment attach to the quantity for the material. The bid is based on the per ton of steel placed in the structure.

The estimate of materials involves extensive calculation to include allowances for waste, short ends, or losses. After these calculations are completed, often with the aid of software or data catalogs on weights, allowances, and the like, costs are determined. The types of losses involved are discussed later in this chapter.

For the purpose of construction-cost analysis, materials are divided into two categories

- Direct materials
- Indirect materials

Direct materials are important to construction costs. Our initial definition is not particularly precise, but we posit that direct materials appear in the design and can be "touched." Direct material cost is the cost of material used in the design. This includes the innumerable materials that are found in buildings, dams, highways, industrial plants, and homes—from the complex to the simple. The nuances of the definition depend partly on the application, whether by an owner, contractor, or subcontractor. Those distinctions are covered in Chapters 7 and 8. *Direct material cost* includes costs of delivery, storage, sales and other taxes, and losses incurred in the construction.

The cost should be significant enough to warrant the effort of estimating it. Some direct material, because of the difficulty of estimating, may be analyzed as indirect, i.e., evaluated as overhead, which is the method used for indirect materials. For accuracy, direct costs are preferred to overhead methods of analysis.

The major portion of the quantity take-off deals with direct materials. Direct materials are further subdivided into

- Raw
- Standard commercial or bulk
- Subcontract

Raw materials include constructed, fabricated, or processed materials in a condition that will receive direct labor work, which converts it to another state of the design. For worked material it is necessary to have designs and planning and then to add direct labor.

Standard commercial materials are a class of materials normally not converted. Rather they are accepted in a ready state (as is, for example, concrete used by a subcontractor). Standard commercial materials may be a significant proportion of total material cost and are considered separate from constructed raw materials. Sometimes these are called *bulk* materials. Standard materials may be charged with out-of-pocket expenses for procurement, freight, receiving, handling, inspection, and installation.

Standard commercial parts are costed by either engineering or buying. Engineering may provide a bill-of-materials listing of the standard parts to the buying department, which will price the parts from catalogs or from quotations or using software and the internet, and return the information to engineering. Sometimes when a short lead time for the bid prevails, estimating may compile the standard part costs. But information from catalogs or the internet can be faulty, what with frequent price changes and negotiated price-volume breaks between the company's buying agents and seller.

For some jobs incidental standard materials are estimated as a percentage correlated to the direct materials; for example, nails relating to board feet of lumber. This happens whenever the standard purchased materials are insignificant.

Subcontract items are assemblies, intermediate materials, or equipment produced by a supplier according to designs, specifications, or directions applicable only to the design being estimated. Subcontract materials are discussed in greater detail in Chapters 7 and 8.

If direct materials are touchable, then it would follow that a contrasting logic is necessary for indirect materials. We say that *indirect materials* are "not touchable as they do not appear in the end item." Examples of indirect materials are brooms, ladders, and so on.

Some materials can be classified as either indirect or direct. Convenience of the costing dictates whether it is simpler to organize some material costs as direct or indirect. Indirect materials are then costed by overhead practices.

There are exceptions to all cost definitions. "Wooden forms" for concrete forming, while they do not appear in the end item, are classified for the sake of the method of work as a direct material.

Qualitatively the cost for direct material is found by using

$$\text{cost of direct material} = \text{quantity take-off} \times \text{cost/quantity} - \text{salvage value} \qquad (3.1)$$

Cost finding includes historical and measured information. In some cases, material costs are uncovered by using company records, though for other materials the engineer must seek out the basic costs. In other cases, suppliers will be called for this information. Catalogs, internet, and subcontractors divulge cost information, too.

Equation (3.1) is straightforward, except that accuracy in finding the direct material cost depends on quality of the components in the equation. *Salvage* is a recovered material having a credit or debit applied against the direct material cost of the design that is being "taken off."

Several questions arise concerning the practice of uncovering direct material cost. When are the values assessed? Are there practices that differ between subcontractors, contractors, suppliers, and owners? What effect do losses have on direct material cost? What is the tolerance of the cost, if any? Because of the great number of materials, is there a way to organize this task? These questions are answered in the next several sections.

3.2 QUANTITY TAKE-OFF

Determination of the construction-material quantity is divided into three parts:

- Take-off
- Shape and unit of measure
- Contemporary practice

Shape implies mass or area or length or count, or one of the many engineering dimensional units. "Count" means the number of discrete units—for instance 23 studs of a particular size. Equations for the mensuration of special shapes are found in the end papers of this text. Simple geometry is the starting point for finding shape. But contemporary CAD software is helpful here, as many programs are able to give shape amounts.

The *unit of measure* is the dimensional unit that is expected with the pricing information. For instance, if structural steel is the object, and if the pricing information will be quoted in dollars per ton by the steel supplier, then the customary English "tons" are desired. If shape is pounds, then $/lbm is the appropriate cost rate. The cost per unit shape is

PICTURE LESSON
Concrete Arch Bridge

The picture, dated 1866, shows spectators standing by the Stockton & Darlington Railway's Hownes Gill Viaduct as a train passes over it. A wagon way alongside is under construction. The viaduct was opened in 1858. The picture shows the bridge's fine aesthetic appearance. Arguably, the bridge material was concrete only and did not include reinforcing iron bars imbedded in it.

Before the early periods of increased demand by the railroads, bridges would carry horse-drawn roadway traffic only. With the advent of railways, bridges were required to carry much heavier and more concentrated loads. Decks sagged, repeated rhythmic vibrations were troublesome for the early bridge engineers, and some bridges failed.

Portland cement, the basis of today's concrete, was invented in 1824. The underlying idea is that concrete takes compression, and steel reinforcing is incorporated to resist tension stresses that concrete cannot withstand. Concrete is well adapted to construction of beams and arch bridges. Arch-type bridges were more rigid and better suited to heavy railway traffic than suspension bridges. The foundation supports the weight of the bridge and the traffic on it, but the arch (as in suspension bridges, too) calls on the ground and recognizes that there are thrusts outward on the abutments. Thus, an arch is in a state of compression and thrusting outward. This type relies on the ground to resist a horizontal force. Arches may be semicircular, elliptical, or pointed.

in compatible dimensions. Finding this shape quantity is called *taking off,* a popular and useful term in construction-cost analysis.

The bidder determines the amount or count of the material during material take-off as required by the design. Essentially, the design gives the theoretical amount. The take-off could be a bill of materials, or it could be a separate listing indicating material specification, size, weight, length, and so on. Not all drawings provide a listing of the materials that is separately identified in columnar format. Very frequently the only indication is a drawing note leading to the item—for example, "No. 6 steel." Contemporary CAD software indicates the amount in a variety of ways. Sometimes the designer will provide the standard binary or computer files to bidders for the downloading of the quantities.

The preparation of the contractor's estimate is an important step in identifying, through interpretation of the drawings and specifications, the direct labor work, along with the direct material, that will be required. This step is called the *quantity take-off* or quantity survey. This coupling between the material and direct labor work is described in Chapters 6 and 7, which introduces the unit quantity model.

Sometimes these take-off quantities are termed a *quantity survey*, and the person is called a *quantity surveyor*, a term found in Europe and Canada. In those countries the quantity surveyor is a licensed professional and is frequently in the employ of the owner, architect, or consulting office. When the owner releases the drawings and quantities for bidding, the contractors are using identical quantities, which is unlike the situation in the United States where the quantity surveyor analog is not found[1] and where, if the bid is unsuccessful, the quantity survey is wasted effort.

Computer-aided design and drafting[2] is a normal part of take-off of materials. Further, the computerized software gives dimensions, even though they may be unstated on the drawing. The advantages are too numerous to repeat here, and the reader is aware of them. But what might be unknown to students living in the twenty-first century is that once upon a time a large effort was necessary to find the take-off dimensions and count and the identity of the materials, because that information was not readily available on the drawings. "Blueprint reading" and the engineer's triangular scale were a big part of taking off dimensions.[3] Quantity take-off that in those days required weeks is now done in minutes. Then a two-dimensional orthographic view (but with three views: front, top, and end) was the model and rolls of drawings were common for the take-off procedure. A hand-held planimeter was used to measure lengths by moving it along a line on the drawing, or dimensional stackups would be found and differences noted for the specialized components.

Now, besides the standard orthographic views, the monitor shows three-dimensional isometric or virtual reality with dimensions. The planimeter is a software feature hooked to the mouse, and lengths, areas, mass, and the take-off shapes are retrieved. Other software enhancements for quantity take-off are available, which aid the speedy process.

Engineer/architects may provide quantities, but seldom do they assume responsibility for the accuracy or completeness. If they do provide this information, it is with "payment scheduling" in mind. Prime contractors, when they perform take-offs, do it with the expectation that their work force is installing the quantities. Nor do general contractors take

---

[1] With the possible exception of some highway departments, where the practice is found.

[2] This is known as CAD or CADD, where "design" or "drafting" is implied. In any case the advantages remain the same irrespective of the mnemonic letters.

[3] Ask a senior citizen who was involved in the construction field about drawings and the use of triangular scales and slide rules to find these "take-off measures." There is a history here that should not be neglected in your professional growth.

off all quantities; instead they rely on the subcontractors to perform this take-off. Reinforcing and structural steel are examples.

There is a system approach to taking off quantities. Procedures, accuracy, consistency, and knowledge of construction and materials are considerations in this work. The student will want to think how these objectives are met. Various take-off forms exist, but paperless cost analysis and estimating use the monitor and intranet for this activity. We will bypass the routine showing of the many paper forms for this work.

It is important to realize that the quantity take-off will couple to construction work. There is a linkage between direct material and direct labor.

3.3   MATERIAL CALCULATIONS

Other steps follow the material take-off described above. First, the amounts are increased for losses that occur in delivery and construction. It is said that more material must be purchased than is used. Losses are lumped into:

- Waste
- Shrinkage
- Scrap

There is *waste* in most construction jobs. A simple picture illustrates. When concrete is ordered, a small surplus is added to take care of possible shortages. In wood framing, joists, studs, and plywood are bought in standard sizes, but they are cut to fit where the drop off is waste. Standard sizes often are the reason for waste. For an 11-ft room where a 12-ft piece is the only full stock available, waste = 9%. Precutting to exact length by a supplier may be a more costly alternate.

The list of places where waste is found is very long. Exceptions may be such things as windows for homes, precast concrete panels, structural steel, and so on, where there may be no waste.

Shrinkage is the loss of materials because of theft or physical deterioration. Originally, shrinkage dealt with volumetric reduction of lumber owing to drying, but now we mean the economic effects owing to deterioration of materials because of aging, oxidation, rusting, chemical reaction, and natural spoilage.

A polymer raw material may have a limited shelf life, and if it is not molded and cured before the onset of chemical aging, there is economic, not physical, shrinkage. Rusting of ferrous materials is a common form of aging. Food commodities have serious problems with shrinkage losses, both economic and in the reduction of flavor or moisture. Misplacement, vandalism, and theft are other reasons for shrinkage.

Scrap is material that is faulty because of human mistake. Mislocation of drill holes for rivets on a column or girder is a supplier mistake. Mislocation caused by a designer's error is an engineering mistake. Quality requirements may uncover defective installation of work on the job site, and scrap results.

The material take-off gives a "theoretical" amount, and it is increased in the following way.

$$S_a = S_t(1 + L_1 + L_2 + L_3) \tag{3.2}$$

where S_a = actual shape in units of area, board feet, length, mass, volume, count, etc.

          S_t = theoretical shape required for design in units of area, length, etc.

L_1 = loss due to waste, decimal

L_2 = loss due to shrinkage, decimal

L_3 = loss due to scrap, decimal

Those losses are determined as a percentage and added to the theoretical amount. Also, material wastage is found from historical records or known from experience. A great deal of engineering effort is concerned with the reduction or elimination of waste. Despite this effort, 1% to 25% is often appropriate as an allowance for waste.

Once the amount of material is known, the procedures continue in order to find

- Efficiency of material usage
- Cost per unit of material
- Salvage

Having discussed shape, cost per unit shape, and salvage, we can give the cost for direct material as:

$$C_{dm} = S_a C_{ms} - V_s \tag{3.3}$$

where C_{dm} = cost of direct materials, dollars per unit

C_{ms} = material cost in compatible units to S_a, dollar per unit

V_s = salvage cost, dollars per unit

S_a and C_{ms} are stated in compatible dimensional units. In some cases there is a salvage cost, and waste and scrap are sold for a credit to the job. But more often there are residual costs entailed by removal and landfill requirements that add to the material unit cost.

The efficiency of conversion of raw materials is found, such as

$$E_s = \frac{S_t}{S_a} \times 100 \tag{3.4}$$

where E_s = shape yield, percent

If we consider N_p, the number of pieces constructed, then we have total or lot cost, depending on the nature of N_p. The final term in Eq. (3.3) deals with salvage. In many cases, scrap and waste are sold to a junk dealer or returned to the original supplier, who then credits the contractor. In some cases the credit is significant enough to subtract from the total unit cost, though in most construction it is not worth the effort. Some waste and scrap even add to the cost because of disposal problems.

$$E_m = \frac{N_p S_a C_{ms} - N_s V_s}{N_p S_a C_{ms}} \times 100 \tag{3.5}$$

where E_m = material cost yield, percent

N_s = number of salvage units

Measurement of materials includes not only finding dimensions and calculating quantities, but also the description of the material and the work, so that it can be properly reasoned out, identified, and priced. An item's description and dimensions cannot be separated, since they are parts of the same whole—one is meaningless without the other. It is this topic that we consider now.

3.4 PRICING MEASURED QUANTITIES

With the quantities determined, material cost analysis continues to the second step: finding the cost per unit shape. Determining an accurate cost value is not simple, and engineering classification, accounting systems, vendors, and professional, consulting, or association advice can make it bewildering. The key points to consider are:

- Variation of materials
- 80/20 rule
- Computer and database lookup and entry

Pricing of material is different from measuring, as it requires information from several sources. Costs of materials, labor, and equipment are needed. Information from past jobs about material waste, labor productivity, overhead costs, and other cost information and data obtainable by job costing are crucial to ongoing successful future operations.

The material items vary widely, ranging from the ordinary ones to complex and specialized items that have limited application. The specialized items are often subject to wide price variation. Greater care is required in estimating those values. *Engineering equipment*, such as reactors, pipe stills, large heat exchangers, and so on, are specified in detail and are tailor-made for the project. These items are fabricated in shops by vendors who specialize in manufacturing this line of equipment. Perhaps, there are only a few active shops, and the shop load is sensitive to demand, and the shop's price is coupled to this business activity.

Bulk materials include lumber, piping, concrete, steel, electrical wiring, and so on, and these are the ordinary materials for custom engineering or standard designs. Often they are required in large quantities (thus the term bulk) and, since they are openly available on the market, and are standard, and are produced in significant amounts, these materials are sold more or less "off the shelf." Because they are broadly available, their price behavior tends to be more stable.

Price behavior varies from the simple to the sublime. There are prices that are found by a telephone call and prices that are contractually dependent. Instant inquiry methods are straightforward. However, the task is complicated by the massive nature of the data collection and the range of data from accurate and timely to inaccurate and outdated. Further, the task is international as well as local. A typical chemical plant needs something of the order of 10^5 pieces of cost data, a daunting requirement. When geographic location is a consideration, literally millions of pieces of data are helpful. Internet sources,[4] too, can provide useful information.

The popular, but unproved, 80/20[5] rule says that about 80% of the total costs of a project come from about 20% of the number of items in the cost estimate. Therefore, it seems reasonable that only the costs of the critical minority of major items need to be based on current quotations and about 80% can be taken from the estimator's database. The 20% of

---

[4] The student may want to visit *www.ThomasRegister.com*, *www.Sweets.com*, or *www.HomeDepot.com*, for catalogs of information. These commercial sources have hot links to the supplier and allow making orders for purchase.

[5] Attributed to Vilfredo Pareto, 1848–1923, an Italian economist, this ratio is called Pareto's law.

PICTURE LESSON
Eiffel Tower

The Eiffel Tower, completed in 1889 to commemorate the centennial of the French Revolution, uses wrought iron, the ferrous material of the industrial revolution. Designed by Gustave Eiffel, a structural engineer and bridge builder, this plan was selected from 700 other designs. Eiffel also designed the iron skeleton of the Statue of Liberty.

This ferrous material contains less than 0.1% carbon, but it has 1%–3% finely divided slag distributed uniformly throughout the metal. In a photomicrograph of wrought iron, the slag is seen as long streaks or globs and is separate from the iron. It is called wrought because it starts out as a red-hot high-temperature pasty mass containing slag from the molten pig iron and is hammered (or "wrought") and mixed until the slag is uniformly distributed. The mass is then rolled or hammered to the desired shape.

Wrought iron during the 1830s was being commercially produced as plate in good quality to satisfy demand for steam boilers, among other applications. It was superseding cast iron for structural work. The newly improved material was malleable and ductile and much stronger in tension than cast iron. Wrought iron has advantages of resistance to corrosion, high ductility, and an ability to hold protective coatings. Wrought iron preceded the production of construction steel by more than 100 years.

Sections were prefabricated off site, a novel development that Eiffel used for his bridges. Moreover, the sections could be riveted instead of having to be bolted. Following the raising by cranes, the section's rivet holes had to align where placed, and some 2.5 million rivets locked 12,000 pieces together. Hydraulic jacks inside the base of each of the four legs allowed raising or lowering of the structure for minute alignment.

the items will need detailed analysis. Another unproved fable indicates that an error margin of $\pm 7\%$ is about the practical accuracy of estimates. Firm prices for minor items are not usually sought until a bid becomes a contract. Consider a simple situation—cost of concrete. In a market that is competitive, a cubic yard may cost $73.50, and if this is entered in the spreadsheet, then at the time of the work that cost will be certainly something else, unless a contract is agreed to between the seller and buyer at the time of the estimate and quote.

There are basically two approaches to the dilemma—(1) gather the specific information as the cost analysis is being prepared, or (2) use older data and refresh it as necessary with the aid of indexes or other adjustments. This older information is sometimes called "standard" or "base" cost. For simple construction, cost encyclopedias are available, which collect and divulge the data. For more complex designs, older data are preserved in some fashion. Feedback of historical information is the key to this step.

Technical specifications of construction materials differ from those of manufacturing materials (e.g., AISI 1035 vs A36). Further, a construction firm will not normally construct a design from a replenished inventory, though a manufacturing firm will reorders materials on a periodic basis to restock supplies. A contractor will not order materials until a particular bid is accepted. But these differences in classification and material ordering policies are superficial for cost-analysis purposes.

Materials informally bought or items that are widely sold and are produced by a number of manufacturers have an ongoing market price. In this case market competition establishes reasonable prices. Engineers use those market values as they pick a cost.

The "quotation cost" contract is the most widely used means of obtaining costs for specialized items. The price of the material is established by a supplier. The delivery price is considered fixed, subject to guarantees of the mutually-agreed-to contract. Quotations can be solicited orally if insufficient time is available, or if the purchases are relatively small in value. Written solicitations should generally be tendered where special specifications are involved, where a large quantity is included in a single proposed procurement, or when telephone quotations are not considered economical or legal. Involved in the wording of the tender quotation are statements about the design and specifications, terms or conditions, delivery date, and price. The obligation to contract at fair and reasonable prices does not diminish as we move down the scale from multimillion-dollar for system acquisition to the dime and quarter prices for nuts and bolts.

The "quotation or price in effect" (QPE) contract is a collaborative legal agreement between the buyer and seller. As usually established, the contract allows for adjustment to the original price should the seller incur material costs more than those promised. This is called an *escalator* [6] clause. If the seller's material costs fall below those which were estimated in the contract, then the buyer agrees to the original price and adjustment is unnecessary. If at the time of delivery of the design, the seller can prove that the material costs incurred to the seller escalated above the original estimated value, then the buyer will make up the difference by use of a formula or through negotiation. The quote or price-in-effect method appears to avoid the troublesome problem of making accurate material analysis and forecasts. However, competition may not allow the luxury of this contract.

Quotation or price-in-effect contracts are used whenever periods of inflation or deflation are important. Big-ticket cost and long-lead-time components and equipment are the things that QPE considers. See Chapter 11 for additional details on QPE contracts.

If the bid is accepted and a contract signed, then there is the possibility that the contractor costs are not completed until at least the main work activities are planned. For

---

[6] Another description of escalation defines it as the ratio of completed cost to the low bid value.

example, the total costs of the form work may be as much as a third of the total costs of a concrete structure. It may not be justified in the estimating phase to choose the amount of the form work material or the number of times it can be recycled until job operations are planned, and this is such that the original estimate, while not invalidated in total, will need a review to assess how those costs were handled.

Engineering needs assistance in performing estimates and cost analysis. A *purchasing associate* often has to obtain prices of materials, equipment, and services; however, this may relate to a few major items whose costs are most significant. A purchasing associate may have buyers and expediters as assistants Once the contract is signed, it is an *expediter* who is responsible for timely procurement and delivery of materials, supplies, and equipment for the company's work and for its operation. In larger companies, purchasing may be a separate department. The essential functions include expediting to assure that materials are on site when required and inventory control of materials purchased. With smaller contractors and subcontractors, a superintendent or foreman, or owner, or contract manager may do purchasing.

Most materials are delivered directly to the job site. Off-site inventory of construction materials is usually discouraged.

Computers[7] handle pricing of measured quantities, resource management, inventory control, and purchasing. Some estimating programs automatically generate bills of materials prepared from the contract documents and the estimate. Information for this software may be provided by the software firm on a periodic updated basis, or it may be derived from past data through a job analysis of won work. Most suppliers of estimating software include a cost database, normally of unit prices, for use with the project. The database may use national averages, or it may be based on selected cities. The computer links architectural-engineering-contractor standards on the web. Once connected by the browser to a web site, hot links will launch the web browser to industry information.

Pricing the quantities of the work generally completes the estimate, which then leads to the considerations for the bid, which may lead to the contract. Those topics are covered in later chapters.

3.5 CODING AND SPECIFICATIONS OF MATERIALS AND WORK— A NECESSARY COMPUTER CHASE

The purpose of coding is to identify items of material and work that are used in estimates, cost accounts, and general and job accounting. This is a "paper chase," a topic students shudder at.

Specifications, on the other hand, are additions to the drawings that affect the choice and cost of materials, design, and construction. Specifications are standard test methods, practices, guides, classification, and terminology.

3.5.1 Coding

Many categories of work and materials are involved in project construction. For each project a special estimate of labor, materials, engineering, and so on, is necessary. It makes sense to keep some sort of order and to contrive a scheme that segregates those descriptions and costs. An art of classification, known as *coding,* gives these advantages:

---

[7]See the References for software that is used to assist the pricing of quantity take-off.

PICTURE LESSON
Saltash Bridge

The most famous of the early wrought-iron spans was the Saltash or Royal Albert Bridge built by I. K. Brunel in 1855–59. The picture shows that the bridge carried a single track of railway.

This was Brunel's last and greatest bridge. Although the massive wrought-iron tubes that form its upper chord look strange to us today, it achieved a marked advance in appearance and magnitude. Brunel used prestressing for the Saltash bridge.

The bridge combined both arch and suspension spanning features. A composite span, known as a lenticular truss, combined elements of three major bridge types—arch, suspension, and, on the approaches, beam. Twin tubular arches of wrought iron, each spanning 455 feet, are connected by ends to link chains. There are two main spans and 17 shorter spans.

Each 1060-ton span was fabricated on shore and floated into position by two twin pontoons. The pontoons were partially flooded and towed out as the tide ebbed. The span was then lifted 100 feet into position by hydraulic jacks, the masonry of the piers being built up beneath the truss after each 3-foot lift.

- Estimate preparation
- Cost control and assurance
- Data management

Those codes are the glue between engineering, estimating, buying, scheduling, and cost accounting and controlling. The estimate is more useful if it is integrated with other information. Further, the code is a dictionary for the language of the cost records, and it becomes important in data management. Old data are recorded in some fashion, manipulated, verified, and reissued for new estimates, in a round-robin fashion. Cost forecasting and cost reporting are two sides of the same coin.

While a code's purpose makes overall sense, its preparation needs specific language. For example, a code system needs to satisfy these points:

- Tied to execution of project
- Detailed
- Descriptive
- Encourages project job control

The code is linked to the overall manner in which the job is executed and the way that costs are collected. Most codes are determined after the company studies its construction business and how it pays its bills. Some companies recognize that construction work is project oriented, and to achieve a goal of maximum profit the project must be individually accounted. One project may be a winner while another is losing money.

It is necessary that the code be detailed. The control of costs requires a level of detail to allow the examination of records and permit the influence of engineering to control cost overrun. A caveat, however: detail for detail sake is avoided. A computer chase can be counterproductive.

Misunderstandings are commonplace and perhaps unavoidable (like the occasional burp). However, code books allow broader study for those who are interested. The code description discusses what is included and what is excluded.

Selecting an existing cost-code system or preparing a new one is not to be taken lightly. The nuts and bolts of a cost code are a series of alphabetic, numeric, or alphanumeric symbols, sometimes with dots and dashes thrown into this primordial soup. Commercial coding systems can be bought, and they are compatible with Windows. Specialized trade associations have as their mission the improvement of coding schemes and standard specifications. Many companies develop their own code. In some construction firms, cost systems have a structured sequence corresponding to the order of appearance of the various trades or types of construction processes typical of the company's construction activity.

There is no better source than the experience of past projects for information and data for new projects. Most contracting companies specialize in one kind of work. Often there are similarities, and most building contracts contain the following: clearing and excavating, concrete footings and foundation walls, concrete slabs, concrete or steel columns and beams, steelwork, carpentry, plumbing, and electrical work. The exterior wall in one project may be of similar specification as that in another, and the material and labor costs may be similar. Some work items resemble those in other projects.

Building contractors are interested in accounts that describe the cost of framing and casting structural concrete as used in building frames. Heavy construction contractors are interested in earth-work-related accounts such as grading, ditching, clearing and grubbing, and machine excavation. The more utilitarian the work, the more it tends to be commonplace.

The codes allow cost data filing, and the engineer is able to draw on facts obtained from other jobs in making a new estimate. Data filing refers to the standup metal files or the hard drives. The codes encourage cost control and prompt reporting of potential overruns or underruns.

Final cost analysis of the job compares actual costs and times spent on work activities with those of the estimate. Job tickets, computer retrieval, and task records are sought.

The project's actual results are applied to correction of productivity and unit prices. This is the information for the new estimate. This experiential information and data, and after some statistical juggling and indexing for future period, will be the grist for future estimates and bids.

A software/computer system is the means for handling these procedures. Often it the field personnel who are able to correct coding entries by their job-site observations, which may be at odds with estimates and other office information. With ubiquitous computer hookups, correction of data is a simple entry. The ongoing renewal of estimates is important.

The coding structure determines the level of detail for the project and the time and effort required to manage the system. Several structures of accounts are possible for a project. One is needed to separate costs and operations, like a work breakdown structure (WBS, see Chapter 8). A second structure can be used to analyze the costs and efficiency of different operations performed on the project. Another structure is used for managing resources and for requisitioning and reporting. These differing possibilities need to be refined into one scheme that meets the several objectives.

For smart business it is important that the project be divided into the same identified units and elementary work classifications. The engineer when preparing the quantity take-off identifies each work and material item by its code designation. Throughout the construction progress, a continuous record is maintained of the actual costs of construction of these work items. From estimate to project completion, the same cost codes and WBS numbers apply.

Even though companies have developed their own coding system, there are North American coding systems. There are agencies that push various approaches, such as the American Road Builders Association, and the Construction Specification Institute, which emphasizes building-oriented accounts.

Those coding structures are arranged in various ways. There are levels that give specificity to the identification. The refinement and detail depend on the number of levels. The more numerous the levels, the greater the detail.

One system for organizing and coding construction work has been in existence since the 1970s: MasterFormat system, developed by the Construction Specification Institute[7] and Construction Specifications Canada (CSC). The code has 16 divisions. See Table 3.1. It should be noted that the CSI has ongoing development of specifications.

The MasterFormat is the most widely adopted standard. It is a closed system and, of its 16 standard divisions, only some are useful to a particular company. It incorporates an

TABLE 3.1   MasterFormat

	Title		*Title*
1	General Requirements	9	Finishes
2	Site Construction	10	Specialties
3	Concrete	11	Equipment
4	Masonry	12	Furnishings
5	Metals	13	Special Construction
6	Wood & Plastic	14	Conveying Systems
7	Thermal & Moisture Protection	15	Mechanical
8	Doors & Windows	16	Electrical

[7] *www.CSINET.org*

TABLE 3.2   Subdivision for Concrete

	Title		Title
03100	Concrete formwork	03400	Precast concrete
03200	Concrete reinforcement	03500	Cementitious decks
03250	Concrete accessories	03600	Grout
03300	Cast-in place concrete	03700	Concrete restoration and cleaning
03370	Concrete curing	03800	Mass concrete

organizational format for project manuals, bidding, and contract forms. It is accepted by the U.S. federal agencies and most state and local governments. Estimating encyclopedias have adopted the system (see Chapter 7).

MasterFormat uses only numbers. MasterFormat also contains broad-scope headings (such as 03300 Cast in place concrete) and subsidiary medium-scope headings that use the last two digits for the final identification (for example, 03310 structural concrete).

Structural concrete is represented by the number 03330 in the MasterFormat code, but because of the wide variety of concrete placing, it is necessary to have more identifications for placing concrete. (How many methods for the placing of concrete can you identify?) But MasterFormat allocates only nine unassigned digits under 03310—structural concrete (from 03311 to 03319). No more numbers may be added. See Table 3.2 for the subdivision for concrete.

There are no widely adopted codes for industrial, heavy engineering, and plant estimating that are comparable to the MasterFormat. Most industrial and plant contractors develop their own specialized database for material and labor costs, which are mostly expressed in unit hours, man months, or man years. Heavy engineering construction estimating for dams, roads, tunnels, and technical plant construction, etc., generally does not adopt the man hour for functional operations that is typically found for building construction. Sometimes, the unit of measure is the man year or man month.

3.5.2   Specifications

Specifications are the written items of work that complement the construction drawings. The drawings show what is to be built, and the specifications describe how the project is to be constructed and what results are expected. Historically, specifications have referred to specific statements concerning technical requirements of the project, such as

- Materials
- Workmanship
- Operating characteristics or performance

Specifications also include the bidding and contract documents.

Material standards are sponsored by a variety of technical societies and the federal government. These specifications have been written by specialists and are sometimes called consensus standards, because there is wide public and technical agreement. Standards of this type are accepted as authoritative.

Materials have quality characteristics that are frequently specified by the American Society for Testing Materials[8] and are routinely understood and incorporated into engineering specifications. These written specification statements can often be downloaded from the web. A good technical or a large general library will have written specifications, and sometimes copies are available as CDs.

ASTM issues numerous specifications pertaining to the physical, chemical, electrical, thermal, performance, acoustical, and other properties of a wide variety of materials, including almost all those used in construction. ASTM standards are employed widely. For example, specifications are used as quality references for Portland cement, reinforcing steel, asphalt, and many other materials.

Workmanship deals with the contractor's operations rather than the materials used. Specifications relate to the quality of this workmanship. Consider the situation where the owner and designer provide plans, drawings, and specifications, and the builder follows those instructions. If a quality fault is found, and the owner demands restitution in some legal way, the question can be asked: "Who has failed, the designer or the builder?" Broadly speaking, if the builder follows defective or inadequate plans and specifications furnished by the owner, architect, or engineer, the builder is not chargeable for damages or loss that can be traced to such defects. Should the contract documents specify exactly how the work is to be done, then the engineer/designer has largely assumed responsibility for securing the desired result. This example illustrates the connection between the drawings and specifications.

Specifications are abundantly found with materials, and there are attempts to write specifications that instruct designers and engineers about criteria for building design. This is an example of a performance specification. As a method, performance specifying means requiring the result rather than prescribing the means of accomplishing it. A material-specification approach is "Construct the building using the following materials and products in the following manner...." A performance specification is stated in the following way: "Design and construct the building to give the following performance...." Product and material characteristics are probably measurable. Even seemingly intangible characteristics, such as durability or maintainability, can be defined with scientific procedures, albeit the cost-effective compliance of the characteristics can be nettlesome to test.

SUMMARY

Discussion in Chapter 2 and 3 has dealt with direct labor and direct material, which constitute the majority of project costs. The direct costs are traceable, those which the contractor considers economically feasible to isolate, e.g., costs of concrete, forming, or laborer's wage, which are segregated by the contractor.

The engineer begins by calculating the final exact quantity or shape required for a design. To this quantity the engineer adds to compensate for losses of waste, scrap, and shrink-

---

[8] There are many national and international standards association which are available on the internet. American Standards and Testing Materials, www.ASTM.org; The American National Standards Institute, www.ANSI.org; The American Association of State Highway and Transportation Officials, www.AASHTO.org; American Concrete Institute, www.ACI-INT.org; The American Institute of Steel Construction, www.AISC.org; the American Water Works Association, www.AWWA.org; the American Welding Society, www.AWS.org; American Concrete Pavement Association, www.PAVEMENT.com, and The Construction Specification Institute, www.CSINET.org.

age. It is possible to find the direct material cost, once the cost of the material is referenced to a shape dimension. Contractual arrangements affect the method by which the material cost rate is found. Pricing of the take-off is the initial step of cost analysis.

QUESTIONS

1. Give an explanation about the following terms:

Commodities	Take-off
Engineering materials	Indirect materials
Direct materials	Bulk materials
Waste	Efficiency of conversion
Escalation	Coding
MasterFormat	Specifications

2. Discuss the advantages and disadvantages of each contractor doing a material take-off.
3. Why bother to classify construction information?
4. State and explain the principles for the measurement of materials, using examples.
5. Devise a cost code suitable for cast-in-place concrete, form work, and concrete reinforcement.
6. Explain the point, "A quantity survey interprets the drawing."
7. Give examples of direct and indirect materials.
8. What is the most important loss in construction? How should this loss be considered if you were the cost analyst? Or the owner?
9. How do you strike a balance between too much or too little coding?
10. Devise a code for recreational construction.
11. Compare the "claims" made by websites that divulge material cost data.

PROBLEMS

3.1 (a) Find the board feet for a job that requires 2500 joists that are 2 × 10 in. × 18 ft.
(b) Also, find the board feet for 14,880 lineal feet of 2 × 4 in.
(*Hint:* One board foot measure is a wood member 1 in. thick by 12 in. wide by 1 ft long. The measure of board foot is defined as shown below.)

$$\text{bft} = \frac{t \times w \times 1}{12}$$

where t = nominal thickness, in.

w = nominal width, in.

l = length, ft

3.2 Find the board feet for 100 ft^2 of partition shown by Fig. P3.2. Studs, plates, and fire stops are 2 × 4 in. (*Hint:* Assume 5% waste.)

	x	y	Fire Stops	Studs	Bottom Plate	Top Plate
(a) 8 ft 4 in.	12 ft	6	7	1	2
(b) 7 ft 2 in.	14 ft	8	9	1	2

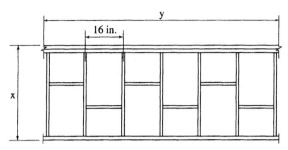

Figure P3.2

3.3 Find the cost for the 8-ft partition shown in Fig P3.3. (*Hints*: Let the cost be $1.50/bft. Include 5% waste. Work the problem only to the width of the king stud.)

Window Size, A	Header Size	Stud
(a) 8/0 × 3/0	4 × 8 in.	2 × 4 in.
(b) 8/0 × 5/0	4 × 10 in.	2 × 6 in.

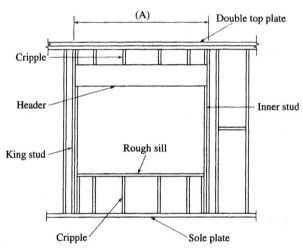

Figure P3.3

3.4 A contractor has a choice to buy precut lumber or buy standard lengths and use carpenters who will make cuts. The project requires 36,000 bft, and there is a waste of 4%. Standard lumber costs $1.50 per bft. Precut lumber costs 20% more. Labor time for cutting is 0.1 man hour per cut. Gross hourly wage cost for carpenters is $25/hr. The number of pieces and the cuts are given as

No. of Pieces Required	No. of Cuts
3000	0
2000	1
1000	2

Find the preferred method— precut or standard length of lumber.

3.5 Find the material quantities and the total cost for a foundation that has a centerline distance of 392 ft. What is the cost per 100 ft? Examine Fig. P3.5 for information. The schedule for the materials is given as

Material	Specification	Frequency of Use	Waste, %	Rate	Unit Cost
Concrete	5.5 sack Portland cement	1/1	4		$68.50/yd³
Forms	2 × 12 in. × length	3/1	7		$150/100 bft
Stakes	Oak, 2 × 2 × 24 in.	4/1	15	4 ft O.C.	$0.65/ea
Spreader ties	1 × 3 in. × 3 ft	2/1	16	4 ft O.C.	$150/100 bft
Nails	2½ in. common	1/1	10	1 lb/100 bft	$1.00/lb

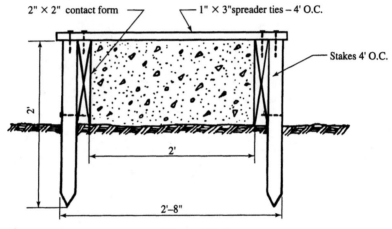

Figure P3.5

3.6 Determine the quantity of excavation and the amount of material required for back filling using Fig. P3.6. How much excess of excavated material is removed from the site if the swell factor is 15%? (*Hint:* The centerline length of a rectangular residential footing is 165 ft.)

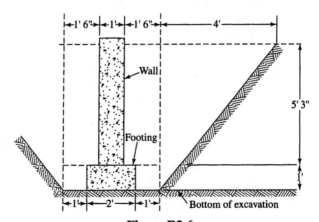

Figure P3.6

3.7 A foundation plan for a residence is given by Fig. P3.7. Find the volume in yd³ if waste is assumed to be 1%. Discuss various methods for doing these volumetric calculations. The cost of 4000-psi concrete is $78.50/yd³. What is the bid? (*Hints*: Dimensions are in ft. In doing any "mensuration" calculations for construction quantities, an important purpose is to minimize the errors of simple arithmetic, and outside of having CAD provide the yd³, many of quantity take-off procedures require a calculator. There are rules of arithmetic, even for simple problems.)

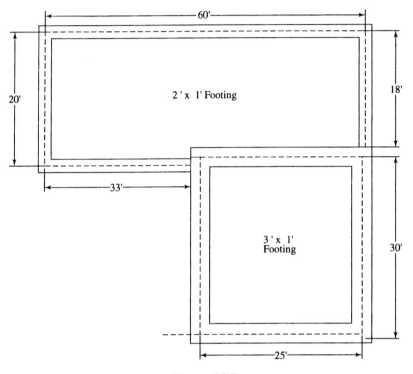

Figure P3.7

3.8 Find the material requirement and cost for the concrete step design shown as Fig. P3.8. There are 10 risers. (*Hints*: Cost of 3,000 psi concrete is $73.50/yd³. Assume a 5% waste.)

   (a) 4 ft wide.

   (b) 6 ft wide.

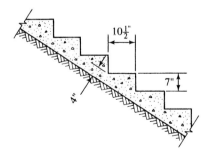

Figure P3.8

3.9 A CAE program determines total requirement of steel reinforcing bars for a concrete design as 17.1 ton. The purchased amount is 17.63 ton. Find the shape yield. If the average price for reinforcing bar is $0.903/lb, and the debit salvage of waste is $0.104/lb, then what is the material cost yield? (*Hint*: Consider the difference as loss and returned for salvage and as a credit to the cost of the subcontractor.)

3.10 A detail sketch of a shop-fabricated truss is given by Fig. P3.10.

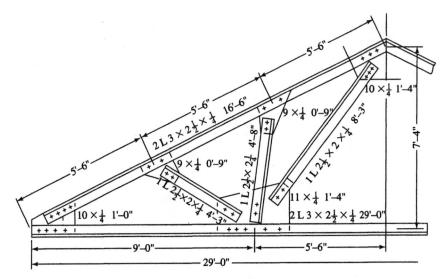

Figure P3.10

A summary bill of material for the steel truss is

Component	Size	lb/ft
Structural steel	$3 \times 2\frac{1}{2} \times \frac{1}{4}$	4.5
	$2\frac{1}{2} \times 2 \times \frac{1}{4}$	3.62
Gussets	Weight, lb	
$10 \times \frac{1}{4}$	8.5	
$11 \times \frac{1}{4}$	9.4	
$9 \times \frac{1}{4}$	7.7	
10×14	8.5	
$4 \times 3 \times 5/16$	7.2	
$\frac{1}{4}$ in. O.D. rivet		

The costs for finished and fabricated trusses are given as structural steel, $10/lb; gussets $8/lb; and installed rivets, $5 each. Flatbed truck delivery to the job site is $1000. Find the total cost for the truss.

3.11 A simplified diagram of a shop welded and fabricated A36 steel truss is given by Fig. P3.11. A bill of materials is provided as

Item	lb/ft
$3\frac{1}{2} \times 3 \times \frac{1}{4}$	5.4
$4 \times 4 \times 5/16$	8.2
$2\frac{1}{2} \times 2 \times \frac{1}{4}$	3.6
$2\frac{1}{2} \times 2 \times \frac{1}{4}$	4.1

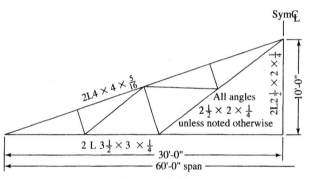

Figure P3.11

Other details, such as gussets and plates, add 20% to the weight of the truss. The costs for finished and fabricated trusses are given as structural steel material, $10/lb; and the material and welding of gussets as $15/lb. Delivery to the job is $1000. Find the cost of the fabricated truss. (*Hint*: Scale any unknown lengths by comparing to known lengths.)

3.12 Various formulas modeling the weight of ferrous metal trusses have been available for many years.[9] Consider the following approximation:

$$W = \frac{P}{45}\left(1 + \frac{S}{5\sqrt{A}}\right)SA$$

where W = weight of truss of horizontal projection of roof portion supported by one truss, lb per ft^2

S = span, ft

P = truss loading, lb per ft^2

Reconsider the design in Prob. 3.11, and let $P = 60$ lb/ft^2 and $A = 10$ ft. The cost for finished and fabricated trusses is given as structural steel material, $10/lb. Transport cost of the truss to the project location is $1000. Find the cost of a purchased truss delivered to the job site. Discuss the change of truss design over the years, and pay attention to the improvement in material properties.

3.13 Determine the direct material cost for the steel structure given by Fig. P3.13. The cost per ton is $1350.

---

[9] Of interest to the author is one by Professor Milo Ketchum, civil engineer and Dean of the College of Engineering, University of Colorado, during the 1910s period, and for whom a campus building is named.

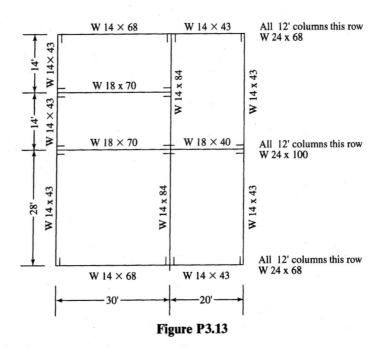

Figure P3.13

3.14 A central steam power plant for 320-MW electrical load is designed. Among the significant hardware systems is steam piping. The major high-pressure and high-temperature piping from the boiler header outlet to steam turbine inlet is 1027 ft long, as given by a CAD take-off. The pipe is either prefabricated in a pipe shop or field welded. Previous history for the installed cost of the material from a pipe fabricator/subcontractor indicates the following:

Element	Pipe Shop	Field Fabricated
Waste, %	4.2	11.6
Shrinkage, %	0.2	1.3
Finished cost	$711.25/ft	$1136.43/ft
Percentage of CAD length	69%	31%

What actual amount of pipe is necessary for the electric plant? Find the direct material cost for the pipe.

3.15 Direct material elements of chemical process plants are often given by *cost estimating relationships*. These CERs are developed by statistical methods. For example, consider a CER for a coal-fired central-steam electric plant, where the cost of screw conveyors for the handling of crushed bituminous coal from the ball mills to the boiler is estimated. The direct material cost for a 12-in. internal-diameter conveyer is given by the linear CER:

$$C = 224{,}250 + 2034L$$

where C = cost of screw-conveyor handling system, dollars

L = length of conveyor, ft

The conveyor runs from the discharge exit of the ball mill to the boiler inlet auger and is 717 ft long. Find the direct material cost for the conveyor. What is the percentage of fixed cost?

3.16 Direct material equipment of process plants is often estimated by log-log curves where a family of materials is available for the same general design. Notice Fig. P3.16 for estimates of horizontal-tube heat exchangers, where there is a choice of tube material ranging from high-carbon cold-rolled steel to Monel metal. The entry *cost driver* is area of the tube surface.

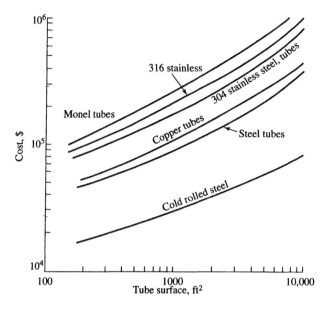

Figure P3.16

A design flow sheet specifies the area of a tube heat exchanger for a fluids-processing plant as 1250 ft². What is the cost ratio in "orders of magnitude" of Monel to cold-rolled steel material, if the only difference in the heat-exchanger design is the tube material?

3.17 Early estimates are used for *budget definition* of indirect material cost in construction of chemical process plants. Notice Fig. P 3.17, which is an example of published information. This is an *upper and lower limit chart*. The cost driver is steam capacity, lb/hr, and the *y*-axis dependent variable is plant indirect material cost. The centerline is the best value, but upper and lower numbers are suggested, which aid early budgeting for maximum and minimum values. Find the upper, likely, and lower limit indirect material cost for capacities of 50,000 and 3,000,000 lb/hr of steam flow.

   The compiler of the graph shows parallel lines that bound the centerline. What is the approximate tolerance of the boundaries at the lower and upper ends of the graph? Discuss the nature of these tolerances.

3.18 Pressure vessels for chemical process plants may be field constructed, in addition to the usual practice of prefabrication. Early estimating of vessels use expressions for the shell weight, given by

$$t = \frac{PD_m}{2SE} + C_a$$

where t = wall thickness of vessel, in.

   P = pressure, psi

   D_m = average of inside and outside vessel diameter, in.

   E = joint efficiency, dimensionless decimal

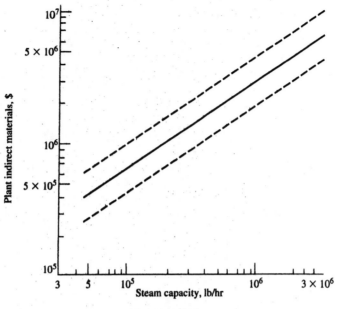

Figure P3.17

S = allowable maximum working stress, psi

C_a = corrosion allowance for additional wall thickness, in.

$$\text{Vessel cost} = K\rho CtL$$

where ρ = material density, lb/in.^{3}

K = material, dollars per lb

C = mean circumferance, in.

L = vessel length, in.

Find the cost of a vessel. The following information is determined from engineering flow-sheet calculations: C_a = 1/16 in., S = 10,000 psi, D_m = 220 in., E = 1 for a welded structure, P = 25 psi, ρ = 0.265 lb/in.^{3}, K = \$2.45/lb for AISI 6120 chromium-vanadium steel that is rolled to the vessel diameter, and L = 18 ft. For this type of field welding, a waste allowance of 20% is anticipated. Find the direct material cost of the shell and ends for this vessel. (*Hints*: The top and bottom ends will be likely elliptical in shape. This material cost excludes welding fabrication and erection labor.)

PRACTICAL APPLICATION

This module continues the internet development for construction-cost analysis and estimating. The application involves *ftp*, or *file transfer protocol*, which means to retrieve a file. The essential commands are "open"—to connect to a remote computer; "get"—to retrieve a file from the computer; and "bye"—terminate the connection. And the immediate purpose of this lesson is to obtain information that your team uses in a construction-cost study. The study can focus on a variety of objectives. For example, there may be current information that aids the cost analysis of a job that the team proposes to evaluate during the academic term. Indexes from commercial or government sources can be beneficial in a study. Current information that guides and colors judgment may be obtained.

Chat rooms or bulletin boards that enhance the location and divulging and sharing of information can be a helpful asset. The possibilities are limitless.

Free downloadable software, such as WS_ftp, allows you to use the windows graphic interface to obtain the files.

Your instructor will aid the definition of the study, its subject, size, and focus.

CASE STUDY: CHEMICAL PROCESS PLANT

A *budget* estimate is required for a pipe still design using high-pressure high-temperature steam flow as the medium. Figure C3.1A gives a simplified sketch of the plant. A pipe still is a first-stage key-process plant in a refinery that distills the crude oil into heating oil, gasoline, and so on. A variety of correlated data give a *rough order-of-magnitude* (ROM) investment estimate of the plant cost. The steam rate is 2×10^5 lb/hr, and the heat exchanger has 300 ft^2 of corrosion-resistant austenitic 304 stainless steel tube area. Find the direct and indirect material cost for the plant. (*Hint*: Use Fig. C3.1B and other nearby graphs for information.)

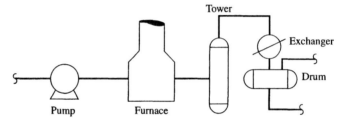

Figure C3.1A

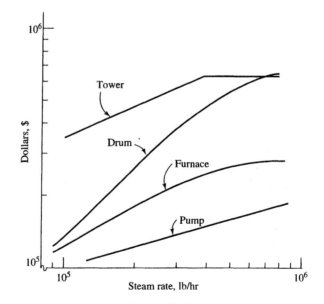

Figure C3.1B

CHAPTER 4

Accounting Analysis

Accounting is the means of analyzing the money transactions of business. Accountants prepare balance sheets, statements of income, and information to aid the control of cost, and these facts are essential for the determination of overhead. Although cost accounting data may have been carefully collected and arranged to suit the primary purposes of accounting, these raw data are incompatible with construction analysis, estimating, and pricing.

Accounting specialties are general, public, auditing, tax, government, and cost. Cost and tax accounting are more important to engineering than are the others. Cost accounting emphasizes accounting for costs, particularly the cost of using productive assets. Tax accounting, because of myriad laws and practices, includes the preparation of tax returns and, importantly, the consideration of the tax consequences on business transactions.

The relationship between engineering and accounting is an active one, for both deal with much of the same information. On the one hand, the engineer deals with costs and designs before the spending of money; and on the other, cost accounting records the cash flow facts. Roughly, engineering looks ahead and accounting looks back, and both are necessary for successful construction cost analysis.

This chapter exposes the student to basic accounting fundamentals. Business transactions, the income and balance-sheet statements, and overhead are the significant lessons for learning.

4.1 BUSINESS TRANSACTIONS

A transaction is an exchange of *wants*. Construction work is given or received and a value, right, or service, collectively referred to as wants, is returned.

The transaction is composed of two elements that are reported in a financial record. This duality leads to *double-entry bookkeeping*, a practice several centuries old.[1] Tested and found true, principles of double-entry bookkeeping have changed little, even though the overall growth of business has complicated professional accounting. The essential practices show similarity to the earliest commercial records.

In double-entry bookkeeping the business transactions are collected in records called *accounts*. The simplest form is the T-account. The recording of a business transaction is an

---

[1] The beginning of rational economic analysis is sometimes linked to the work of the Italian Benedetto Cotrugli, who published in 1458 the first known work on double-entry bookkeeping. The revolutionary nature of double-entry bookkeeping paved the way for understanding business transactions aimed at achieving profit, then creating a commingling of commerce and profit.

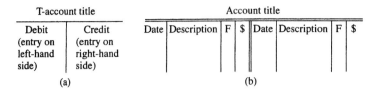

Figure 4.1 Forms for the entry of debit and credit: (a) T-account, and (b) columnar style.

entry, as shown by Fig. 4.1(a). A *T-account* is a graphical presentation of a transaction. An entry on the left side of the account is called a *debit* (Dr.), and an entry on the right side is a *credit* (Cr.). The terms debit and credit, when used for the bookkeeping of transactions, have several meanings. For example, debits increase assets and expense accounts. Credits increase capital, liability, and revenue accounts. Double-entry accounting is basic to balance-sheet and income-statement presentation.

The T-account is used for textbook illustrations. In practice, a more complete description supplies columns for additional data, as shown in Fig. 4.1(b). The columns provide space for the date of each entry, description, folio (F), or a cross reference to another record. If the current status of the account is desired, then it is only necessary to total the debit and credit entries and show the balance on the larger side. This summing and finding the larger amount is called *footing*. This columnar arrangement is suited to computer data processing.

Vouchers, invoices, receipts, bills, sales tickets, checks, and the many documents relating to the transaction, are the evidence for the entry. With this evidence the original entry is made to a journal that contains the chronological record of the transactions. The information is summarized in Fig. 4.2. In practice, the journals are files, computer hard drives, floppy disks, or other media, which we collectively call *journals*. A single journal may suffice for entries for a small contractor or owner. For the larger business, however, many types of journals exist, such as cash, sales, purchase, and general journals. This recording in journals is termed *journalizing*.

The transferring of journal entries to appropriate accounts in a ledger is the next step. The *ledger* is a group of accounts. Perhaps one page of the ledger is used for each account. *Posting* is the term applied to the process of transferring the debit and credit items from the journal to the ledger account. As each item is posted, the number of the ledger account is indicated in the journal folio. Similarly, the journal page number is placed in the folio column of the account of the ledger.

The flow from the original recording of business transactions to the eventual development of the balance sheet and income statement is shown in Fig. 4.2. From time to time, the accountant adjusts the accounting of the business to present the financial situation accurately.

4.2 CONVENTIONS

Accounting uses many important conventions, which the student needs to understand. A *money measurement* fundamental requires that business transactions be recorded only with money. This practice expresses many different situations in common units. A fabricating plant can be compared to a construction company. Moreover, designs denominated in var-

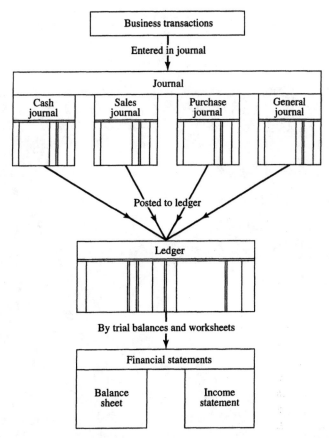

Figure 4.2 Flow of the raw transactions to the balance sheet and the income statement.

ious currencies, such as dollars, euros, or pesos can be algebraically manipulated to guage the company's performance. Because of the importance of the money-measurement fundamental, it is universally adopted, even though it must ignore technology expertise, engineering skill, prominence of design, and contractor reputation. While these are important features, they are difficult to demarcate in money units.

Another fundamental insists that business transactions be recorded in double-entry fashion and that total debits must *equal or balance* total credits. This leads to the important accounting equation, where assets equal liabilities plus owner's equity, or, modeled differently,

$$\text{assets} = \text{liabilities} + \text{owner's equities}$$

$$\text{assets} = \text{liabilities} + \text{net worth} \tag{4.1}$$

Both creditors and owners have claim to equities. The creditor has first rights, leaving the remaining rights to the contractor or owner. Owner's (now meaning owner, contractor, and subcontractor) equities or net worth or proprietorship is the ownership interest in the business assets.

PICTURE LESSON
Thames Tunnel

Tunnels have been used for many centuries. Tunnels under water present special hazards. Whether they pass through silt, sand, gravel, mud, clay, or rock, they entail the risk of caving in and flooding. Floods occurred five times during the 18-year project of the Thames Tunnel. Its construction, late in the nineteenth century, was planned for the purpose of relieving London's traffic. It remained for the London subway system to effectively employ the Thames Tunnel for commercial success.

Marc Isambard Brunel, using an analogy of the mollusk's burrowing technique, employed a basic tunnel shield. This shield made possible the construction of London's Thames Tunnel—called the world's first true soft-ground subaqueous tunnel. Brunel's shield used

12 cast iron frames, each divided into three vertical floors, or cells. The frames were joined together, forming an 80-ton structure that contained 36 cells, each measuring about three by six feet, which was large enough to hold a miner while he excavated. The shield fitted over the tunnel face and temporarily supported the tunnel walls until a brick lining was constructed in place. The shield was advanced by screw jacks.

The production of wrought iron plates for shipbuilding contributed to this tunnel construction. Mechanical machines producing rivets, the simplest of all fasteners, monumentally improved the process. Riveting plays an important role even today. Double-headed red-hot rivets are hammered tight within the hole. Fabricating a clearance hole into a hammered interference fit with the rivet, which fills the hole, is a basic construction process.

Company officers and engineers are standing at the entrance to a large wrought iron structure, which is marked Thames Iron Works and Ship Building Company, Blackwall. Shipbuilding and the production of tunnel structure had things in common. Both ships and tunnel structures would use wrought iron, which could be riveted instead of bolted. The technology of rivets for structural integrity is shown by the close spacing.

Other longer tunnels have been constructed—the Channel Tunnel that goes from England to France, and Japan's 34-mile-long Seikan Tunnel, the world's longest, which was completed in 1988.

A *conservatism* convention encourages the recording of financial data as the lower of possible values. For example, in the evaluation of contractor inventory (materials held by the contractor for sale or construction in the job), we could value the inventory as either the cost to the business for construction or the market value. Conservatism dictates the lower value. In public disclosure for the annual financial report, the statement is made "the cost or market value, whichever is lower" to reflect this convention.

The *consistency* convention states that business transactions, if accounted for in one way, are recorded and accounted for that way in the future. Companies obtain discounts for prompt payments of materials. One firm may use the discounts to reduce the cost of the materials. Another firm may take discounts and record them as income realized on prompt payment. The consistency convention adopts one of those two methods and persists in its use in succeeding periods. This convention discourages a business from manipulating its figures to reflect favorable conditions on one occasion and then, as convenient, changing its approach.

The *going-concern* fundamental implies that the business is operated in a prudent and rational way. This policy, it is assumed, perpetuates the business over an extended period. This going-concern assumption is one justification for the depreciation principles, discussed later.

A *business-entity* fundamental is a simple notion that accounting transactions of a business are for the sake of the business rather than for individuals. If an owner withdraws money from the cash box, then the owner is richer but the business has less. The accountant records the effect of this transaction on the business and ignores the effect on the owner.

A *cost* fundamental recognizes that cost does not necessarily equal value. Though it is possible to list the assets in preparing a balance sheet, their value is determined by several methods. Consider a car for sale. Early in the day a potential customer asks what we will sell it for. Later, a tax assessor asks the same question. Though the answer might honestly differ, the point is that "value" is subjective. Even so, there are two other ways to evaluate business assets. Market value can be obtained, but this depends on the purchaser's needs. It is possible to find several evaluations of the worth of an asset, however, which returns the selection of value to subjective reasoning. We can value assets on the cost of their replacement. A replacement approach leads to a range rather than a single value. Market and replacement value lead to confusing choices, but it is possible to determine the value of the asset as given by the *original* cost. The receipt for the payment of the asset is a document of record and demonstrates cost, albeit not value necessarily. The primary appeal of the cost fundamental is objectivity and expediency. When coupled with the going-business principle, it is assumed that the asset is used to advance the business.

If a *cash basis* is used for accounting, then income is recorded when the cash is received and expense is recognized when cash is paid out. The cash method is used by small contractors and individuals for personal and family records. Cash accounting is not allowed by the Internal Revenue Service for larger contractors or owners.

The *accrual* fundamental is concerned with the majority of business and is generally unfamiliar to us. In accrual, income is recorded when it is earned, whether the payment is received during the period or not. Expenses incurred in earning the income are recorded as expense whether or not payment has been made during that period. The profit-and-loss statement includes incomes earned during the period covered by the statement and expenses concurrent to the same period. Many businesses prefer accrual accounting because it matches revenues to the expenses in a specific period. The time of collection of the income or payment of the expense is not a factor.

Business transactions that affect income are measured by increases or decreases in net worth or services. Moneys received increase net worth and are called revenues. Costs that a business incurs to provide the design, goods, or services decrease net worth and are called expenses. If a transaction is irrelevant to the financial results of the business, then there is discretion as to how and when to record that event. If a piece of paper and pencil were consumed, then, theoretically, the paper and pencil become an expense. However, this penny watching is unwise, because average monthly office-supply expenses are a common-sense way of handling this transaction. A convention of *relevance* permits the accountant to use judgment in handling those expenses.

4.3   CHART OF ACCOUNTS

An account is established for the various business transactions. For instance, a total sales account may lead to a specific customer sales account. Every item of financial information on the balance sheet and income statement has an account. A chart of accounts is necessary for a business. A typical chart or listing of accounts intended for a construction company is provided by Table 4.1.

Formal charts of accounts are grouped into five principal accounts:

- Assets
- Liability
- Net worth

TABLE 4.1   Illustrative chart of accounts for an accrual type of contractor

Assets	*Revenue*

Current Assets
    Notes receivable, customers
    Inventories
    Prepaid insurance, taxes, interest
    Supplies
Fixed assets
    Land
    Buildings, equipment
    Accumulated depreciation
	Sales
	Cash in bank
	Petty cash
	Interest earned
	Dividends received
	Construction costs

Liabilities	*Expenses*

Current liabilities
    Notes payable
    Accrued wages payable
    Accrued interest payable
    Accrued taxes
    Deferred rent income
    Dividends payable
Fixed liabilities
    Mortgage payable
    Bond payable
	Construction costs
	Purchase of materials
	Salaries and wages
	Heat, light, water
	Telephone
	Depreciation
	Direct labor
	Indirect labor
	Insurance
	Repairs and maintenance
	Construction supplies used
	Selling expenses
	    Salaries and commissions
	    Advertising and samples
	General and administrative expenses
	    Salaries
	    Traveling expenses
	    Telephone, postage
	    Supplies

Net Worth	*Taxes, All Kinds*
Capital stock, preferred Capital stock, common Retained earnings	

- Revenue
- Expenses

Assets.   The *assets* of a contractor are those things of dollar value that it owns, such as land, buildings, equipment, or inventory, or intangibles such as trademarks, designs, and patents. Assets may be segregated into current assets, fixed assets, and intangible assets.

Current assets have three inventory accounts representing raw materials, in-process goods, and completed products. Construction work-in-progress inventories, however, are more intangible, in that they generally consist of improvements to someone else's real property, and therefore they are not owned by the construction company.

Fixed assets include office equipment, construction equipment, and buildings, less accumulated depreciation. Land is an asset that does not depreciate. In retail and manufacturing businesses, inventories are owned by the company and are a tangible asset with dollar value. Assets are capable of providing future benefits; otherwise, they are an expense.

As an example of intangible assets, an engineering construction firm designs new methods through its research and development department and obtains patents on the procedures. The cost of research, engineering, and testing leading to the development of a new procedure may be significant and in theory could be treated as an asset. However, under present financial tax accounting, those costs are expenses when incurred. Because many research projects may be underfoot at the same time, cost can be incurred over a period of years. Some firms treat engineering costs as a part of current operating expenses. Patents and copyrights are not recorded as assets if developed within the firm. If they are purchased or acquired through merger, then they are shown as assets.

Liabilities.   The *liabilities* of a contractor are the debts the firm owes. The category of liabilities is frequently broken down into current and long-term debts. Current liabilities are less than or equal to 1 year. Some of the more common items of business liabilities are (1) accounts payable (debts of the firm to creditors for materials and services received), (2) bank loans (amounts the firm owes to banks for money borrowed), and (3) mortgage payable (debt to investors for money loaned to the business on the security of its real estate or equipment).

Net Worth.   The *net worth* of a business is the ownership interest in the firm's net assets. In certain accounting situations, proprietorship and capital are considered to be synonymous with net worth. In a simple case the net worth of a business consists of its capital stock and the retained earnings[2]:

Net Worth	
Capital stock	$45,000
Retained earnings	4,000
Total	$49,000

Broadly speaking, capital stock is the portion of the net worth paid in by the owners. Surplus, another term used within net worth, is that portion that the owners paid for stock, over and above the par value. Retained earnings refer to the accumulated profits and losses of the firm.

A company issues capital stock, which is divided into units of ownership, called *shares*, and the owners of the company are referred to as *shareholders*. The ownership of a shareholder in the net worth of the company is related to the number of shares he or she owns. The surplus of a company increases as the company earns profit and decreases as the company incurs losses or distributes the profits among the shareholders as dividends. If the losses and dividends of a company since inception exceed its profits, then a negative profit or a deficit within retained earnings results.

Income and Expense in Business.   *Revenues* are generated by sales before the deduction of cost. *Expenses* represent costs of doing business. Income is received from the sale of jobs, products, or materials. Expenses include salaries, advertising, power and light, tele-

[2] Generally, this chapter deals with business structures that are incorporated. An "Inc." differs from sale proprietorships or partnerships, which do not issue stock or have retained earnings. These and other differences can be studied in an accounting textbook.

phone, rent, insurance, and interest. The profit-and-loss statement of a contractor is a summary of its incomes and expenses for a stated period. If the statement discloses a net profit or loss, then the change represents an increment or decrement in the retained earnings during the period, arising from the contractor's incomes and expenses, and is carried to the net worth section of the balance sheet.

Gross income is the difference between income and expense. Once taxes are removed, we have *net profits*. Because there are so many types of taxes, our discussion on taxes is simplistic, and we refer to taxes as "all kinds." The interested student is referred to other sources for detailed information.

4.4 STRUCTURE OF ACCOUNTS

An expanded accounting equation with T-accounts is given as

$$\underset{+\,|\,-}{\underline{\text{Assets}}} = \underset{-\,|\,+}{\underline{\text{Liabilities}}} + \underset{-\,|\,+}{\underline{\text{Net Worth}}} + \underset{-\,|\,+}{\underline{\text{Revenue}}} - \underset{+\,|\,-}{\underline{\text{Expenses}}} \qquad (4.2)$$

This is sometimes referred to as the *financial and operating* equation, because it has plus and minus signs for each class of T-accounts. The plus and minus signs show increases and decreases and are summarized as follows:

Debit Indicates:	Credit Indicates:
Asset increase	Asset decrease
Liability decrease	Liability increase
Net worth decrease	Net worth increase
Revenue decrease	Revenue increase
Expense increase	Expense decrease

Now consider an example for General Construction Company. The integrated example flows from the transactions to the noting of the effect on the T-accounts, to a trial balance, and finally to the profit-and-loss and balance-sheet statements. Our simplified approach starts with the initial capitalization of the firm, leading to its first reporting of financial documents.

Each transaction is worded to suggest a dual recording of the transaction. The numbered entry connects to the affected accounts. Consider transaction 1 in Table 4.2. A cash account is an asset account, and capital stock is a net worth account. The effect of the cash on the asset is to increase the T-account by the recording of a debit to the Cash T-account, and net worth is increased with the entry of $50,000 to the credit side of the T-account. Each of the transactions follows the financial and operating equation.

Notice that Table 4.3 is hooked to Table 4.2, showing the T-accounts for the transactions shown in Table 4.2. By using those rules for applying debit and credit, we record the transaction into a T-account. The identity of the account is described further, such as A (= asset). A number of T-accounts are presented in Table 4.3. The numbers in parentheses

TABLE 4.2   Transaction effects for a business called General Construction Co.

Transaction	Accounts Affected	Type of Account	On Account	Effect of Transaction Is Recorded by a Debit of	Effect of Transaction Is Recorded by a Credit of
1. General Construction is incorporated and $50,000 paid for capital stock	Cash	Asset	Increase	$50,000	
	Capital stock	Net worth	Increase		$50,000
2. Business buys materials from S. W. Specthrie on account, $10,000	Inventory	Asset	Increase	10,000	
	Acct payable	Liability	Increase		10,000
3. Pay monthly rent on shop, $1500	Rent	Expense	Increase	1,500	
	Cash	Asset	Decrease		1,500
4. Pay S. W. Specthrie on account, $4,000	Acct. payable	Liability	Decrease	4,000	
	Cash	Asset	Decrease		4,000
5. Sell to P. Hall on account, $15,000	Acct. receivable	Asset	Increase	15,000	
	Sales	Income	Increase	15,000	
6. Pay salaries, $2,850	Salaries	Expense	Increase	2,850	
	Cash	Asset	Decrease		2,850
7. Retire for cash $5,000 of capital stock	Capital stock	Net worth	Decrease	5,000	
	Cash	Asset	Decrease		5,000
8. Collect $2,000 from P. Hall	Cash	Asset	Increase	2,000	
	Acct. receivable	Asset	Decrease		2,000
9. Buy equipment for $3,000 cash	Equipment	Asset	Increase	3,000	
	Cash	Asset	Decrease		3,000
10. Receive $500 rebate on month's rent	Cash	Asset	Increase	500	
	Rent	Expense	Decrease		500
11. P. Hall returns $4,000 of material for credit	Sales	Income	Decrease	4,000	
	Acct. receivable	Asset	Decrease		4,000
12. Pay advertising bill $800	Advertising	Expense	Increase	800	
	Cash	Asset	Decrease		800
13. Buy on credit $12,000 computer from Englewood and Co.	Computer	Asset	Increase	12,000	
	Acct. payable	Liability	Increase		12,000
14. Pay insurance premium, $500	Insurance	Expense	Increase	500	
	Cash	Asset	Decrease		500
15. Take depreciation charge, $600	Depre.	Expense	Increase	600	
	Accum. depre.	Asset contra	Increase		600
16. Pay taxes $1,250	Taxes	Expense	Increase	1,250	
	Cash	Asset	Decrease		1,250
Total				$113,000	$130,000

next to the dollars amount in Table 4.3 relate to the transaction numbers shown by Table 4.2. For example, the Cash T-account has a debit of $50,000 shown with a preceding 1. This indicates the transaction presented in Table 4.2. The transaction deals with the founding of

TABLE 4.3 T-accounts for Table 4.2.

Cash (A)				Accumulated Depreciation (A)				Insurance (E)			
(1)	$50,000	$1,500	(3)			($600)	(15)	(14)	$500		
(8)	2,000	4,000	(4)								
(10)	500	2,850	(6)								
		5,000	(7)								
		3,000	(9)								
		800	(12)								
		500	(14)								
		1,250	(16)								

Inventory (A)			Accounts Payable (L)				Depreciation Expense (E)		
(2)	$10,000		(4)	$4,000	$10,000	(2)	(15)	$600	
					12,000	(13)			

Accounts Receivable (A)				Sales (I)				Rent (E)			
(5)	$15,000	$2,000	(8)	(11)	$4,000	$15,000	(5)	(3)	$1,500	$500	(10)
		4,000	(11)								

Computer (A)		Salaries (E)		Capital Stock (NW)			
(13)	$12,000	(6)	$2,850	(7)	$5,000	$50,000	(1)

Equipment (A)		Advertising (E)		Taxes	
(9)	$3,000	(12)	$800	(16)	$1,250

General Construction, and $50,000 is paid for capital stock. The dual entry is in the Capital Stock T-account, where $50,000 is recorded as a credit, and is noted with a 1 next to the $50,000.

In Table 4.2, one transaction affects two accounts—for example, the transaction of the founding of General Construction. Both cash and capital stock are influenced, and those accounts are identified into asset and net worth types. The effect of the transaction is to either increase or decrease the account, which is further explained by a recording of either a debit or a credit.

Because of the dual effect of a transaction, the record of the transaction must balance, and for every debit there must be a credit. It is not necessary that there be the same number of debit and credit items in any T-account. Table 4.3 is a record of transactions that affect the amounts of assets, liabilities, net worth, income, and expenses of a business. Any business transaction may be recorded in terms of equal debit and credit elements on the basis of the increase or decrease effect. Purchases are considered an expense because they are an offsetting cost to income sales. That portion of purchases remaining unsold at the close of the period is termed inventory and is classed as an asset.

An open account has either a debit or a credit balance. A closed account has debit and credit of equal amount and, therefore, has no balance.

	Rent	
	8,000	$2,000
		6,000
Balance	0	0

	Capital Stock	
	5,000	$50,000
Balance		$45,000

The Rent account is a closed account, because the sum of the debits equals the sum of the credits. This is not the case for Capital Stock, which is open.

Each journal entry provides for equal debits and credits, which are posted to the ledger accounts. If the posting is accurate, then the ledger must have equal debits and credits. Additionally, the sum of the debit ledger account balances must equal the sum of the credit ledger account balances. This equality is periodically tested by a *trial balance*, which is a list of the open ledger accounts as of a stated date. The trial balance shows the debit or credit balance of each account.

The account groups in the trial balance are broadly divided into those used to prepare the balance sheet and the income and expense statement. A few accounts contain both balance-sheet and profit-statement elements and are called *mixed*. Those are separated into the two components during worksheet analysis.

Note that for the Cash T-account in Table 4.3, the cash balance is a debit of $33,600, which is entered in the trial balance under Dr. and the Cash account. In a similar way all T-accounts are footed and their debit or credit balance entered under the trial balance column in Table 4.4

TABLE 4.4 Trial balance for a business called General Construction Co.

Account	Trial Balance		Income Statement		Balance Sheet	
	Dr	Cr	Dr	Cr	Dr	Cr
Cash	$33,600				$33,600	
Inventory	10,000				10,000	
Account receivable	9,000				9,000	
Computer	12,000				12,000	
Equipment	3,000				3,000	
Accumulated depre.	(600)				(600)	
Accounts payable		$18,000				$18,000
Sales		11,000		$11,000		
Salaries	2,850		$2,850			
Advertising	800		800			
Insurance	500		500			
Depreciation expense	600		600			
Rent	1,000		1,000			
Capital stock		45,000				45,000
Taxes	1,250		1,250			
Total	$74,000	$74,000	$7,000	$11,000	$67,000	$63,000
Profit to retained earnings						4,000
						$67,000

The periodic trial balance of the ledger provides reasonable proof of the arithmetic accuracy of journalizing, posting, and ledger account balancing. The ledger lists account balances from which the balance sheet and income statement are later prepared. In most businesses the trial balance is performed after the end of the month.

A worksheet example is given in Table 4.4, which continues the development given in Tables 4.2 and 4.3. Various account titles are listed, and the balances from the ledger are posted in the Trial Balance column as either Debit or Credit. Those entries are carried over to either the income-statement or balance-sheet columns.

Generally, a worksheet adjusts the accounts for accrued expenses, accrued incomes, deferred expenses, depreciation, and bad debts. Once those adjusting entries are disposed, the next step concludes the trial balance by using a worksheet. Note that in Table 4.4 the first trial balance shows the effect of the ledger accounts. Those, in turn, are separated and extended horizontally into profit-and-loss and balance-sheet entries. The worksheet must balance. If it does not, then an error of some kind is indicated, and we must find it.

4.5 BALANCE-SHEET STATEMENT

The *balance sheet* is a tabular presentation of the important accounting equation [Eq. (4.1)] and is a summary of the assets, liabilities, and net worth at a point in time. The information for the balance sheet is obtained from the worksheet. A balance sheet is shown for XYZ Construction Company and repeats the accounting equation terms. Balance sheets have standard forms, especially giving the title of the firm, title "Balance Sheet," and the date of the evaluation.

The General Construction Co. balance sheet is given in Table 4.5. Important points about the balance sheet are the length of time covered, handling of the accumulated depreciation, and the asset and liability groups disclosed. Note that in Table 4.5 the closing date is the end of June. The balance sheet does not provide any hint what the assets, liabilities, and net worth were for any date prior to or subsequent to June 30.

XYZ Construction Co.
Balance Sheet
May 31, 20xx

Assets =		Liabilities	
Cash	$15,000	Bank loan	$15,000
Inventory	10,000	Mortgage	5,000
Land	15,000		
Building and equipment	40,000		
		+ Net worth	
		Capital stock	45,000
		Retained earnings	5,000
	$80,000		$80,000

TABLE 4.5   Balance Sheet for General Construction Co.

General Construction Co.
Balance Sheet
June 30, 20xx

Assets

Current assets		
Cash		$33,600
Accounts receivable		9,000
Inventory		10,000
Fixed assets		
Equipment	$3,000	
Less accumulated depreciation	(600)	
Computer	12,000	
Subtotal fixed assets		14,400
Total assets		$67,000

Liabililities and Net Worth

Current liabilities		$18,000
Net worth		
Capital stock	$45,000	
Retained earnings	4,000	
		49,000
Total liabilities and net worth		$67,000

Balance-sheet assets are not valued on the same basis. Cash, customer receivables, and inventories are valued at cost or net realizable cash value according to the conservatism convention. Land is valued at the amount originally paid for it, and depreciable fixed assets are valued at original cost less the accumulated depreciation. The liabilities are valued at the cash amount required to liquidate at the time of their maturity date. The net worth is a conglomerate value, because it represents the difference between assets and total liabilities.

4.6   PROFIT-AND-LOSS STATEMENT

The statement of earnings of the company, known either as the *profit-and-loss* or *income-and-expense* statement, is a summary of its incomes and expenses for a stated period of time. The net profit or loss represents the net change in net worth during the reporting period arising from business incomes and expenses.

Definition of Profit.   *Profit* represents the excess of revenue over cost. It is an accounting approximation of the earnings of a company (either owner, contractor, or subcontractor) after taxes, cash and accrued expenses (representing costs of doing business), and certain tax-deductible noncash expenses, such as depreciation, which are deducted. Sometimes it is termed *net profit. Loss* represents the excess of cost over bid, such as a construction job costing $1,800,000 while the bid is $1,700,000, which results in a loss of $100,000.

The following example describes the effect on business net worth of profit or loss:

JD Construction
Balance Sheet
May 31, 20xx

Assets =		Liabilities	
Equipment	$400,000	Bank loan	$100,000
Job A material	800,000	Accounts payable	200,000
Job B material	600,000		
		+ Net worth	
		Capital stock	1,500,000
	$1,800,000		$1,800,000

If JD Construction sells Job A material for $1,000,000 cash, its balance sheet changes to

JD Construction
Balance Sheet
June 30, 20xx

Assets =		Liabilities	
Equipment	$400,000	Bank loan	$100,000
Cash	1,000,000	Accounts payable	200,000
Job B material	600,000		
		+ Net worth	
		Capital stock	1,500,000
		Retained earning	200,000
	$2,000,000		$2,000,000

Net worth increases $200,000. If the business sells the asset Job B material for $500,000 cash, then its balance sheet looks like this:

JD Construction
Balance Sheet
July 31, 20xx

Assets =		Liabilities	
Cash	$1,500,000	Bank loan	$100,000
Equipment	400,000	Accounts payable	200,000
		+ Net worth	
		Capital stock	1,500,000
		Retained earning	100,000
	$1,900,000		$1,900,000

TABLE 4.6 Profit-and-loss statement for General Construction Co.

General Construction Co.
Profit-and-Loss Statement
June 30, 20xx

Income	
Project income	$11,000

Expenses		
Salaries	$2,850	
Rent	1,000	
Advertising	800	
Insurance	500	
Depreciation	600	
Total		5,750
Gross profits		$5,250
Taxes @ 23.8%		1,250
Profit (to retained earnings)		$4,000

The $600,000 Job B material is replaced by $500,000 cash, and the net assets and the net worth are decreased $100,000. Notice that profits increase the net worth because profits increase the net assets, and losses decrease the net worth because losses decrease the net assets.

Continuing on with the worksheet and balance sheet, as developed previously for General Construction Company, its profit-and-loss (P&L) statement is given in Table 4.6.

These P&L statements should be studied for their heading, income and expense groupings, and length of time and dates covered. Certainly, profits depend on the time of earnings. Net sales measure the net revenue from sales, and allowances for sales returns, freight out, and sales discounts are deducted from gross sales. Cost of goods sold covers the expense of the products sold to the customer. If the cost of goods sold is a gross value, then freight in and purchase discounts may reduce the value.

Operating expenses list recurring usual and necessary costs for conducting the business. Miscellaneous income and expense arise from interest and discounts and other small items of revenue and expense that are unrelated to the major business thrust. Except for depreciation the income statement items result from current-period transactions. _Depreciation_ is an _allowable noncash tax expense_ and reduces total income. Administrative expenses are found in almost all companies and cover the cost of managing the company; they may also include heat, power, rent, insurance, accounting, engineering, and legal costs. Income taxes, or the provision for income taxes, are an item reducing business income and are identified separately.

4.7 PERCENTAGE-OF-COMPLETION METHOD

There are differences in the methods of presenting the important income and balance-sheet statements between construction, retail, commercial, and manufacturing types of companies. In the retail business, and to some extent in manufacturing, a product is sold for a price between a buyer and a seller, and there is a known cost and title transfers at the time of sale. Preparation of the financial statements is clearer for these firms than for construction companies. In construction, particularly those firms having long-term contracts, elements of the income statement are based on *current-value* estimate of jobs in progress at the income-statement date. Retainage, an amount held back by an owner to assure contract compliance, is another difference in accounting of construction firms.

Revenues and costs for fixed-price contracts are identified by the percentage-of-completion method. As one illustration, this means that revenues and costs are measured by the labor hours incurred to date expressed as a percentage of the estimated total labor hours. A simple example for determining progress of revenues and costs is "units of output," such as the number of cubic yards of concrete in place. Management considers expended labor hours to be a good measure of progress on these contracts. Further, revenues from cost-plus-fee contracts are recognized on the basis of costs incurred during the period plus the fee earned. A typical line for the assets side of the balance sheet for a construction company would be "costs and estimated earnings in excess of billings on uncompleted contracts." Methods to recognize work-in-progress are amplified in Chapter 9.

Financial statements are used by analysts to evaluate performance of general business, where ratios, such as the current ratio (current assets divided by current liabilities), are determined. Perhaps the most important indicator of a construction company's financial position is the amount of uncompleted work and the estimated profit in that backlog.

The ownership structure of contractors, subcontractors, and owners has a plethora of styles: privately held or closely held firms, partnerships, family partnerships, open corporations, publicly held corporations, etc. There are differences of stock ownership. Dividends are customarily declared for public firms that are traded on any market exchange. But family construction firms may forego the payment of dividends, choosing instead to reinvest the retained earnings in the business. Indeed, the financial statements will vary depending on the type of ownership of these firms.

It is important that a firm competently prepare income and balance-sheet statements for bonding and surety insurance. Further, for any contractor who desires to receive maximum credit opportunity, the preparation of lucid financial statements is mandatory.

Thus it is seen that a construction firm will need a different perspective in the preparation and understanding of financial statements. The *percentage-of-completion* income statement together with a balance sheet is an IRS-approved method for long-term contracts. Residences or residential development construction may not qualify for the method. A sample of the method is given in the chapter Case Study. The student will want to study those statements.

Stockholders, partners, and others who have a proprietary interest in the construction company use the financial statements to learn about the company's operation. These reports are influenced by many factors. One such factor is depreciation, an important one for the capital-intensive firms who use construction firms in their business. Students need to understand the importance of these statements and appreciate how they are determined.

4.8 DEPRECIATION

This discussion of depreciation presents the generalized mechanics of depreciation calculations and some of the thinking in cost analysis. Engineering construction is a capital-intensive business and, thus, depreciation becomes important.

Ambiguity exists about depreciation because of the vagueness of the terminology. Terms such as surtax, credits, accelerated write-off, inflation or recession, obsolescence, and new technology confuse any discussion about depreciation. Various accounting terms such as contra-depreciation, reserves for depreciation, allowances for depreciation, and amortization and retirement enlarge the confusion.

Depreciation is an accounting charge that provides for recovery of the capital that purchased the physical assets. It is the process of allocating an amount of money over the recovery life of a tangible capital asset in a systematic manner. It is cost, as of a certain time, not value, that changes with time and is allocated and recovered. There are no interest charges or any recognition of a changing dollar value due to inflation or deflation.

The depreciation charge is not a cash outlay. The actual cash outlay takes place when the asset is acquired. Depreciation charges are the assignments of that initial cost over the recovery life of the asset and do not involve a periodic disbursement of cash. When the rate at which the asset is depreciated increases, the depreciation charge does not increase the outflow of cash. In fact, the opposite reaction happens, because the depreciation charges reduce taxable income and the outflow of cash for taxes. The initial investment is a prepaid operating cost that is expensed or allocated to an operating expense account, typically in overhead.

Depreciation is used for fixed assets, not current spending or expenses. Examples of typical fixed assets include equipment, buildings, land, computers, trucks, cars, and furniture.

Initial costs are undertaken to acquire assets that contribute to revenue over long periods. The cost of construction equipment, for example, remains a positive factor in generating revenue provided that the equipment is used in the construction of a sellable service or job. Because federal laws require income measurement, it is necessary that an appropriate portion of the cost of the equipment be charged to or matched with each dollar of revenue resulting from the income generated by its activity. The amount of such cost, matched with revenue, during any one period is the estimated amount of the cost that expires during the period. Thus, a fixed asset, which will not last forever, has its useful value exhausted over a period of time.

The factors contributing to the decline in utility are considered in the categories of

- Physical wear and tear resulting from ordinary usage or exposure to weather
- Functional factors such as inadequacy and obsolescence
- Governmental and political actions

Physical factors are commonplace, such as the wear and tear and corrosion that impairs efficiency and safety. Owing to technological progress, obsolescence is a frequent situation, where the fixed asset is retired not because it is worn out but because it is outmoded. The superior efficiency of an asset of a later design is one reason that compels new equipment designs for the market. Sometimes alterations of the design through research and development techniques make for immediate obsolescence. There is the instance of inadequacy where the asset, although neither worn out nor obsolete, is unable to meet the demands made upon it. An electric power company installs a hydroelectric generator and, in the course of several years, finds that the demands of an expanding community placed on the generator exceed the ability of the water and hydroelectric power to meet peak loads. Governmental or other forces may prevent operation (e.g., loss of raw material source or a new law prohibiting waste disposal are possibilities).

The government alters the tax laws[3] and frequently affects the depreciation. Changes in the laws governing income taxes and codes are enacted periodically, about every other year. In recent decades the federal government coined the term *modified accelerated cost recovery system* (MACRS). Useful life, recovery period, and salvage value are terms that are carefully defined.

Engineering estimates are required for the value of the asset, erection (if necessary), and operating capital and costs for consideration of new projects. In every project there are certain costs that can be tax-expensed immediately, such as expenditures for non-physical assets, some physical assets of extremely short life, and certain installation and startup expenses.

The question of *life* is an important matter. Some contractors are concerned about economic life with little regard for physical life, while others, public utilities for example, are restricted to earning a specified amount on capital invested, and life takes on another ramification. The economic-life estimate is affected by tax laws and the actions of Congress.

Book value is ordinarily taken to mean the original cost of the asset less any amounts that have been charged as depreciation. Book value should not be confused with salvage.

Before calculating depreciation it is necessary to understand property or asset classification. A depreciable property is used in the business or held for the obtaining of revenue and has a useful life longer than 1 year. The property will wear out, decay, become obsolete, or lose value from natural causes.

Depreciable property may be tangible or intangible, personal or real. *Tangible property* is seen or touched, such as buildings and equipment. Designs, patents, and copyrights are examples of intangible property. *Intangible property* can be amortized if its useful life can be found. *Real property* is land and generally anything attached to, growing on, or erected on the land. Land, which has an indeterminate life, cannot be depreciated. Personal property, which does not include real estate, includes machinery or vehicles, for example. Engineering deals with tangible personal property and tangible real property.

Tax laws affect the practice of depreciation. The Internal Revenue Service (IRS) provides various definitions. *Recovery property* is subject to the allowance for depreciation. A recovery period, a prescribed length of time, is designated for recovery property.

- Three-year property has a life of 4 years or less and is used in connection with construction and tooling.
- Five-year property excludes 3-, 10-, or 15-year public utility property. Examples include machinery and equipment not used in research and development, autos, and light trucks.
- Ten-year property is public utility property with a class life of more than 16 years but less than 20 years. An example could be railroad tank cars.
- Fifteen-year property includes buildings not otherwise designated as 5- or 10-year property.

The amount of annual depreciation charge is calculated by several means. A method called *accelerated cost recovery* is defined as follows:

$$D_j = P(j) \times P \tag{4.3}$$

---

[3] The student may be interested in the web home page of the Internal Revenue Service, where many current details are given. The address is *www.IRS.gov.*

where D_j = depreciation in jth year for specified property, dollars

$P(j)$ = percentage for year j for specified property class

P = cost of asset, dollars

TABLE 4.6   Illustrative schedule of Accelerated Cost Recovery factors.

	Percentage Depreciation			
Recovery Year	3-Year	5-Year	10-Year	15-Year Public Utility
1	33	20	10	7
2	45	32	18	12
3	22	24	16	12
4		16	14	11
5		8	12	10
6			10	9
7			8	8
8			6	7
9			4	6
10			2	5
11				4
12				3
13				3
14				2
15				1

Typical values of $P(j)$ are given for years 3, 5, 10, and 15. See Table 4.6. Note that the sum of each column is 100% and that values are not constant year to year. If percentage values were equal (i.e., $33\frac{1}{3}\%$ for the 3-year property class), then the depreciation is straight line. The method in Eq. (4.3) disregards any expected salvage value.

IRS codes are connected with the use of this table. For instance, if an asset is commissioned any time during the first year of a 3-year recovery period, the entire first-year depreciation is not permitted the year the asset is removed from service. If the property is held for a period at least as long as the associated recovery period, then the asset value will be entirely depreciated. If, on the other hand, the asset is disposed of prior to the period, then the asset will not be entirely depreciated. If the asset is held for the recovery period or longer, then it is possible for the book value to be less than the anticipated salvage value.

Consider a 3-year recovery property having a cost of $100,000. The property will be sold at the end of the fourth year.

Year	Cost	Percentage Depreciation, P(j)	Book Value at Year Beginning	Yearly Deprecation, D$_j$
0	$100,000		$100,000	
1		33	100,000	$33,000
2		45	67,000	45,000
3		22	22,000	22,000

A firm may choose an alternative method to determine depreciation by using a longer recovery period.

Recovery Period	Optional Recovery Period, $N(k)$
3	3, 5, or 12
5	5, 12, or 25
10	10, 25 or 35
15	15, 35, or 45

The percentage for each and every year is

$$P(j) = \frac{1}{N(k)} \tag{4.4}$$

where $N(k)$ = recovery period, years

For Eq. (4.4) we have the *straight-line* method, where salvage value is assumed as zero. Assume a $100,000 asset, which for physical wear-and-tear reasons we choose to depreciate over a 5-year life.

Year	Cost	Straight Line, %, $P(j)$	Book Value at Year Beginning	Yearly Depreciation, D_j
0	$100,000		$100,000	
1		20	100,000	$20,000
2		20	80,000	20,000
3		20	60,000	20,000
4		20	40,000	20,000
5		20	20,000	20,000

In view of its worldwide popularity and for general understanding, a *salvage value* is often associated with straight-line depreciation.

$$D_j = \frac{1}{N(k)}(P - F_s) \tag{4.5}$$

where F_s = future salvage value of investment, dollars

Reconsider the investment of $100,000 with a $10,000 salvage at the end of 5 years.

Year	Cost Less Salvage	Straight-Line Depreciation, %	Book Value at Year Beginning	Yearly Depreciation, D_j
0	$90,000		$100,000	
1		20	100,000	$18,000
2		20	82,000	18,000
3		20	64,000	18,000
4		20	46,000	18,000
5		20	28,000	18,000

Another ad hoc method, sometimes used for depreciation accounting, is the *sum-of-the-years' digits* and provides a declining periodic depreciation charge over an estimated life that is achieved by applying a smaller fraction recursively each year to the cost less its salvage value. In the recursive relationship the numerator of the changing fraction is the number of remaining years of life, and the denominator is the sum of the digits representing the years of life. With an asset having an estimated life of 5 years, the denominator of the fraction is $15(= 1 + 2 + 3 + 4 + 5)$. For the first year the numerator is 5, for the second 4, and so forth.

$$D_{sd} = \frac{2}{N}\left(\frac{N + 1 - k}{N + 1}\right)(P - F_s) \tag{4.6}$$

where D_{sd} = sum-of-the-years'-digits depreciation charges, dollars
  k = current year
  N = life defined for depreciation, years

Reconsider the investment of $100,000 with a $10,000 salvage at the end of 5 years.

Year	Cost Less Salvage	Rate	Book Value at Year Beginning	Yearly Depreciation, D_{sd}
0	$90,000		$100,000	
1		5/15	100,000	$30,000
2		4/15	70,000	24,000
3		3/15	46,000	18,000
4		2/15	28,000	12,000
5		1/15	16,000	6,000

Still another method of depreciation is based on the premise that an asset wears out exclusively as demands are placed on it. Called the *units-of-production* method, the computation is given as

$$D_{up} = \frac{1}{\Sigma N(i)}(P - F_s)N(i) \tag{4.7}$$

where D_{up} = unit-of-production depreciation charge, dollars
    $N(i)$ = units for ith year

This method has the advantage that expense varies directly with operation activity. Retirement in those cases tends to be a function of use. An estimate of total lifetime production is necessary. The example involving $100,000 cost and $10,000 salvage value is estimated to have 200 units of output over the 5-year period.

Year	Cost Less Salvage	Units of Production, $N(i)$	Book Value at Year Beginning	Yearly Depreciation, D_{up}
0	$90,000		$100,000	
1		15	100,000	$6,750
2		45	93,250	20,250
3		50	73,000	22,500
4		55	50,500	24,750
5		35	25,750	15,750
		200		

The advantages of accelerated methods of depreciation are compatible with the logic that the earning power of an asset is created during its early service rather than later, where upkeep costs tend to increase progressively with age. Accelerated methods offer a measure of protection against unanticipated contingency such as excessive maintenance, and they return the investment more quickly and simultaneously, decreasing the book value at the same rate. A high book value would tend to deter the disposing of unsuitable equipment even when the need for replacement is pressing. Rapid reduction of book values, provided the owner or contractor overlooks the tax benefits from capital loss, leaves the owner or contractor more free to dispose of inefficient and unsatisfactory equipment.

Depreciation of costs is collected under the control of general ledger accounts, such as accumulated depreciation or allowance for depreciation. Periodic fixed charges resulting from those accounts are analyzed and charged against the equipment under the proper cost classification through the application of worksheet analysis. Distribution of depreciation of machinery and equipment is thus made based on factors such as cost of equipment, rates of depreciation applicable to each unit of equipment, and location of each equipment. If the job will exhaust the equipment depreciation, then naturally the depreciation is charged to rates supporting the job.

An accumulated depreciation account contains the accumulated estimated net decrease in the value of the particular asset account to which it pertains. In most businesses the amount shown in the depreciation account does not appear as cash. If a special fund is set aside specifically for this purpose, then it will be called a sinking fund. To create a fund of this nature suggests that a fund is actually invested outside the company to earn interest. However, interest rates found on the outside are hopefully less than the earning rate enjoyed by the company. Usually, it is wiser to employ the money for operations. The amount equal to the depreciation will appear as other assets, such as working capital, raw materials, or materials in storage. When it becomes necessary to buy new equipment or replacements, management must convert physical assets into cash (unless sufficient cash is on hand), or use existing profit to pay for the new equipment.

The appearance of accumulated depreciation on the balance sheet, as with other assets, represents capital retained in the business, ostensibly for the ultimate replacement of the capital asset being depreciated. Accumulated depreciation is also known as a "contra" account, whose balance reduces the value of an asset, which is shown at cost. The value of the contra account is enclosed in parentheses, indicating a negative value. For example, in Tables 4.2, 4.3, 4.4, and 4.5 the contra account is ($600).

Factors in any depreciation model are subject to estimation: salvage value and, particularly, life. If the estimates prove faulty, then it is possible to retire an investment before its capital is recovered. In those circumstances where net income received is less than the amount invested, an unrecovered balance remains. This unrecovered balance is referred to as *sunk cost*, which is the difference between the amount invested in an asset and the net worth recovered by services and income resulting from the employment of the asset.

As an illustration of sunk cost, consider a capital investment of $5000 to be recovered in 5 years with a residual $1000 salvage, or $4000 depreciation. On the basis of straight-line depreciation, the amount invested and to be recovered per year will be $800. As a result of excessive use, the equipment is sold after 3 years for $1400, and has actually consumed $3600 in 3 years or $1200 per year on the average. The sunk cost is equal to the difference in the actual depreciation and the depreciation charge, in this case $3600 − $2400 = $1200. Stated yet another way, sunk cost is the estimated depreciation value (or book value) minus the realized salvage of the asset. Sunk cost is unaffected by decisions of the future and must be faced realistically.

From an accounting point of view, costs are often identified as to their behavior, either as a function of time or as a function of the object of the construction or service performed. This classification is known as fixed, semifixed, and variable. We now turn to this topic.

4.9   FIXED, SEMIFIXED, AND VARIABLE COSTS

We have defined costs in a number of ways. In another interpretation, costs are fixed, semifixed, or variable.

Fixed costs are incurred regardless of the differences in levels of construction activity. Generally, they do not change in the period as a result of more or less construction activity. The decision to build a plant in a city under a tax system makes the costs of carrying the depreciation and taxes on the plant uncontrollable and fixed. The governmental agency controls the taxes. Only the closing down or removal of the plant can affect these continuing charges against the property. Assuming the ongoing-business convention, other fixed costs are salaries of the president, engineers, and superintendents, who are employed the full year regardless of whether or not there are contractor jobs available.

Sometimes fixed costs step up or down as operations vary beyond a certain level, and these are called *semifixed costs*. Sometimes they are known as programmed fixed costs, because they are forecast ahead according to a level of business activity. Once the decision is made to commit to a level of construction activity, say build two houses instead of one during a fiscal period, these costs usually remain fixed unless significant revisions are made in schedules during the period covered by plan. The prefix "semi" clarifies their difference from fixed costs.

Workers' compensation, social security taxes, unemployment taxes, energy rates, major building or process alterations, and employee training are in the noncontrollable portion of semifixed costs. Some of the block tax rates are set by law, while others are the responsibility of management. Within the semifixed category of costs, there are a number of controllable items such as maintenance, overtime pay, supervision, and power. It is possible that these costs will vary with levels of construction activity, but not with the same sensitivity as "pure" variable costs.

Costs that vary directly with changes in activity are called *variable costs*. Variable costs occur only because of activity. Examples are direct material and direct labor used in the construction process. These costs do not exist unless there are jobs. We have examined examples of variable costs in Chapters 2 and 3.

A variable cost for one contractor may be fixed for another. For example, one firm hires and fires job superintendents as a function of work load, while another firm may retain the superintendent on its payroll irrespective of business. Sometimes variable costs are assumed to be linear, although nonlinear behavior is possible.

Theoretically, variable costs in the very short term can be classified as fixed, because of the inability to terminate the variable costs. And fixed costs in the long term can be viewed as variable, because of the ability to liquidate those charges. So the length of the period has an influence on the nature of costs. Costs, in an accounting sense, are often identified as to behavior, either as a function of time or as a function of the object of the construction or service performed. The behavior of fixed, semifixed, and variable costs is illustrated in Fig. 4.3(a).

PICTURE LESSON
Pile Driving

Piles are driven to form the posts for a trestle. Rings, or iron bands, are placed around the chamfered heads of the spruce logs to prevent them from brooming and splitting while being driven. Work clothing typical of Pacific Northwest construction crews during the 1891–1892 era included hats, jackets, and vests. Hand tools include the adz, lumberman's double-edge axe, pliers, and 16-lb sledge.

Early contract specifications for work of the picture lesson required a price per lineal foot for piles delivered to the work site, and another price per lineal foot for piles driven. This is identical to practice today.

The tower is moved and secured for each pile. The pile driver shown in the picture is a free-fall design. The hammer is raised by block and tackle with horse teams. Two teams are used alternately. After the horse team raises the hammerhead, the hoisting claw is yanked loose and the hammerhead falls upon the pile with gravity force. The upright timbers that guide the hammer are called ways, and a common weight for the hammer is 2000 to 3000 lb. The free-fall hammer was much used in remote areas. A portable steam engine used a claw-and-clutch design to grasp, raise, and release the hammer and was available only near the railroad tracks. Steam-hammer blows could be applied upon the pile several times faster than with horse teams.

Productivity of the construction crew using horse teams for driving 15-ft piles into stiff dirt and rock, according to Halbert Gillette, an information pioneer who provided estimating data for all kinds of construction work during the early part of the twentieth century, is 21 piles per day. Fully loaded cost for material, labor, and rented horse teams is $0.85 per pile.

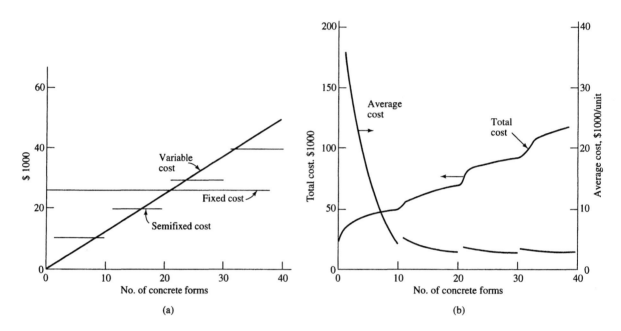

Figure 4.3 (a) Fixed, semifixed, and variable costs, and (b) total and average cost.

These costs can be modeled in the following way:

$$C_t = C_f + C_{sf} + nC_v \qquad (4.8)$$

where C_t = total cost at n units or activity, dollars
   C_f = fixed cost, dollars for period
   C_{sf} = semifixed cost for level of n or activity for period, dollars
   n = number of units or level
   C_v = variable cost per unit, dollars/unit

$$C_a = \left(\frac{C_t}{n}\right) \qquad (4.9)$$

where C_a = average cost, dollars/unit

The cost per unit, C_a, can be thought of as *average cost*, since it considers fixed, semifixed, and variable cost per unit. Further, if the range of the activity variable n is reasonably stable and known, we can combine the semifixed costs with the fixed costs. For this relationship n is used, indicating that volume or quantity is implied. Applications, such as yd^3, ft^3, number of concrete forms, etc., are possible.

Consider the example of the framing of a dam face with constructed wooden and steel frames, where we have the following illustration:

Type of Cost	Calculation	Cost, $
Variable		
Direct labor	10 man hours each @ $25/hr = $250	
Direct material	$1000 each	
Total		$1250/frame
Fixed		$25,000 for the period
Semifixed		$10,000 for units $1 \leq n < 10$
		$20,000 for units $10 \leq n < 20$
		$30,000 for units $20 \leq n < 30$
		$40,000 for units $30 \leq n \leq 40$

Semifixed and variable costs are not defined for $n = 0$. Fixed costs are necessary before any frame construction can begin, and thus are charged at $n = 0$. Semifixed costs are planned according to the level of forms, presuming that additional costs are necessary. These additional semifixed costs could be portable cranes and equipment to move materials. Refer to Fig. 4.3(b) for the relationships. The curves are herky-jerky because of the way in which the semifixed discontinuous costs are defined. In other applications the cost figures will be smooth. Development of a fixed, semifixed, and variable cost equation for construction equipment is deferred to Chapter 7.

A contractor's fixed and semifixed costs, if they are for several simultaneous jobs, cannot be spread costwise into one construction job. To do so may create inaccuracies. Some costs are *periodic* (time based); others are *object* based, such as direct materials or direct labor; and some are *joint*. For accounting analysis a new means of looking at joint costs is required. Because of difficulties in treating fixed, semifixed, and joint costs, we now consider the important overhead procedures.

4.10 OVERHEAD

Contractor's overhead makes up an important part of the construction price. Overhead is a real cost of running the business. Overhead applies to the owner, general contractor, and subcontractors. It is necessary to be familiar with these practices in order to convert raw construction cost and build it up into a contractor's bid cost.

Overhead is an *estimate* of the contractor's cost that cannot be clearly associated with particular jobs, projects, or systems and must be prorated among all work on some *arbitrary* basis. The key to this puzzle is the way in which these indirect expenses are collected, allocated, and charged to individual estimates. Direct costs, such as direct labor and direct materials, present little if any allocation problem. Those costs do not exist if there is no job.

The underestimating or overestimating of overhead is serious in view of the proportion to total job cost. As an illustration, consider equipments A and B, both having identical labor rates of $30 per hour for operation. Equipment A is initially worth $150,000 while Equipment B is worth only $15,000. If an average *burden* rate of 200%, which is based on direct labor dollars, is used, Equipment A and B would each cost, on a equipment-hour basis, $60 per hour. However, this is false hour costing, because the investment in machine A is 10 times that in machine B. Proper cost allocation of overhead to handle discrepancies of this sort is necessary. Years ago, equipment investment per worker was lower, and it was common that overhead rates were uniformly distributed over the direct labor base. In recent decades the ratio of fixed cost to variable cost has risen, and the expediency of overhead distribution by a single rate is misleading. Traditional cost-accounting systems were developed in the early part of the twentieth century for labor-intensive work. At that time, overhead cost was not so major a component of construction cost.

Overhead is allocated to a designated base. A base for distribution of construction overhead may be direct labor dollars or hours or combined with materials, equipment hours, or subcontractor costs. There is an arbitrary choice on how the overhead is costed to a job.

Overhead costs become complicated whenever the costs have a jointness or commonness with different levels of variability. *Joint costs*, or costs incurred together, are depreciation, insurance, taxes, and maintenance and are dependent on each other. For example, these items are tied to equipment value. As equipment value increases, so do depreciation, insurance, and so on. Accounting joint costs are handled by the overhead process.

Overhead for construction projects is of two types: office and job. Overhead costs exclude direct labor and direct and subcontract materials. Items appearing in overhead must not be included in those estimates.

Office overhead includes general business expenses, such as home office rent, office insurance, heat, light, supplies, furniture, telephone, legal expenses, donations, travel, advertising, bidding expenses, and salaries of the executives and office employees. Those charges are incurred for the benefit of the owner's or contractor's overall business. In office overhead, the final cost objective is multiple (i.e., several projects), and those costs cannot be isolated as specific estimating amounts for any one job. Though variety is possible in the calculation, one approach would use

$$R_o = \frac{C_o}{C_p} \times 100 \qquad (4.10)$$

where R_o = office overhead rate on basis of direct costs, percent

   C_o = overhead charges summed for office activity of contractor, dollars

   C_p = cost of direct labor and direct and subcontract materials, dollars

Job overhead pertains to the project. If a firm has only one field project in mind, then a consolidation of office and job overhead is a possible convenience. On the other hand, improved accuracy becomes apparent if separate job overhead is found for one or all projects. Job expenses are those costs that do not become an integral part of the construction but are directly chargeable to the contract and must be separated in the accounting journals from overhead expenses, which are general.

Each project requires a special analysis to determine its own items. Typical items of job overhead are listed below.

Permits and fees	Insurance
Performance bond	Electricity at job location
Job office expense	Water at job location

Office salaries	Barricades
Cost clerk	Badges
Timekeeper	Survey
Supplies	Parking areas
Superintendent	First aid
Depreciation	Storage and protection

The office salaries are connected to the specific job and are not incurred for the total business. Those overhead charges do not include mandatory contributions for direct labor, such as FICA, workers' compensation insurance, unemployment insurance, or any of the union-contract contributions. Employee-related costs are included in the gross hourly cost of direct labor, as is discussed in Chapter 2.

The job overhead rate is found using

$$R_j = \frac{C_j}{C_p} \times 100 \tag{4.11}$$

where R_j = job overhead rate on basis of direct cost, percent
   C_j = overhead charges summed for project, dollars
   C_p = cost of direct labor and subcontract materials for project, dollars

Technically, the dimensions on Eqs. (4.10) and (4.11) are not a percentage. Both the numerator and denominator terms are distinctively different. The student will want to consider these dimensions.

As a rule of thumb, if the base C_p increases, then the overhead rate R_j should be decreasing. Conversely, if the base is decreasing, then the overhead rate generally will be increasing. Comparison of rates between projects may make no sense, remembering that the rate is a ratio of two cost sums. A low rate may indicate that a project is labor intensive and the direct labor base is high. The overhead account is less because it includes little or no charges for depreciation or rental of equipment.

The overhead is applied to the estimate of the job by using

$$C_{op} = C'_p(R_o + R_j) \tag{4.12}$$

where C_{op} = overhead charged to future project, dollars
   C'_p = cost of future direct labor and direct and subcontract materials estimated for project, dollars

This formula, Eq. (4.12), is used in the application stage, where the overhead is *applied or posted* to the job.

Consider this simple example. Last year a contractor had annual construction direct costs of $6,000,000 and a general office overhead of $C_o = $240,000. Using Eq. (4.10) we find $R_o = 4\%$. Additionally, actual cost tickets for several projects were reviewed, and $C_j = $310,000 was uncovered. Then $R_j = 5.17\%$. Though the rates are guided by historical patterns, those computations are for future periods, specifically budgeted for the duration of the project. It is necessary that both the base C_p, C_o, and C_j be forecast and computed for identical periods. We apply these rates for future estimated costs, provided that they are similar to those historical costs, and thus the use of Eq. (4.12).

An important new concept in overhead accounting is *activity-based costing*, which attempts to find variables that demonstrate cause and effect for overhead charges. One simple approach identifies two types of cost drivers: volume related and nonvolume related. *Volume*-related cost drivers have a direct positive relationship with construction quantity. An example would be direct labor and equipment hours. More construction quantity means more direct labor and equipment hours.

Nonvolume variables obviously increase the cost of the project, as managing a few or many subcontractors leads to a difference in overhead costs. Similarly, if there are few or many engineering-drawing changes for the project, we can expect differences in the cost of overhead.

Volume-Related Variables	*Nonvolume Variables*
Direct labor hours	No. of subcontractors
Machine hours	No. of architect/engineering drawings
Direct labor costs	No. of engineering drawing changes per job
Construction volume	No. of job sites for one contract
	No. of different materials
	No. of schedule changes
	Amount of rework
	Strikes and downtime
	Weather

The student will want to study the references for more information on activity-based costing.

4.11 JOB-ORDER AND PROCESS-COST PROCEDURES

Job-reporting procedures provide historical cost information and are known as *job order* or *process cost*. The completion of daily or weekly labor, material, or equipment time and cost reports is the task of field personnel, those knowing what is going on at the job site. Frequently, this may mean the owner (for a small contractor), foreman, timekeeper (for a larger firm), or cost engineer (where specifications and designs require compliance reporting).

These reports are returned via hand-carry, fax, email, or telecommunication and computer methods to a central office, where they are processed. We need not concern ourselves here with the various technology involved in these intricate schemes, other than to mention that they include personal names, pictures, and badge numbers, credit-card identification with magnetic strips providing information about individuals or equipment, bar coding of materials, holograms, laptop computer, etc.

Purposes for these reports include:

- Payroll and timekeeping
- Development of standards, both time and cost, for future work and estimates
- Cost, budget, schedule, IRS requirements, and legal verification of work-in-place
- Monitoring progress for productivity

There are accounting procedures to tally material, labor, and burden costs to control accounts. In job construction, the significant activities of construction are assigned order numbers. This is a convenient way to collect the material requisitions and labor job tickets. A typical job-cost ledger describes the costing points to which the work refers. A job order may cover the construction of one unit or a number of identical units. Payroll and timekeeping are subsidiary bookkeeping and accounting activities that initiate with these forms.

Sometimes these records can be queried to determine standards, both labor or equipment, since the records will have hours or days paid and the quantities for work completed. This historical information can be reused for time and cost estimating for similar and future designs.

We have seen that progress payments by the owner to a contractor are guided by work and material that is placed. These records often constitute legal verification for progress payments.

The exact information needed to control job progress varies with the jobs and the people who manage them. There are many "forms" used for the paper chase of job-order or process costing. Space does not permit their illustration here.

SUMMARY

Cost accounting is important to the performance of diverse engineering functions. Accountants provide overhead rates, some historical costs, and budgeting data. The engineer reciprocates with manpower and material estimates for the designs. The estimate in many situations will serve as a forecast of a profit-and-loss statement for projects. Interestingly, the accountant provides the verification of the P & L after the project has closed out. Thus, there is mutual dependence between accounting and engineering.

The engineer is more curious about balance sheets, profit-and-loss statements, and overhead rate determination than about the intimate details of the structure of accounts and the bookkeeping maneuvers. We have stressed this viewpoint of the accounting fundamentals. The chapter problems reinforce the learning of the important financial documents.

Accounting costs, if purely historical and derived from job-order or process-cost procedures, are usually inadequate as estimates of future costs. The design may have changed or costs may have increased or decreased. It is essential for intelligent cost analysis to have a continuous flow of reliable information from all activities of the organization. Accounting is an important contributor of old information. The next chapter deals with the future speculation involving labor, materials, and other accounting data.

QUESTIONS

1. Define the following terms:

Asset	Account balance	Accounting
Liability	Depreciable value	Income
Net worth	Book value	Expenses
Accounting equation	Service life	Overhead
Fixed costs	Trial balance	P&L
Capital stock	Semifixed cost	Variable cost
Surplus	Open account	Activity-based costing
Debit	Overhead	Job costs
Credit	Accrual	T-account

2. List 10 kinds of liabilities for a business with which you are familiar.

3. Why do expense accounts normally have debit balances and income accounts normally have credit balances?

4. What is the purpose of a trial balance? Into what main groups are trial-balance accounts divided?

5. Distinguish between the cash and accrual bases of accounting.

6. What are the functions of special and general journals?

7. Why is equitable distribution of cost essential throughout the organization to cost finding, analysis, and prediction?

8. Define overhead. What is the essential and philosophical purpose of overhead?
9. What is the difference between job and office overhead?
10. Prescribe and contrast several methods for the distribution of indirect cost.
11. Study the website of Construction Financial Management Association, www.CFMA.org. What are the topics that this organization considers important?

PROBLEMS

4.1 Evaluate the effects of transactions by constructing a daily balance sheet showing an *asset* side and a *liabilities and equities* side. The business is called John Smith, HVAC. January 1: John Smith starts a HVAC business installing products for the residential construction market. Smith deposits $10,000 of his own money in a bank account that he has opened in the name of the business, creating capital stock. January 2: The business borrows $5000 from a bank, giving a note, thereby increasing the assets and cash, and the business incurs a liability to the bank. January 3: The business buys inventory in the amount of $10,000, paying cash. January 4: The firm sells material for $300 that cost $200.

(*Hints*: Balance sheets follow a formal business style. Create a heading for each day's activities using the name of the business, type of business document, and date. For each transaction there are two entries.)

4.2 For the following ledger T-accounts from Weichman Contractor, set up a balance sheet for the end of March. (*Hint*: Note that for each T-account entry there are two dates, amounts, and titles of transaction that are cross identified.)

Cash

March	1	Capital	$1000	March	3	Rent	$200
	10	Consulting fee	250		20	Salaries	350
	25	J. A. Wilson on acct.	500				

Customers (Accts. Receivable)

| March | 10 | J.A. Wilson | $1200 | March | 25 | Cash on acct. | $500 |

Supplies on Hand

| March | 5 | Accts. Payable | $360 | March | 31 | Supplies used in March | $110 |

Equipment

| March | 4 | Notes payable | $3200 | | | | |

Accounts Payable

| March | 5 | Supplies | $360 | | | | |

Notes Payable

| March | 4 | Equipment | $3200 | | | | |

Weichman Construction Co. Capital

March	3	Cash	$200	March	1	Cash investment	$1000
	20	Salaries	350		10	Consulting	250
	31	Supplies used	110		10	J. A. Wilson	1200

4.3 By using the following ledger giving the T-accounts, construct a profit-and-loss statement and a balance-sheet statement for the month of June. The name of the business is Precision Construction. (*Hints*: Determine the Profit-and-Loss Statement, noticing the location of the gross profit into the capital T-account. Ignore taxes.)

Equipment							
June	1	Balance	$7210				

Accts. Payable							
June	2	Cash Meyer Co.	$350	June	1	Balance	$360
					6	Supplies	600
					28	Misc. exp.	40

Notes Payable							
				June	1	Balance	$3200

Capital							
				June	1	Balance	$1790
					30		1570

Income							
June	30	To P&L	$2500	June	4	Accts. Pro. A. B. Jones	$2200
					15	Cash I. N. Smith	300

Lease Expenses							
June	1	Cash	$200	June	30	To P&L	$200

Misc. Office Expenses							
June	10	Telephone cash	$60	June	30	To P&L	$100
		Elec.	40				

Salaries							
June	20	Cash	$350	June	30	To P&L	$350

Supplies Expense							
June	30	Supplies used	$280	June	30	To P&L	$280

4.4 Construct a year-end balance sheet for Construction Management Inc.:

Account Title	$
Retained earnings	610,000
Cash	150,000
Outstanding debt	450,000
Raw materials	90,000
Finished goods	50,000
Current liabilities	40,000
Stock ownership	400,000
Fixed assets	1,100,000
In-process materials	110,000

(*Hint*: Identify the account with the three divisions of the balance-sheet statement before writing the balance sheet.)

4.5 Prepare the profit-and-loss statement and find the retained earnings at the end of the period by using the following account balances of Billerbeck Construction Supplies, Inc. for the year ended December 31. BCS is a publicly held corporation.

Account Title	$	Account Title	$
Dividends paid	45,000	Retained earnings, Jan 1	90,000
Sales	700,000	Rent	70,000
Sales returns	40,000	Salaries	130,000
Inventory, January 1	120,000	Interest earned	2,000
Purchases	270,000	Sales discounts	10,000
Purchase returns	20,000	Interest expense	5,000
Inventory, September 30	160,000	Taxes, all kinds	47,000

4.6 Evaluate the effects of each transaction by constructing a balance sheet showing an assets side and a liabilities + net worth side.

Transaction
(a) Eastwood Construction is organized with a capital stock of $250,000 cash for the entire stock.
(b) Bought from Culpepper Co. on credit $100,000 of construction material.
(c) Borrowed $80,000 cash from First National Bank.
(d) Paid Culpepper $30,000 on account.
(e) Returned $10,000 of defective material to Culpepper.
(f) Loaned $50,000 to Robert Gondring.
(g) Paid $20,000 cash for building site.
(h) Erected a building at a cost of $120,000 cash.
(i) Borrowed $70,000 from Friendly Insurance, giving a mortgage for collateral.

(*Hints*: Find the T-account balances for these titles: Cash, Accounts Receivable, Material, Land & Building, Accounts Payable, Notes Payable, Mortgage, and Capital Stock. Then determine the balance sheet.)

4.7 Evaluate the effect of each transaction by constructing a balance sheet showing an assets side and a liabilities + net worth side.

Transaction
(a) Samuel Specthrie establishes SS Construction, paying $250,000 cash for the entire capital stock.
(b) Paid $50,000 cash for a building and storage site.
(c) Erected building costing $200,000, paying $50,000 cash and issuing a $150,000 first mortgage for the balance
(d) Borrowed $60,000 cash from First National Bank.
(e) Bought furniture, costing $15,000, on a open account from Wood Furniture Co.
(f) Purchased $50,000 of tools from Universal Tool on credit.
(g) Bought $30,000 of computer equipment from Byte Co. for cash.
(h) Returned $15,000 of faulty tools to Universal Tool.
(i) Paid $25,000 in reduction of bank loan.
(j) Bought $40,000 of U.S. Treasury bonds for cash.

4.8 Patrick Lyell Construction started the year with the following balances:

Account Title	Balance as of January 1, $
Cash	100,000
Inventory	100,000
Construction equipment	400,000
Accounts payable	50,000
Net worth	550,000

Transactions during the year were limited to the following: paid $100,000 for labor; purchased $150,000 worth of materials; noted equipment depreciation of $50,000; added one apartment unit costing $300,000 to Lyell; sold apartment unit for $600,000 cash; and purchased new equipment costing $200,000.

Make an end-of-year income statement and balance sheet. (*Hints*: Neglect income taxes. Accounts payable and inventory at the end of the year were the same as at the beginning of the year.)

4.9 An engineering construction firm is chartered to provide consulting services, design, and a limited range of construction services. Its designs have developed into registered patents. Some designs are licensed by subcontractors, the firm not choosing to use them itself; it does receive income from that business arrangement. These accounts and their balance are summarized below.

Find the income statement and balance sheet for a family-owned firm titled "Eastwood Engineering Services," Second Quarter. (*Hints*: These values are in units of $1000, but ignore that. For each account title, first identify if it is an asset, liability, net worth, income, expense, or tax. Each account will have a spot in either the income or balance-sheet statement.)

Account Title	$	Account Title	$
Cash on hand	3	Receivables from projects	55
Fees received	73	Royalties on patents	40
Bonds owned	12	Equipment book value	58
Sale of design	7	Interest on owned securities	2
Salary expense	28	Office rent	1
Equipment lease	6	Travel	4
Utilities paid	5	Supplies expense	9
Patent assets	11	Accounts payable	30
Capital stock	25	Retained earning balance at start of quarter	27
Taxes, all kinds	12		

4.10 The following data are the assets, liabilities, incomes, and expenses of the business Warren Andrews, Design-Build Inc. The firm is employee owned.

Prepare an income and expense statement for the 6-month period ending June 30. Prepare a balance sheet at June 30. (*Hints*: Remember to have the right headings for the documents. It is necessary to prepare the income and expense statement first because the profit amount is needed to complete the balance sheet.)

Account Title	$	Account Title	$
Cash in bank	65,000	Office fixtures owned	50,000
Income from fees	150,000	Bank loan	100,000
Interest from owned securities	5,000	Accounts payable	70,000
Rental income on owned properties	30,000	Interest expense on bank loan	1,500
Staff salary expense	120,000	Receivables from clients	90,000
Traveling expense	7,500	Automobiles owned	40,000
Telephone expense	2,500	Capital stock	425,000
CAD supplies expense	8,500	Surplus, January 1	75,000
Real estate owned	250,000	Securities owned	200,000
		Taxes	20,000

4.11 What follow are data for assets and liabilities at December 31, and incomes and expenses for the business year of Wilmer Hergenrader, P.E., Construction Consultant, Inc. By using this information, prepare an income statement and a balance sheet.

Account Title	$	Account Title	$
Cash in bank	80,000	Furniture and fixtures	30,000
Fees earned	420,000	Rental income on building lease	26,000
Interest received from bonds	2.500	Bonds owned	120,000
Land	60,000	Traveling expense	35,000
Receivables from clients	50,000	Taxes paid	15,000
Staff salaries	270,000	Office expense	10,000
Buildings	180,000	Telephone expense	5,000
Bank loan payable	70,000	Accounts payable	25,000
Capital stock	275,000	Automobiles owned	20,000
Surplus, January 1	65,000	Association dues expenses	2,500
Interest expense on bank load	4,000	Donations to charity	2,000

4.12 The following data, expressed in units of 10^4, are the closed ledger accounts showing assets, liabilities, income, and expenses of Blue River Builders, Inc. Blue River is a family-owned firm and does not declare dividends. Prepare an income statement and balance sheet for the year.

Account Title	10^4	Account Title	10^4
Cash in bank	1560	Construction fixtures owned	1000
Project income	3000	Bank loan	2000
Interest income, owned securities	100	Accounts payable	1400
Rental income	600	Interest expense on bank loan	50
Salary and wage expense	2400	Receivable from contracts	1800
Office expense	200	Equipment owned	800
Materials expense	150	Capital stock	8300
Buildings owned	5000	Retained earnings, January 1	1700
Taxes, all kinds	140	Securities owned	4000

4.13 A heavy-duty truck is purchased for $35,302. It will cost its owner $108,694 by the time it has been driven 10 years and 100,000 miles. Besides depreciation to zero, this sum includes $20,792 for repairs and maintenance; $16,768 for gas and oil; $14,472 for garaging and parking; $10,800 for insurance; and $10,560 for taxes.

Find the depreciation cost per mile. What are the yearly and per-mile costs? What are the percentages for the elements of ownership?

4.14 A Chevrolet 1500 work truck's value and upkeep costs are found to follow this schedule:

End of Year	Drop in Value, %	Drop in Value $	Upkeep Costs, $
1	28	$4904	$412
2	21	3600	540
3	15	2700	1076
4	11	2000	1372
5	9	1504	1288
6	6	1036	1388
7	4	756	1828
8	3	484	968
9	2	340	1216
10	1	192	308

(a) How many years does it take for a truck to depreciate two-thirds and three-fourths of its value?

(b) If the truck is driven 13,500 miles per year, then what are the operating yearly costs in dollars per mile? Plot those operating costs. When is the advantageous time to trade a truck, assuming that the chief criterion is per-mile economy?

(c) Discuss: If the immediate cost of repairing an old truck is less than first-year depreciation on a new one, is it the best policy to buy a truck and drive it until it is ready to be junked? (*Hint*: How does "prestige" influence your decision to keep an old truck rather than buying a new one? Does it influence your decision if the truck is self-owned or owned by the company?)

4.15 Construction equipment has a capital cost of $43,000, salvage value of $3000, and an asset life of 10 years. Compute the depreciation expense for the first 3 years under

(a) Accelerated cost recovery   (b) Straight line   (c) Sum-of-the-years'-digits.

4.16 Contractor A purchases construction equipment that costs $250,000. Economic life is estimated as 5 years with a salvage value of $15,000. This company chooses the accelerated cost recovery method for depreciation. Contractor B, in competition with company A, buys the same equipment at identical cost. The management of company B uses straight-line depreciation.

Determine the yearly depreciation charges and the end-of-year book value for companies A and B. Discuss the two methods under this competitive situation, and comment on the importance of the depreciation method for cost estimating.

4.17 A construction company will buy a tire-mounted excavating machine for a delivered price of $250,000. Life is 6 years or 12,000 hours for total rated dipper capacity of 144,000 yd^3. Salvage is expected to be nil. At the end of the life the machine will be retired to secondary and emergency service.

Determine the depreciation charge and book value for accelerated cost recovery, straight-line, sum-of-the-years'-digit methods, and production-units method. Plot "book value" for the four methods and discuss the merits of book value.

4.18 General Construction has the following 10^5 balances on its accounts for the year to date, June 30:

Income	*$11,000*
Assets (other than fixed assets)	64,600
Capital stock	45,000
Expenses (other than depreciation)	5,150
Liabilities	18,000
Equipment (at beginning of period)	3,000
Taxes	1,250

The depreciation for the period is $600. Prepare the income statement and the balance sheet for the period. Comment on the importance of depreciation to the cash flow of the business.

4.19 Work is estimated to have a fixed cost of $25, a variable cost of $1/unit, and a semifixed cost as given by

No. of Units	Semifixed Cost, $
1–10	$5
11–20	10
21–30	15

Find the total cost for 15 units and for 30 units. What is the average cost for these quantities?

4.20 A cost schedule of electrical rates is given as

Energy	Cost
First 50 kWh or less per month	$25.00
Next 50 kWh per month	$0.15 per kWh
Next 100 kWh per month	$0.09 per kWh
Next 800 kWh per month	$0.06 per kWh
Over 1000 kWh per month	$0.25 per kWh

What is the cost for 500 kWh? For 1250 kWh? Find the average cost per kWh. Discuss how "block" rates relate to fixed, semifixed, and variable cost concepts. Because this is a power company committed only to rural and farm service, how is it discouraging new business to commercial users?

4.21 A contractor specializes in concrete airport/road and commercial parking-surface paving. The engineer determines monthly fixed and variable costs as $10,000 and $2.50/ft^2, respectively. Semifixed costs are estimated as

Construction Rate, 1000 ft^2	Monthly Semifixed Costs, $
0–5,000	$40,000
5,000–9,000	60,000
9,000–16,000	140,000
16,000–20,000	200,000

Draw two plots, total cost versus construction rate, and cost per ft^2 versus units of area. If the engineer anticipates that the concrete paving rate will be in the range of 12,000–14,000 per 1000 ft^2 for the forthcoming summer months, what simplifications can you make to the cost analysis?

4.22 A large international engineering-construction firm has the following historical costs, expressed in $10^6, for the last 4 years:

Year	C_p	C_o	C_j
1	$240	$12	7.0
2	320	14	5.6
3	280	13	8.0
4	300	15	7.6

Calculate office and job overhead rates, and determine the rate for year 5. If in year 5 a project is estimated to have direct labor and material cost of $150 million, what is the overhead amount?

4.23 Historical costs for one project have provided the following data:

Direct material: $7,500,000
Direct labor:
   100 employees at $25 per hour for 525 hours each
   500 employees at $20 per hour for 875 hours each
   200 employees at $15 per hour for 1050 hours each
Subcontract costs: $2,500,000
Equipment: $600,000

General expenses include office rent, $250,000; insurance, $160,000; furniture and supplies, $150,000; telephone and computers, $50,000; and salaries, $500,000. Project expenses include permits, $25,000; superintendence, $100,000; storage and protection, $75,000; and other project expenses, $400,000.

Find the office and job overhead rate. (*Hint*: Consider equipment as project overhead expense.)

MORE DIFFICULT PROBLEMS

4.24 Analyze these transactions following the approach of Tables 4.2, Transaction Effects of a Business; Table 4.3, T-Accounts; Table 4.5, Balance Sheet; and Table 4.6, Profit-and-Loss Statement.

	Transactions	Account Affected
1.	Founded AJAX Construction paying $300,000 cash for capital stock	Cash Capital stock
2.	Bought material from Contractor Supply on credit for the amount $150,000	Inventory Accounts payable
3.	Paid rent for month, $4000	Rent Cash
4.	Sold material to M. Meyers for $35,000 cash	Cash Sales
5.	Paid $50,000 to Contractor Supply on account	Accounts payable Cash
6.	Borrowed $80,000 cash from First National Bank	Cash Note payable
7.	Retired $60,000 of capital stock	Capital stock Cash
8.	Sold material on account to K. Wilson for $60,000	Accounts receivable Sales
9.	Returned $20,000 material to Contractor Supply	Accounts payable Inventory
10.	Paid salaries and wages, $3000	Salaries Cash
11.	Paid taxes of $5000 from cash	Taxes Cash

4.25 A contractor acquires truck-mounted concrete-pumping equipment. Owing to the initial and operating costs, the contractor is in doubt as to the preferred method for charging depreciation. The contractor considers three methods: straight line, MACRS, and units of production.

The facts are as follows: initial cost, $1,460,000; useful life, 10 years; and a 20% salvage value. Production output starts at 200 yd^3/hr for a billing year of 2000 hr. It is estimated that the pumping rate will decline 10 yd^3/hr each year.

For the data, plot the book value of the asset against time for the methods and select the best choice. Discuss the merits of these allocation schemes. (*Hint:* The best choice is the minimum book value as time proceeds, thus minimizing taxable income.)

4.26 A contractor acquires truck-mounted concrete-pumping equipment. Her billing rate for pumping of concrete is dependent on a fixed cost, number of yd^3 pumped, round-trip miles distant from home base, and the cleaning of the equipment, which is conditioned on the yd^3 pumped. The equation is as follows:

$$\text{Cost per job} = 125 + 1 \times \text{No. of yd}^3 + 15 \times \text{No. of miles}$$

$$+ \begin{bmatrix} \text{yd}^3 \text{ pumped} & \$ \\ 15 - 25 & 56 \\ 25 - 59 & 104 \\ 60 \text{ or more} & 144 \end{bmatrix}$$

Overhead costs of 11% are additional.

A job is 5 one-way miles distant from the equipment's home location and requires 55 yd^3.

Find the job cost. (*Hint*: Notice the two types of variable costs and the semifixed costs. This cost is exclusive of the concrete material.)

4.27 Find the full cost of a job with the following information: Direct materials, $175,000; labor hours for project = 3800 hours, and labor wage without fringes and mandatory contributions = $25 per hour; fringes and mandatory contributions = 30%; office overhead rate = 10% on basis of full direct labor and material costs; permit and fees = $2500, bonds = $5000; job expenses and salaries = $18,000, supplies = $1500, and telephone = $1200; insurance = $15,000; utilities = $8000; and surveys = $28,000. What is the project overhead rate?

PRACTICAL APPLICATION

The understanding of financial management of construction is important. Indeed, if the student aspires to leadership in this business, then it is critical that he or she comprehend the important business documents, such as the income statement and the balance sheet. This practical application asks you to find a current annual report of a business and summarize the financial position of the company.

Your instructor will suggest several businesses that are able to respond to your questions about financial management of construction. As most construction businesses are "closely held," finding these annual reports can be challenging. There may be a business department at your campus, and they may have access to these resources. Then, too, *The Wall Street Journal* lists companies that are willing to provide annual reports. Web sites are candidates for requesting annual reports. For instance, visit the web address of Construction Financial Management Association, *www.CFMA.org*, and survey the topics that this professional organization considers to be important.

Your assignment is to prepare a written summary of a construction business, where the information is gleaned from their annual business report. Then give an oral report to your class. Your instructor will amplify these directions.

CASE STUDY: PERCENTAGE COMPLETION CONSTRUCTION, INC.

You are given an employment opportunity with a construction firm, and they provide you an annual report. Annual reports have three parts: prose, graphs, and tables. It is the tables that state business strengths or weaknesses. Basically, the income statement tells you how the company did this year in comparison to last year, and the balance sheet tells you how strong the finances are by indicating what the company owns and what it owes as of a certain date.

Conduct an analysis of Percentage Completion Construction, Inc. What ratios are useful? From another viewpoint, would you buy stock of this company, and would you work for this company? Prepare a one-page summary of this company, stating your opinion and giving an analysis of the financial documents.

Percentage Completion Construction, Inc.
Balance Sheet
December 31, 20xx

	Assets	
	This Year	Last Year
Cash	$265,000	$220,000
Certificates of deposit	40,000	
Contract receivables	3,800,000	3,300,000
Costs and estimated earnings in excess of billings on uncompleted contracts	80,000	100,000
Inventory	90,000	95,000
Prepaid charges and other assets	120,000	80,000
Property and equipment, net of accumulated depreciation	950,000	1,100,000
Total assets	$5,345,000	$4,895,000

	Liabilities	
	This Year	Last Year
Notes payable	$470,000	$580,000
Lease obligations payable	190,000	250,000
Accounts payable	2,540,000	2,600,000
Billings in excess of costs and estimated earnings on uncompleted contracts	207,600	563,350
Subtotal Liabilities	$3,407,600	$3,993,350
Shareholders' Equity		
Common stock—$1 par value, 300,000 outstanding shares	$300,000	$300,000
Retained earnings	1,637,400	601,650
Subtotal shareholders' equity	$1,937,400	$901,650
Total liabilities and shareholders' equity	$5,345,000	$4,895,000

Percentage Completion Construction, Inc.
Income Statement
December 31, 20xx

	This Year	Last Year
Contract revenues earned	$22,500,000	$16,000,000
Cost of revenues earned	20,000,000	14,500,000
Gross profit	2,500,000	1,500,000
Selling, general, and administrative expense	900,000	800,000
Income from operations	1,600,000	700,000
Other income (expense)		
Gain on sale of equipment	10,000	5,000
Interest expense	(60,000)	(70,000)
Income before taxes	1,550,000	635,000
Provision for taxes	364,250	133,350
Net income per share	1,185,750	501,650
$3.95 (this year)		
$1.67 (last year)		
Retained earnings, beginning of Year	601,650	250,000
Less dividends paid ($0.50 per share)	150,000	150,000
Retained earnings, end of year	$1,637,400	$601,650

CHAPTER 5

Forecasting

Despite all statements to the contrary, "emotional estimating" or "guesstimating" has not disappeared from the construction estimating scene. Nor has its substitute, estimation by formula and mathematical models, been universally nominated as a replacement. Somewhere between those extremes is a preferred course of action. In this chapter we consider the ways to enhance construction cost analysis by graphical and analytical techniques.

Business forecasting comprises the prediction of prices of material, availability and cost of labor, market demand, and costs of construction. Business forecasts influence every major construction decision. The usual approach to business forecasting involves extrapolation of past data into the future by using linear or nonlinear curves or mathematical relationships. Future demand is expected to follow some pattern of growth and decline. Most business forecasts are made for the short-run period of up to two years. Medium-term forecasts cover two to five years, and long-term forecasts are for more than five years.

In this chapter we consider statistical methods and indexing. Our intention is to be basic and to point out the special connections to construction. Forecasting does not mean estimating. Forecasting depends on statistical or mathematical methods and is a small but important part of the professional field of construction cost analysis. Students who have learned something of the statistical approach to construction estimating have found that cost advice is sharpened with a clearer understanding of these methods.

Computers are the normal way to handle forecasting, and software programs are available in such vast variety that we make no mention of them, except to remind the reader that the effective use of these conveniences begins with a knowledge of fundamentals. The present chapter focuses on some of these fundamentals.

5.1 GRAPHIC ANALYSIS OF DATA

Descriptive statistics are concerned with methods for collecting, organizing, analyzing, and summarizing, presenting visual or graphical images, drawing conclusions, and making decisions. In a more restricted sense, the word "statistic" connotes the data or the numerical measures derived from the data (for example, the average).

The data gathered for descriptive statistics and graphical presentations may be either discrete or continuous. They may be the result of a series of observations taken over time or another controllable or noncontrollable variable. Raw data communicate little information. An improved way to communicate information develops a frequency distribution, which compacts the data into viewable records.

A frequency distribution begins with the collection of observations into a tabular arrangement by intervals. For instance, a survey is made of the price for a roll of heavy 100 × 4 ft plastic underlayment.

Price interval, ($/roll)	Number of Observations	Relative Frequency	Cumulative Frequency
12.35–12.75	1	0.003	0.003
12.75–13.15	6	0.019	0.022
13.15–13.55	33	0.102	0.124
13.55–13.95	51	0.157	0.281
13.95–14.35	121	0.373	0.654
14.35–14.75	50	0.154	0.808
14.75–15.15	44	0.136	0.944
15.15–15.65	13	0.040	0.984
15.65–16.05	5	0.016	1.000
	324	1.000	

The price range of $12.35–$12.75 is called an interval, and its end numbers are called limits. The size of the interval is 0.40(= 12.75 − 12.35). The first midpoint is 12.55 $[= (12.35 + 12.75)/2]$. Relative frequency is the number of observations for each interval divided by all observations. Graphical representations of relative-frequency distributions are called histograms. Figure 5.1, which plots percentage observations against price ($/roll),

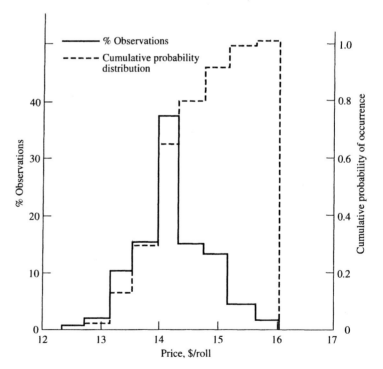

Figure 5.1 Relative-frequency curve and cumulative-probability-of-occurrence curve for data.

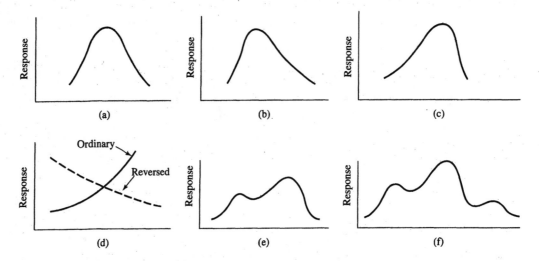

Figure 5.2 Several types of frequency curves obtainable from analysis of data: (a) symmetrical; (b) skewed to the right; (c) skewed to the left; (d) J shaped: (e) bimodal; and (f) multimodal.

is a histogram. If the relative frequencies are consecutively summed, then a cumulative frequency results. That is shown by the dashed line and right-hand scale of Fig. 5.1. If those data are considered representative of the parent population, then the term cumulative probability of occurrence can be used.

If the midpoints of various histogram cells are joined by a line and smoothed, the frequency curves will appear as in Fig. 5.2, which presents several types of frequency curves obtainable from analysis of data. A common title for the y axis would be "percentage of observations" or "count." Those graphical plots can be constructed similar to Fig. 5.1.

Curve construction calls for a "trained eye." For instance, if a wealth of data exists, then the bimodal or multimodal plot may be evident. But without an abundance of data, graphical conclusions of this sort are seldom found. Mathematical analyses are preferred for this and for other reasons. After the graphical plot is concluded, we calculate a measure of central tendency, such as mean, median, or mode, and a measure of dispersion, such as the standard deviation or range.

An average is a number typical or representative of a set of data. Sometimes called the arithmetic mean, it is found as

$$\bar{x} = \frac{x_1 + x_2 + \cdots + x_n}{n} = \frac{\sum\limits_{i=1}^{n} x_i}{n} = \frac{\Sigma x}{n} \tag{5.1}$$

The median is found by first ranking the data in ascending or descending order of magnitude. The middle value for an odd number of data, or the mean of the two middle values if the set number is even, is the median. The mode of data is the value that occurs with greatest frequency. The mode may not exist or may not be unique. The set, $-1, 0, 2, 4, 6, 6,$

7, and 8, has the mean 4, median 5, and mode 6. The set, $-1, 0, 2$, and 4, has no mode. The set, $-1, 0, 0, 4, 6, 6, 7$, and 8, is bimodal with the modes 0 and 6.

The degree to which numerical data spread about a mean value is called the dispersion of the data. Various measures of dispersion are available, the most common being the range and standard deviation. The range of a sample of data is the difference between the largest and smallest numbers in a set. The standard deviation of a set of n numbers, $x_1, \ldots, x_n$, is denoted by s and is defined by

$$S = \sqrt{\frac{\sum_{1}^{n} (x_i - \bar{x})^2}{n - 1}} \tag{5.2}$$

The standard deviation is determined relative to the mean of the sample. The variance of a set of data is defined as the square of the standard deviation, or s^2. The set $(-1, 0, 2, 4)$ has the range of $5, s = 2.22$, and $s^2 = 4.92$.

Plotting of data is an important step in graphical analysis. Despite the ease with which data can be mathematically and statistically analyzed with computers, plotting of two-variable data, y versus x, for example, is useful to gain a "feeling" from the actual data. Mathematical relationships, void of a visual or real-life experience, lead to misjudgment in estimating. Very frequently you need to see the plot to understand what is going on. Knowledge, skill, and practice derived from direct observation of or participation in the statistical analysis help to develop judgment that is so necessary for estimating.

A graph is a pictorial presentation of the relationship between variables. Most graphs have two variables, where we presume for construction cost estimating that x is the independent or controlled variable and y is the dependent variable. Once raw data are gathered, the next steps select the axes and divisions, locate the points, and draw the straight line (preferably). A rule of thumb to follow in plotting by eye is to have half of the points above the line and half below it, excluding those points that lie on the line.

With the line drawn we find the graphical straight line equation by using

$$y = a + bx \tag{5.3}$$

where y = dependent variable
           a = intercept value along the y axis at $x = 0$
           b = slope, or the length of the rise divided by the length of the run
           x = independent or control variable

Measurements of a and b are determined from the graph, because the intercept and slope can be measured from the line drawn through the points. This step, frequently overlooked in cost analysis, improves the gaining of experience that is necessary in construction cost estimating.

Both arithmetic and logarithmic axes are popular in cost analysis and are described by Fig. 5.3. Trial plots should attempt different groupings of the axes, such as semilog or log log, in addition to arithmetic axis. That plot that best "straightens out" scattered data is selected. In Fig. 5.4(a), the axes are arithmetic, and slope b is calculated from the right triangle. The line is extended to the y axis at $x = 0$, and at this point the value a is measured from

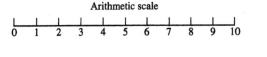

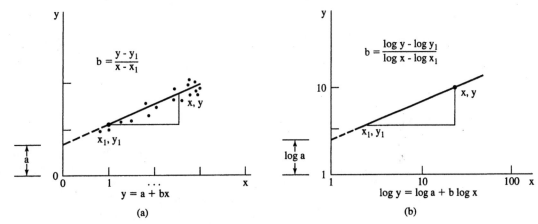

Figure 5.4 Plotting straight lines on (a) arithmetic and (b) logarithmic scales.

the graph. The logarithmic copy of Eq. (5.3) is given by Fig. 5.4(b). The slope is found by using the points (x, y) and (x_1, y_1) after the line is plotted by using the rule of thumb.

Figure 5.5 is an example of a linear plot. Note that the points are left unerased on the graph and indicate the variability that may exist with time or cost data. At the nominal 10-in. pipe size, the variability of man hours is four times from a minimum to maximum, not an unusual event for cost data. Frequently, a graphical line may show a sharp slope change that is not evident in mathematical analysis. Graphical plotting shows this knee jerk in data. When possible, graphical plots are important to give this feel.

Graphical plots and equations will vary between two analysts, but their mathematical equations should be identical. Initially, graphical plots are used to screen and aid the selection of the best model that fits the data as a straight line, and then the method of least squares analyzes data for their equation. The simple formula $y = a + bx$ is often sufficient for practical day-to-day estimating work, but when the line appears nonlinear or is

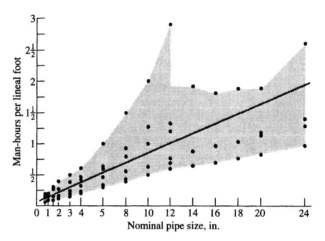

Figure 5.5 Illustrative example of man hours to handle and install standard pipe.

logarithmic straight, curves other than those found by using arithmetic scales are used. Attention turns to mathematical analysis after the plot and the graphical equations are concluded.

5.2 LEAST-SQUARES AND REGRESSION

The descriptive statistical methods are concerned with a single variable and its frequency variation. But many problems in the construction estimating field involve several variables. Simple methods for dealing with data associated with two or three variables will be explained here. Those are known as regression models and are important tools.

In regression, and on the basis of sample data, we want to find the value of a dependent variable y corresponding to a given value of a variable x. This is determined from a least-squares equation that fits the sample data. The resulting curve is called a regression curve of y on x, because y is determined from a corresponding value of x. If the variable x is time, then the data show the values of y at various times, and the equation is known as a time series. A regression line or a curve y on x or the response function on time is frequently called a trend line and is used for prediction and forecasting. Thus, regression refers to average relationship between variables.

5.2.1 Least Squares

The notion of fitting a curve to a set of sufficient points is essentially the problem of finding the parameters of the curve. The best-known method is that of least squares. Since the desired curve or equation is to be used for estimating or predicting purposes, the curve or equation should be so modeled as to make the errors of estimation small. An error of estimation means the difference between an observed value and the corresponding fitted curve value for the specific value of x. It will not do to require that the sum of these differences or errors to be as small as possible. It is a requirement that the sum of the absolute values of the errors be as small as possible. However, sums of absolute values are not

PICTURE LESSON
Wooden Trestle

The importance of neighborhood materials is evident in this picture, where a railroad company uses the immediate timbers for construction of the standard-gauge railway trestle—a braced framework serving as a support of timbers, piles, or steelwork for carrying a road or railroad over a depression.

Washington State loggers depended on getting their product of large old-growth timber to market, and the numerous logging companies encouraged the buildling of railroads. Crossing the streams, rivers, and tributaries meant that many thousands of bridges were built during any decade of railroad construction. Cost of a logging railway, used for the one purpose of bringing product to market, was low. As an example, for an average grade of 3.3%, heavily wooded terrain, with the presence of roots, railway cost about $9100 per mile.

Saw mills near the construction site provided all framed timber. Horse and steam power was used for hoisting and moving. Creosoting was unlikely, and unseasoned timbers were the rule rather than the exception. Wooden trestles, while inexpensive to build, were expensive to maintain. But replacement of wooden trestles was unnecessary, as the forest was extensively harvested, and the railway was allowed to deteriorate.

Although it has valuable properties, timber is not a suitable material for permanent bridgework, and it is not used in large works of importance today.

convenient mathematically. The difficulty is avoided by requiring that the sum of the squares of the errors be minimized. If this procedure is followed, the values of parameters give what is known as the best curve in the sense of least-squares difference.

The principle of least squares states that if y is a linear function of an independent variable x, then the most probable position of line $y = a + bx$ will exist whenever the sum of squares of deviations of all points (x_i, y_i) from the line is a minimum. Those deviations are measured in the direction of the y axis. The underlying assumption is that x is either free of error (a controlled assignment) or subject to negligible error. The value of y is the observed or measured quantity, subject to errors that have to be "eliminated" by this method of least squares. The value y is a random-variable value from the y-population values corresponding to a given x. For each value x_i we are interested in corresponding y_i.

If a very large number of experiments were made, a histogram and eventually a normal curve, or Gaussian curve, or bell-shaped curve, could be constructed. The bell-shaped curve, shown as Fig. 5.6(a), has the remarkable property that as an experiment is repeated under similar circumstances, the response variable will vary as a normal curve. If natural processes are conducted in the experiment, 68.27% of the data fall within $\pm 1\sigma$, 95.45% within $\pm 2\sigma$, and 99.73 within $\pm 3\sigma$, as is evident in Fig. 5.6 The area under the normal curve is 1.

Suppose that our observations consist of pairs of values as x_i, y_i, which are assumed to give a linear plot as in Fig. 5.7. The symbol $\bar{y}_i$ ("y-bar") is the average value resulting from the x_i controlled variable. For instance, if the value of $x_i = 5$ and the experiment were repeated 10 times, $\bar{y}_i$ would be the mean value of 10 observations. The bell-shaped curve is positioned with respect to the y-axis value and is considered a random variable. Our problem is to uncover a and b for what is known as "a best fit." For a general point i on the line, $y_i - (a + bx_i) = 0$, but if an error ε_i exists, then $y_i - (a + bx_i) = \varepsilon_i$. For n observations we have n equations of

$$y_i - (a + bx_i) = \varepsilon_i \tag{5.4}$$

where ε = difference between actual observation and regression value.

Summing, we can write the sum of squares of these residuals as

$$\sum_{i=1}^{n} \varepsilon^2 = \sum_{i=1}^{n} \left[y_i - (a + bx_i) \right]^2 \tag{5.5}$$

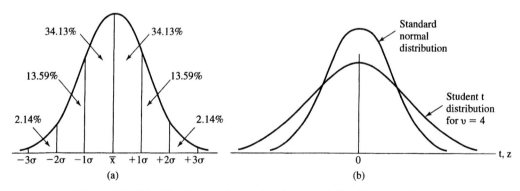

Figure 5.6 Distributions: (a) standard normal distribution showing proportion of area in standard-deviation zones, and (b) normal and Student t distributions compared.

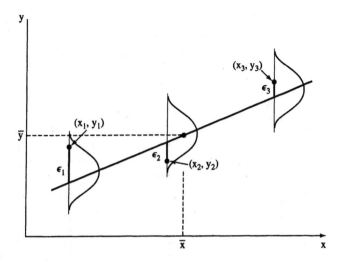

Figure 5.7 Regression line of $y = a + bx$ showing deviations from regressed values.

For a minimum, we insist on

$$\frac{\partial \Sigma \varepsilon^2}{\partial a} = 0, \quad \frac{\partial \Sigma \varepsilon^2}{\partial b} = 0 \tag{5.6}$$

or

$$n\Sigma xy - \Sigma x\Sigma y = n\Sigma(x - \bar{x})(y - \bar{y})$$
$$= n(\Sigma xy - \bar{x}\Sigma y + \bar{x}\Sigma y - \bar{y}\Sigma x) \tag{5.7}$$

and solving those two normal equations for a and b we have

$$a = \frac{\Sigma x^2 \Sigma y - \Sigma x \Sigma xy}{n\Sigma x - (\Sigma x)^2}$$

and

$$b = \frac{n\Sigma xy - \Sigma x\Sigma y}{n\Sigma x^2 - (\Sigma x)^2} \tag{5.8}$$

Those a and b values are substituted into $y = a + bx$. This least-squares equation passes through $(\bar{x}, \bar{y})$, which is the coordinate mean of all observations. The calculations of those coefficients are handled with a computer or calculator when serious numerical problems are given.

Table 5.1 is an example of a tabular form in determining a and b regression values for the equation $y = a + bx$. If Y = index and X = year for the index, the equation of the least-squares line is $Y = 84.875 + 2.389X$. The four left columns of Table 5.1 are necessary for regression. The columns x and y are the original data, and x^2 and xy are required computations. Equation 5.8 uses the total of those columns for finding constants a and b.

TABLE 5.1   Calculation of Regression Coefficients

Year, x	Index, y	x^2	xy	$\hat{y}$	$\varepsilon = y - \hat{y}$	ε^2
0	87	0	0	84.875	2.125	4.516
1	89	1	89	87.264	1.736	3.014
2	90	4	180	89.653	0.347	0.120
3	92	9	276	92.042	−0.042	0.002
4	93	16	372	94.431	−1.431	2.047
5	99	25	495	96.820	2.180	4.752
6	97	36	582	99.209	−2.209	4.879
7	100	49	700	101.598	−1.598	2.554
8	101	64	808	103.987	−2.987	8.922
9	106	81	954	106.376	−0.376	0.141
10	106	100	1,060	108.765	−2.765	7.645
11	109	121	1,199	111.154	−2.154	4.640
12	115	144	1,380	113.543	1.457	2.123
13	118	169	1,534	115.932	2.068	4.277
14	122	196	1,708	118.321	3.679	13.535
105	1524	1015	11,337			63.167

For $Y = a + bx$, the constants

$$a = \frac{\Sigma y \Sigma x^2 - \Sigma x \Sigma xy}{n\Sigma x^2 - (\Sigma x)^2} = \frac{(1524)(1015) - (105)(11{,}337)}{15(1015) - (105)^2} = 84.875$$

$$b = \frac{n\Sigma xy - \Sigma x \Sigma y}{n\Sigma x^2 - (\Sigma x)^2} = \frac{15(11{,}337) - (105)(1524)}{15(1015) - (105)^2} = 2.389$$

$\hat{Y} = 84.875 + 2.389X$ or if X is year, then index $= 84.875 + 2.389X$. The forecast mean value for year $x_i = 15$ is $Y = 84.875 + 2.389(15) = 120.71$.

$$S_y = \left(\frac{\Sigma \varepsilon^2}{n - 2}\right)^{1/2} = \left(\frac{63.167}{13}\right)^{1/2} = 2.204$$

$$S_{y_i} = S_y\left[\frac{1}{n} + \frac{(x_i - \bar{x})^2}{\Sigma(x - \bar{x})^2}\right]^{1/2} = 2.204\left[\frac{1}{15} + \frac{(15 - 7)^2}{280}\right]^{1/2} = 1.198$$

For degrees of freedom $= 13$ and a 5% level of significance, $t = 2.160$. The confidence interval is $120.71 \pm 2.160(1.198) = (118.122, 123.298)$. For a single estimated value y_i for $x_i = 15$,

$$S_{y_i} = S_y\left[1 + \frac{1}{n} + \frac{(x_i - \bar{x})^2}{\Sigma(x - \bar{x})^2}\right]^{1/2} = 2.204\left(1 + \frac{1}{15} + \frac{64}{280}\right)^{1/2} = 2.508$$

The prediction interval is $120.71 \pm 2.160(2.508) = (115.293, 126.127)$. The confidence interval for the slope and intercept is

$$S_b = \frac{S_y}{[\Sigma(x - \bar{x})^2]^{1/2}} = \frac{2.204}{(280)^{1/2}} = 0.132$$

slope interval $= 2.389 \pm 2.160(0.132) = (2.103, 2.674)$

$$S_a = S_y\left[\frac{1}{n} + \frac{x^2}{\Sigma(x - \bar{x})^2}\right]^{1/2} = 2.204\left(\frac{1}{15} + \frac{49}{280}\right)^{1/2} = 1.083$$

intercept interval $= 84.875 \pm 2.160(1.083) = (82.536, 87.214)$

The limitations of the method must be pointed out. The method of least squares is applicable when the observed values of y_i correspond to assigned (or error-free) values of x_i. The error in y_i (expressed as a variance of y) is assumed to be independent of the level of x. If inferences are to be made about regression, it is also necessary that the values of y_i corresponding to a given x_i be distributed normally, as presented in Fig. 5.7 with the mean of the distribution satisfying the regression equation. The variance of the values of y_i for any given value of x must be independent of the magnitude of x.

Though the evidence of this statement can be statistically shown, experience shows that only a comparatively small number of the distributions met within cost analysis can be described by normal distributions. Cost data are limited at the zero end. Distributions influenced by business, engineering, and human factors are generally skewed. Despite those drawbacks, the least-squares method is widely used. Imperfection, apparently, does not reduce popularity.

5.2.2 Confidence Limits for Regression Values and Prediction Limits for Individual Values

If variations around the universe regression are random, then the method of least squares permits the computation of sampling errors and provides for determining the reliability of the estimate of the dependent variable from the fitted line. Confidence limits for regression values can be constructed through the extension of simple statistics. The confidence limits for individual regression values and for the straight line are quadratic in form around the sample line of regression. The confidence band for the slope is fan shaped with the apex at the mean, and the confidence limits for the intercept are parallel lines.

The variance of an estimate permits the forming of confidence limits of the estimate. The approach, similar to the variance of a sample, in this case reckons the deviations from a line instead of a mean. The variance of y, estimated by the regression line, is the sum of squares of deviations divided by the number of degrees of freedom available for calculating the regression line, or

$$s_y^2 = \frac{\Sigma \varepsilon_i^2}{\nu} \tag{5.9}$$

where s_y^2 = variance around the regression line

ε_i is as defined previously by Eq. (5.4). Only two bits of information are required to determine the regression line: means $(\bar{x}, \bar{y})$ and either slope b or intercept a. With n as the number of paired observations, ν is defined as

$$\nu = n - 2 \tag{5.10}$$

where ν = degrees of freedom, number

Also,

$$s_y^2 = \frac{\Sigma \varepsilon_i^2}{n - 2} \tag{5.11}$$

Define

$$s_{\bar{y}}^2 = \frac{s_y^2}{n} \tag{5.12}$$

where $s_{\bar{y}}^2$ = variance of the mean value of y, or $\bar{y}$

We can now write the confidence limits for $\bar{y}$. A table of the Student t distribution and of the values of t corresponding to various values of the probability (level of significance) and a given number of degrees of freedom ν is found in Appendix 2. The Student t distribution is similar in shape to the normalized standard variable (compare Appendixes 1 and 2), except that the t distribution is flatter, i.e., has area and probability farther in the tails, both right and left. See Fig 5.6(b). With this t value we state that the true value of y lies within the interval

$$\bar{y} \pm t s_{\bar{y}} \tag{5.13}$$

The probability of being wrong is equal to the level of significance of the value of t. Because the regression line must pass through the mean, an error in the value of $\bar{y}$ leads to a constant error in y for all points on the line. The line is then moved up or down without change in slope.

If limits for an individual value are desired, then a different approach must be asserted. An individual value is not the same as the mean value. An index value, for instance, is a single calculated number and would lead to a future individual value, not a future mean value.

The statement that usually describes the limits for individual values goes like this: If we use the sample line of regression to estimate a particular value for y, then we add to the error of the sample line of the regression some measure of the possible deviation of the individual value from the regression value. For individual values, a new set of parabolic loci may be viewed as prediction limits. Figure 5.8 presents the prediction loci for individual values as well as the confidence loci for average values. Note that the prediction limits for y get wider as x deviates up and down from its mean, both positively and negatively. This implies that predictions of the dependent variable are subject to the least error when the independent variable is near its mean and are subject to the greatest error when the independent variable is distant from its mean.

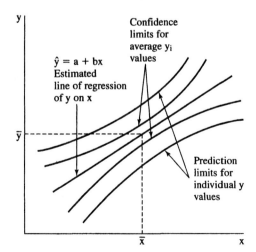

Figure 5.8 Confidence bands for average values and prediction limits for individual values.

If we require an estimate of the confidence limits corresponding to any x_i, we calculate the limits for $\hat{y}$ (called "y-hat" and, strictly speaking, a statistical estimate). The variance of the estimate of this mean value is

$$s^2_{y_{\bar{i}}} = s^2_y\left[\frac{1}{n} + \frac{(x_i - \bar{x})^2}{\Sigma(x - \bar{x})^2}\right] \tag{5.14}$$

where $s^2_{y_{\bar{i}}}$ = variance of each mean value of y

Notice that a bar is over the subsubscript i. A new $s^2_{y_{\bar{i}}}$ is computed for each x value. The confidence interval for the mean estimated value of $y_{\bar{i}}$ corresponding to specific x_i is

$$y_{\bar{i}} \pm ts_{y_{\bar{i}}} \qquad ' \tag{5.15}$$

The interior confidence limits of Fig. 5.8 correspond to Eq. (5.15). For a predetermined level of significance we are able to predict the limits within which a future mean estimated value of y_i will lie with an appropriate chance of error.

We can find the prediction interval of a single estimated value of $\hat{y}_{\bar{i}}$ using the variance of a single value, which has as its variance

$$s^2_{y_i} = s^2_y\left[1 + \frac{1}{n} + \frac{(x_i - \bar{x})^2}{\Sigma(x - \bar{x})^2}\right] \tag{5.16}$$

where $s^2_{y_i}$ = variance of the individual value of y

This variance is larger than S^2_y because the variance of the single value is equal to the variance of the mean plus the variance of $\hat{y}$ estimated by the line, or

$$s^2_{y_i} = s^2_{y_{\bar{i}}} + s^2_y \tag{5.17}$$

Each x requires a separate value of $S^2_{y_i}$. The prediction interval for a single value is greater, too, or

$$y_i \pm ts_{y_i} \tag{5.18}$$

In Fig. 5.8 the external lines are computed by using Eq. (5.18). The terms confidence and prediction interval have different meanings. A confidence interval deals with an expected average Y value. A prediction interval deals with a single Y value. The prediction interval is greater in magnitude.

The variance of the intercept a is a particular case of the variance of any mean estimated y_{y_i}. If we substitute $x_i = 0$ in Eq. (5.14), then the variance of intercept a is

$$s^2_a = \left[\frac{1}{n} + \frac{\bar{x}^2}{\Sigma(x - \bar{x})^2}\right] \tag{5.19}$$

Its confidence band is given by

$$a \pm ts_a \tag{5.20}$$

Refer to Fig. 5.9(a), which corresponds to Eq. (5.19).

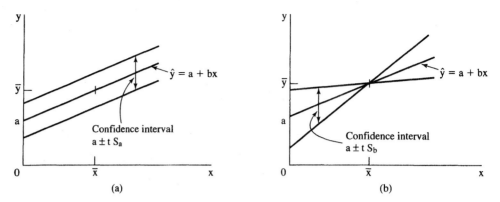

Figure 5.9 (a) Confidence interval for intercept a, and (b) confidence interval for slope b.

The variance of slope b is given as

$$s_b^2 = \frac{s_y^2}{\Sigma(x - \bar{x})^2} \tag{5.21}$$

Its confidence band is given by

$$b \pm ts_b \tag{5.22}$$

Refer to Fig. 5.9(b), which corresponds to Eq. (5.21).

The confidence band for the slope is represented by a double fan-shaped area with the apex at the mean. The number of degrees of freedom is $\nu = n - 2$ for all confidence intervals in this section.

Table 5.1 demonstrates the extended calculation of regression values, namely, the parameters a and b. The $\hat{y}$ column is calculated by using the regression equation $\hat{y} = 84.875 + 2.389x$. For example, if $x = 1$, then $\hat{y} = 87.264$. This calculation is repeated for each observation of x. The $\varepsilon = y - \hat{y}$ column is the error difference between the observed and regression values. The right column is the square of the error. The sum 63.167 is used to calculate S_y. Confidence and prediction intervals are found for $x = 15$ and are given in Table 5.1.

5.2.3 Curvilinear Regression and Transformation

The world of linearity, largely an imaginary one, is a tidy and manageable assumption about which generalizations can be asserted with boldness. The function $y = a + bx$ is a characterization frequently employed for known nonlinear situations. Many prefabricated tools are known that can be used with those assumptions. But computers and a greater awareness are forcing a reevaluation of linearities.

A listing of nonlinear relationships for construction estimating purposes would give the following:

$$y = ae^{bx} \text{ semilog} \tag{5.23}$$

$$y = ab^x \text{ exponential} \tag{5.24}$$

$$y = ax^b \text{ power} \tag{5.25}$$

$$y = a + \frac{b}{x} \text{ reciprocal } x \tag{5.26}$$

$$y = \frac{1}{(a + bx)} \text{ reciprocal} \tag{5.27}$$

$$y = \frac{x}{(a + bx)} \text{ hyperbolic} \tag{5.28}$$

$$y = a + b_1 x + b_2 x^2 + \cdots \text{ polynomial} \tag{5.29}$$

The exponential and power fit, Eqs. (5.24) and (5.25), are used frequently. The regression formulas of Sec. 5.1 can be made to work with those curvilinear models. First, those equations appear like $y = a + bx$ if transformed. As an example, the exponential function $y = ab^x$ transforms to

$$\log y = \log a + x \log b, \text{ which appears as } y = a + bx.$$

The three exponential curves $y = ae^{bx}$, $y = 10^{a+bx}$, and $y = a(10)^{bx}$ are equivalent. For if any two points are chosen and the constants for each curve determined so it passes through these points, then the curves are equal.

Let $y = \log y$, $a = \log a$, $b = \log b$, and $x = x$. This equation gives a straight line when plotted on arithmetic-logarithmic scales. Similar to Eq. (5.8), the values of the parameters of Eq. (5.24) can be solved by using

$$\log a = \frac{\Sigma x^2 \Sigma \log y - \Sigma x \Sigma y \log y}{n\Sigma x^2 - (\Sigma x)^2}$$

$$\log b = \frac{n\Sigma x \log y - \Sigma x \Sigma \log y}{n\Sigma x^2 - (\Sigma x)^2} \tag{5.30}$$

The power function $y = ax^b$ is transformed from a curved line on arithmetic scales to a straight line on log-log scales if we let $y = \log y$, $a = \log a$, and $x = \log x$. Intercept and slope equations can be found by making those substitutions into Eq. (5.8).

$$\log a = \frac{\Sigma(\log x)^2 \Sigma \log y - \Sigma \log x \Sigma (\log x \log y)}{n\Sigma(\log x)^2 - (\Sigma \log x)^2}$$

$$b = \frac{n\Sigma(\log x \log y) - \Sigma \log x \Sigma \log y}{n\Sigma(\log x)^2 - (\Sigma \log x)^2} \tag{5.31}$$

The power equation models the important "learning" concept, useful for estimating construction projects. If we repeat the construction project, then it takes less time or cost to construct what has been constructed earlier, which is known as learning. The principle has been applied to natural-gas-fired power plants (even with different general contractors) to show the declining cost, which reflects the technological learning. Residential construction (for time of total work force), nuclear-fired central steam power plants, and other projects have been successfully modeled by the power equation.

Assume that we have five sets of separate experiences, or the unit number x and the man hours y, to construct that design. Data (unit number, 10^3 hours) are $(10, 510)$, $(30, 210)$, $(100, 190)$, $(150, 125)$, and $(300, 71)$. Thus, the 10th and 300th units require 510,000 and 71,000 hours, respectively. Table 5.2 shows the calculation. The first two columns are the original units, and the four right columns are the transformed calculations. The transformed and original unit equations also give identical results. Let $x = $ 500th unit; then $\log y = 3.1921 - 0.515 \log 500 = 1.8021$, and antilog $y = 63.4$ hours.

Also $y = 1556(500)^{-0.515} = 63.4$ hours (in original units), showing the agreement between the two approaches. We call these approaches "transformed" (in the logarithm units) and "original." Notice the calculation given in Table 5.2.

Each of these nonlinear functions can be statistically evaluated for the fitted data, such as

- Coefficients of the equations
- Listing of calculated estimates
- Standard error of y, $\Sigma \varepsilon_i^2 = (y_i - \hat{y})^2$
- Squared correlation coefficient
- Confidence-limits values for y

Computer programs do these calculations wherever the work is large.

Normally, the best-fit line for regression is judged by the smallest value of the standard error. Other computed values provide an intuitive feel for the accuracy of the correlation. Polynomial regression—that is, where for any x the mean of the distribution of y is given by $a + b_1 x^2 + b_2 x^2 + b_3 x^3 + \cdots + b_p x^p$—is used to obtain approximations whenever the functional form of the regression curve is a mystery. We plot data in a variety of ways, hoping to "straighten out" an arithmetic curve by means of semilog or log-log plots. If samples of paired data straighten out on semilog paper, we conclude that the form is exponential or $y = ae^{bx}$ and we would be tempted to apply least-squares methods that used this form to the data.

Suppose that data, which are coded to facilitate understanding and computation, were obtained as follows, where y, or dollars, is assumed to be related to construction as $y = a + bx^2$.

x	Observed value, y_i	Error-free value, y	Deviation, $y_i - y$
0	1	a	$1 - a$
1	1.4	$a + b$	$1.4 - (a + b)$
2	1.8	$a + 4b$	$1.8 - (I + 4b)$
3	2.2	$a + 9b$	$2.2 - (a + 9b)$

The sum of the squares of the deviations is

$$\Sigma \varepsilon^2 = (1 - a)^2 + (1.4 - a - b)^2 + (1.8 - a - 4b)^2 + (2.2 - a - 9b)^2$$

TABLE 5.2   Least Squares Analysis of Illustrative Cost Data to Find Initial Value and Learning Slope

Given: $y = ax^b$

or

$$\log y = \log a + b \log x$$

which is of the form

$$y = a + bx$$

Let $y = \log y$, $a = \log a$ intercept, $b = b$ slope, $x = \log x$, and $n =$ sample size.

Unit, x	Man Hours, y	$x = \log x$	$y = \log y$	$(\log x)^2$	$\log x \log y$
10	510	1.0000	2.7076	1.0000	2.7076
30	210	1.4771	2.3222	2.1818	3.4301
100	190	2.0000	2.2788	4.0000	4.5576
150	125	2.1761	2.0969	4.7354	4.5631
300	71	2.4771	1.8513	6.1360	4.5859
		9.1303	11.2568	18.0532	19.8443

$$\log a = \frac{\Sigma(\log x)^2 \Sigma \log y - \Sigma \log x \Sigma(\log x \log y)}{n\Sigma(\log x)^2 - (\Sigma \log x)^2}$$

$$\log a = \frac{(18.0532)(11.2568) - (9.1303)(19.8443)}{5(18.0532) - (9.1303)^2} = 3.1921$$

$$b = \frac{n\Sigma(\log x \log y) - \Sigma \log x \Sigma \log y}{n\Sigma(\log x)^2 - (\Sigma \log x)^2}$$

$$= \frac{5(19.8443) - (9.1303)(11.2568)}{5(18.0532) - (9.1303)^2} = -0.515$$

Then

$$\log y = 3.1921 - 0.515 \log x \text{ (in logarithm units)}$$

$$\text{antilog } a = 1556 \text{ (initial value at } x = 1)$$

$$y = 1556x^{-0.515} \text{ in original units}$$

What is the estimate for $x = 350$ units?

In logarithm units:

$$\log y = 3.1921 - 0.515 \log 350 = 1.8819$$

$$\text{antilog } y = 76.19 \text{ hours}$$

In original units:

$$y = 1556(350)^{-0.515} = 76.19 \text{ hours}$$

For a minimum we are required to satisfy

$$\frac{\partial \Sigma \varepsilon^2}{\partial a} = 0, \quad \text{and} \quad \frac{\partial \Sigma \varepsilon^2}{\partial b} = 0 \tag{5.32}$$

or $(1 - a) + (1.4 - a - b) + (1.8 - a - 4b) + (2.2 - a - 9b) = 0$

and $(1.4 - a - b) + 4(1.8 - a - 4b) + 9(2.2 - a - 9b) = 0$

$$4a + 14b = 6.4$$

$$14a + 98b = 28.4$$

Hence, $a = 1.175$ and $b = 0.122$. When substituted back into the general form, the fitted equation becomes $y = 1.175 + 0.122x^2$. A better fit is obtained from an equation of a different form or $y = a + bx + cx^2$.

The previous problem is straightforward, but the application of the method of least squares to nonlinear relations usually requires a good deal of computational effort, made easier with computers/software. In most cases we can "transform or rectify" a nonlinear relation to a straight-line relation. This manipulation simplifies handling of the data and permits a linear graphical presentation that may be more revealing for certain facts. With a rectified straight line, extrapolation is simpler, and the computations of certain other supportive statistics, such as the standard deviation or confidence limits, are simpler.

Now, those previous functions indicated a regression of y on x that was linear in some fashion. Sometimes a clear relationship is not evident, and a general polynomial is selected. A predicting equation of the polynomial form [see Eq. (5.29)] requires a set of data consisting of n points (x_i, y_i). We estimate the coefficients a, b_1, b_2, b_p of the pth-degree polynomial by minimizing

$$\sum_{i=1}^{n} \left[y_i - \left(a + b_1 x + b_2 x^2 + \cdots + b_p x^p \right) \right]^2 \tag{5.33}$$

according to

$$\frac{\partial \Sigma \varepsilon^2}{\partial a}, \frac{\partial \Sigma \varepsilon^2}{\partial b_1}, \frac{\partial \Sigma \varepsilon^2}{\partial b_2}, \cdots, \frac{\partial \Sigma \varepsilon^2}{\partial b_p} = 0 \tag{5.34}$$

which is the least-squares criterion, by minimizing the sum of squares of the vertical distances from the points to the curve. This results in $p + 1$ normal equations of the shape:

$$\Sigma y = na + b_1 \Sigma x + \cdots + b_p \Sigma x^p$$

$$\Sigma xy = a\Sigma x + b_1 \Sigma x^2 + \cdots + b_p \Sigma x^{p+1}$$

$$\vdots \tag{5.35}$$

$$\Sigma x^p y = a\Sigma x^p + b_1 \Sigma x^{p+1} + \cdots + b_p \Sigma x^{2p}$$

where summation super–and subscript notation has been eliminated. We now have $p + 1$ linear equations in $p + 1$ unknowns $a, b_1, \cdots, b_p$.

Consider the following example: Road alignment is restricted to national highway code as specified by the state highway department, and any requirements above the code are incremental costs to the bid price. The data (with x = in., y = cost) are given as follows:

Alignment grade above standard allowance, in.	Incremental Cost, $/ft^2$
0.01	7.00
0.02	8.40
0.03	9.20
0.04	10.10
0.05	10.30
0.20	26.20

Tabulations, very similar to those in Table 5.1, give the following:

$$\Sigma x = 0.35 \qquad \Sigma y = 71.2$$

$$\Sigma x^2 = 0.0455 \qquad \Sigma xy = 6.673$$

$$\Sigma x^3 = 0.008225 \qquad \Sigma x^2 y = 11.0225$$

$$\Sigma x^4 = 0.00016979$$

With those values we are ready to solve the following system of three linear equations,

$$6a + 0.35b_1 + 0.0455b_2 = 71.2$$

$$0.35a + 0.0455b_1 + 0.008225b_2 = 6.673$$

$$0.0455a + 0.008225b_1 + 0.0016979b_2 = 11.0225$$

for which $a = -31.73$, $b_1 = 1660$, and $b_2 = -7025$, and the predicting equation becomes $y = -31.73 + 1660x - 7025x^2$.

In practice it may be difficult to determine the degree of the polynomial to fit data, but it is always possible to find a polynomial of degree at most $n - 1$ that will pass through each of n points. For practical reasons, however, we prefer the lowest degree that describes our data.

5.2.4 Correlation

The methods of regression show the relationship of one dependent variable to an independent variable that can be considered linearly related. There is a closely related measure, called correlation, that tells how well the variables are satisfied by a linear relationship. If the values of the variables satisfy an equation exactly, then the variables are perfectly correlated. With two variables involved, the statistician refers to simple correlation and simple regression. When more than two variables are involved, they are multiple regression and multiple correlation. Only simple correlation is considered in this book.

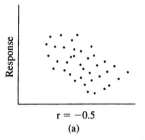

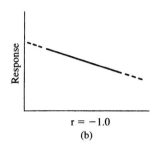

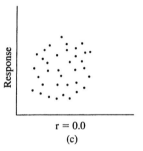

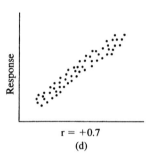

Figure 5.10 Scatter diagrams of data showing levels of correlation coefficients: (a) negative linear correlation; (b) negative linear correlation with points exactly on line; (c) no correlation; and (d) positive linear correlation.

Figure 5.10 indicates the location of points on an arithmetic coordinate system. If all the points in a scatter diagram appear to lie near a line, as in Fig. 5.10(b) or (d), the correlation is presumed to be linear, and a linear equation is appropriate for regression or estimation. If there is no relationship indicated between the variables, as in Fig. 5.10(c), then there is no correlation (i.e., the data are uncorrelated, and any number of straight lines can be drawn through the randomly mixed cloud of observations). In Fig. 5.10(b) the correlation coefficient is negative linear. For Fig. 5.10(d) a positive linear correlation coefficient is found.

With a fitted curve from data it is possible to distinguish between the deviations of the y observations from the regression line and the total variation of the y observations about their mean. A calculated difference between the two variations gives the amount of variation accounted for by regression. The higher this value, the better the fit or correlation. For $y = a + bx$, no correlation exists if $b = 0$, as in Fig. 5.10(c), and the line plots as a horizontal line. Thus, x and y are independent.

There is no correlation of x on y if x is independent of y. These two statements are expressed mathematically as

$$\frac{n\Sigma xy - \Sigma x \Sigma y}{n\Sigma x^2 - (\Sigma x)^2} = 0 \tag{5.36}$$

and

$$\frac{n\Sigma xy - \Sigma x \Sigma y}{n\Sigma y^2 - (\Sigma y)^2} = 0 \tag{5.37}$$

For no correlation the product of Eqs. (5.36) and (5.37), or the product of the slopes, is zero. Conversely, for perfect correlation all the points lie exactly on each of the two regression lines and their lines coincide, or

$$\frac{n\Sigma xy - \Sigma x \Sigma y}{n\Sigma x^2 - (\Sigma x)^2} \times \frac{n\Sigma xy - \Sigma x \Sigma y}{n\Sigma y^2 - (\Sigma y)^2} = 1 \tag{5.38}$$

The square root of this product is called the correlation coefficient, and it is denoted by r:

$$r = \frac{n\Sigma xy - \Sigma x \Sigma y}{\{[n\Sigma x^2 - (\Sigma x)^2][n\Sigma y^2 - (\Sigma y)^2]\}^{1/2}} \tag{5.39}$$

which is a computation equation form.

Correlation is concerned only with the association between variables, and r must lie in the range $0 \le |r| \le 1$. By using the data from Table 5.1 the correlation coefficient may be found as an example. Recollecting that $n = 15$, $\Sigma xy = 11{,}337$, $\Sigma x = 105$, $\Sigma y = 1524$, $\Sigma x^2 = 1015$, and $\Sigma y^2 = 156{,}500$, then

$$r = \frac{(15)(11{,}337) - (105)(1524)}{\{[(15)(1015) - (105)^2][15(156{,}500) - (1524)^2]\}^{1/2}} = +0.98$$

which is a high score for correlation. The magnitude of the correlation coefficient r determines the strength of the relationship, and the sign of r tells us whether the dependent variable tends to increase or decrease with the independent variable. For instance, a positive value indicates that the index is increasing positively with years. If two variables are linearly related, r is a useful measure of the strength of the relationship between them. The value of r will be equal to $+1$ or -1, if and only if the points of the scatter lie perfectly on the straight line, which is unlikely in cost analysis.

The interpretation of the correlation coefficient as a measure of the strength of the linear relationship between two variables is a purely mathematical interpretation and is without any cause-and-effect implications. It is possible to have high correlation coefficients between variables that are only nonsense. Can you think of situations which give high correlation which have no real-life cause and effect?

5.2.5 Multiple Linear Regression (Optional)

It may happen that the method of least squares for estimating one variable by a related variable yields poor success. Although the relationship may be linear, frequently there is no single variable sufficiently related to the dependent variable being estimated to yield good results. The extension to two or more independent variables is natural.

Because linear functions are simple to work with and estimating experience shows that many sets of variables are approximately linearly related, or can be assumed so for a short duration, it is reasonable to estimate the desired variable by means of a linear function of the remaining variables. Problems of multiple regression involve more than two variables but are still treated like those involving two variables. For example, there may be a cost relationship between differential profit (DP), construction investment for high-rise office buildings (I), and market saturation of rental space (MS), which can be described by the equation $DP = a + b_1 I + b_2 MS$, which is called a linear equation in the variables DP, I, and MS. The constants are noted by a, b_1, and b_2. This kind of analysis can explain variations in one dependent variable by adding the effects of two or more independent variables.

This approach is not limited to time-trend problems. Time is a catch-all, which takes into account gradual changes due to different factors, both known and suspected. For three or more variables, a regression plane is a generalization of the regression line for two variables. Multiple linear regression has the form

$$y = a + b_1 x_1 + b_2 x_2 + \cdots + b_k x_k \tag{5.40}$$

where $x_0 = 1$, a is a constant, and $b_1, b_2, \ldots, b_k$ are partial regression coefficients. This is a plane in the $k + 1$ dimension. We do not say that the result so obtained is the best functional relationship. We simply state that, given this assumed function and criterion, we have chosen the best estimate of the parameters.

Regression analysis requires the following assumptions:

- The x_j values are controlled and/or observed without error. Perfection remains a difficult requirement within construction estimating practices, but it is nominally met.
- The regression of y on x_j is linear.
- The deviations $y - [j|x_j]$ are mutually independent.
- Those deviations have the same variance whatever the value of x_j.
- Those deviations are normally distributed.
- The data are taken from a population about which inferences are to be drawn.
- There are no extraneous variables that make the relationship of little real value.

The plane in the $k + 1$ dimension passes through the mean of all observed values, similar to the two-variable case, and

$$\bar{y} = a + b_1 \bar{x}_1 + b_2 \bar{x}_2 + \cdots + b_k \bar{x}_k \tag{5.41}$$

Reworking Eqs. (5.40) and (5.41), we have

$$a = \bar{y} - b_1 \bar{x}_1 - b_2 \bar{x}_2 - \cdots - b_k \bar{x}_k \tag{5.42}$$

$$y - \bar{y} = b_1(x_1 - \bar{x}_1) + b_2(x_2 - \bar{x}_2) + \cdots + b_k(x_k - \bar{x}_k) \tag{5.43}$$

As before, the coefficients are determined by using the method of least squares. To illustrate, consider the case of two independent variables, and we have

$$y = a + b_1 x_1 + b_2 x_2 \tag{5.44}$$

with n sets (y, x_1, x_2) of points at this point. In each set the error is given as

$$\varepsilon = y - (a + b_1 x_1 + b_2 x_2) \tag{5.45}$$

and the sum of squares of errors in the n sets is

$$\Sigma\varepsilon^2 = \Sigma[y - (a + b_1x_1 + b_2x_2)]^2 \qquad (5.46)$$

We minimize $\Sigma\varepsilon^2$ as before, which requires that the partial derivatives of $\Sigma\varepsilon^2$ with respect to a, b_1, and $b_2 = 0$:

$$\frac{\partial(\Sigma\varepsilon^2)}{\partial a} = \Sigma[y - (a + b_1x_1 + b_xx_2)] = 0$$

$$\frac{\partial(\Sigma\varepsilon^2)}{\partial b_1} = \Sigma x_1[y - (a + b_1x_1 + b_2x_2)] = 0 \qquad (5.47)$$

$$\frac{\partial(\Sigma\varepsilon^2)}{\partial b_2} = \Sigma x_2[y - (a + b_1x_1 + b_2x_2)] = 0$$

Subscripts and superscripts for summation notation were dropped for convenience. If we keep x_2 constant, then the graph of y versus x_2 is a straight line with slope b_1. If we keep x_1 constant, then the graph y versus x_2 is linear with slope b_2. Because y varies partially because of variation in x_1 and partially because of variation in x_2, we call b_1 and b_2 the partial regression coefficients of y on x_1 keeping x_2 constant, and of y on x_2 keeping x_1 constant. The normal equations corresponding to the least-squares plane for the y, x_1, and x_2 coordinate systems are

$$\Sigma y = na + b_1\Sigma x_1 + b_2\Sigma x_2$$

$$\Sigma x_1 y = a\Sigma x_1 + b_1\Sigma x_1^2 + b_2\Sigma x_1 x_2 \qquad (5.48)$$

$$\Sigma x_2 y = a\Sigma x_2 + b_1\Sigma x_1 x_2 + b_2\Sigma x_2^2$$

The solution of this system of three simultaneous equations gives the values of a, b_1, and b_2 for Eq. (5.44) and is referred to as y on x_1 and x_2. This is a regression plane, but more complicated regression surfaces can be imagined with four-dimensional, five-dimensional, etc., space.

The problem given earlier as a polynomial is solved again by multiple linear regression. Table 5.3 presents the procedure to distinguish gross construction, $\$10^9$, index of output per man hour, and worker productivity. Those were assumed to be the influencing variables.

5.2.6 Computers and Software

The arithmetic in the simple linear or nonlinear regression equations is digestible; but when we consider many variables, the situation seems impossible. Computers, with spreadsheet and statistical software,[1] are capable of making short work out of massive computations, making it possible to find linear and nonlinear regression equations of 150 variables or more with all the accompanying statistical measures of reliability. This allows the selection of the best equation. But in construction cost estimating, the number of variables is

---

[1] Popular statistical websites are www.SAS.com and www.SPSS.com. Two meta websites are www.MATH.YORKU.CA/SAS/StatResource.html and www.STAT.UFL.EDU/VLIB/Statistics.html.

TABLE 5.3 Multiple Linear Regression

Gross Construction $y(\$10^9)$	Index of Output per Man Hour, x_1	Productivity, x_2	Computations Required in Solution for y on x_1 and x_2
92.6	81.5	1.48	$\Sigma y = 1756.30$
102.0	83.7	1.64	
105.0	86.1	1.74	$\Sigma x_1 = 1520.1000$
111.9	87.6	1.84	$\Sigma x_1^2 = 156,872.75$
103.8	91.2	1.89	$\Sigma x_2 = 32.37$
116.7	96.3	1.96	$\Sigma x_2^2 = 72.01092$
116.4	95.0	2.07	
117.8	100.0	2.20	$\Sigma x_1 y = 180,565.04$
109.7	103.9	2.28	$\Sigma x_2 y = 3860.8860$
121.8	107.2	2.34	$\Sigma x_1 x_2 = 3357.3750$
122.0	108.8	2.44	
122.0	113.1	2.49	
134.1	118.4	2.57	
138.5	121.6	2.67	
142.0	125.7	2.76	

The normal equations are

$$17{,}563.30 = 15a + 1520.10b_1 + 32.37b_2$$
$$180{,}565.04 = 1520.10 + 156{,}872.75b_1 + 3357.3750b_2$$
$$3860.886 = 32.37a + 3357.3750b_1 + 72.0109b_2$$

for which

$$a = 29.1181$$
$$b_1 = 0.7052$$
$$b_2 = 7.658$$

and the multiple linear equation becomes

$$y = 29.1181 + 0.7052x_1 + 7.6458x_2$$

much less. Typically, the fewer the variables that have cause-and-effect strength, the better the nature of the equation.

   The computer has without a doubt created many benefits for construction cost estimating, but unfortunately it has also created many pitfalls. Perhaps, the greatest danger is encountered at the outset when the source and type of data are being selected. Despite the excellence of computation, final results are entirely dependent on the reliability of data used, the interpretation of the computations, and judgment as to reasonableness of the conclusion. Thus, an idea of causation must precede statistics. Computers and computation are the easy part, but the "cause and effect" of the data for construction applications continues to be the biggest challenge.

5.3 TIME SERIES (OPTIONAL)

Escalation is an important element in estimating project costs, especially those with long lead times. While the forecasting of business factors can be handled by regression methods, there are additional methods known as moving averages and smoothing.

PICTURE LESSON
Building the Railroad

The national fascination of the nineteenth century was the railroads. Remember, there was no air or auto travel. Before the industrial revolution (approximately 1850) in the United States, transport had been restricted to walking, horse-drawn vehicles, sailing vessels, and to a lesser degree, canal boats. The Railway Age produced the smoking horses on the iron rails. Virtually everyone agreed that iron roadways were the magic carpets that would promote progress and lead inexorably to a higher standard of living. And it was so.

Railroads were common, even if the trackage was for a few miles. Farming communities would have a tie to the main line. Indeed, if the business was to pick up a small quantity of grain, cattle, or chickens, towns received service.

Construction was intimately connected with railroad expansionism. In the Picture Lesson, construction is shown for the Bollman Truss. Notice the deflection of the approach due to the weight of the train and construction crew. Designed by Baltimore bridge engineer Wendel Bollman, this truss was widely used by railroad companies at that time.

An interesting principle, mathematically proven for the time of 1890s, stated that "the greatest economy is attained when the cost per lineal foot of the superstructure is equal to the cost per lineal foot of the trusses and lateral system." In fairness to that time, other engineers had deduced other equations for optimum economic construction of bridges.

There are business situations where new information is added to the data set on a repeated basis. Periodic observations of labor and material cost, overhead, and construction demand are typical. Data that have periodic additions are referred to as time-series data. The data are refreshed with these costs (or prices), and as the process continues, data that are more senior and perhaps not as important are removed from the data set.

In time-series analysis we are interested in the behavior of these observations. Is the underlying process constant, variable, trend-cycle, seasonal, or regular? Samples of graphs, where time is the X-axis and a response variable (consider this to be any construction-business factor) is the Y-axis, are shown in Fig. 5.11. Look at Fig. 5.11(a). The simplest of the time-series cases is the mean. For the trivial case of a mean that is constant, or nearly so, over the time interval for which the forecast is required, the dependent variable is non-sensitive. In other words, there is no trend, and the future value is the same as it is today. This is the situation for Fig. 5.11(a).

The linear model with a trend, Fig. 5.11(b), is found more widely than the quadratic and exponential models. A trend cycle, or seasonal, suggests a time series to a set of observations taken at specific times. Examples of time series are the monthly home startups for construction.

Movements are generally considered to be cyclical if they recur after constant time intervals. An important example of cyclical movements is the so-called business cycles representing intervals of boom, recession, depression, and business recovery. Seasonal movements refer to identical or nearly identical patterns that a time series appears to follow during corresponding months of successive years. Those events may be illustrated by peak summer construction remodeling activity and lesser demand preceding the Christmas period. Certain of those effects are sometimes superimposed on other effects, Fig. 5.11(f), where a linear and a cyclic pattern are superimposed.

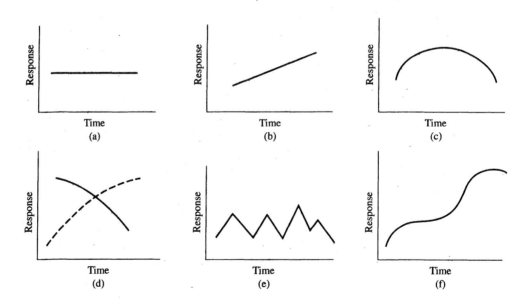

Figure 5.11 Typical time-series models: (a) constant (no trend); (b) linear; (c) quadratic; (d) exponential; (e) cyclic or seasonal; and (f) linear and cyclic.

Cycles may be interpreted in either of two ways: as deviations from established long-term trends or as significant fluctuation due to some time-series effect. In the case where the cycles are interpreted as deviations from a trend, the peaks and troughs are normally referred to as errors from the estimate and are caused by a collection of unknown factors. Again, refer to Fig 5.11(f).

Consider the following data: The most recent value is at the "now" time, and the time 6 years ago is the oldest information.

Years ago	Price, $/100 lb
6	$60.20
5	60.50
4	68.70
3	60.24
2	60.55
1	62.32
Now	75.71

Forecasting calls for professional judgement, and we intuitively believe that the computation of moving average at any single point in time should place no more weight on current observations than on those achieved previously. This is a major logic for the moving average. A reasonable estimate, given by the average price, is $64.03. The forecast for any future observation could be that same value. But this average concludes that the denominator is 7, and that suggests that the average is best analyzed by the total number of observations. But what happens when the next data point is available? Thus a moving average, denominated by a selected value of N, so it is reasoned, better predicts the future value. With this logic, we state the following:

The actual average of N most recent observation, computed at time t, is given by

$$M_a = \frac{x_t + x_{t-1} + \cdots + x_{t-N+1}}{N} \tag{5.49}$$

where M_a = moving average of response variable, price, etc.

$\quad\quad x$ = information data, such as cost, price, yield, etc.

$\quad\quad t$ = time unit, years, quarters, months, etc.

$\quad\quad N$ = selected denominator of group of time units, number

There is a restriction that the number of terms in the numerator be equal to N and x_t be the latest term added. Another arrangement, of course, is to use the most recent three, four, or five observations and divide the sum by 3, 4, or 5. For example, the model can be arranged to another form:

$$M_a = M_{t-1} + \frac{x_t - x_{t-N}}{N} \tag{5.50}$$

for $t, N = 1, 2, 3, \ldots,$ and $t > N$.

Assume a time series of information x_t is gathered as shown below: If a 3-year moving average is desired, then the data would be arranged as follows:

Date, t	Data x_t	3-Year Moving Total	3-Year Moving Average
1	$60.20	...	...
2	60.50	...	...
3	68.70	189.40	$63.13
4	60.24	189.44	63.15
5	60.55	189.49	63.16
6	62.32	187.11	61.04
7	75.71	198.58	66.19

One of the considerations with moving averages is selection of rate of response. The rate of response is controlled by the choice of N of the observations to be averaged. If N is arbitrarily chosen large, then the estimate is stable. If N is selected small, then fluctuations due to random errors or other legitimate causes can be expected. We are able to take advantage of those properties. If the process is considered constant, we may choose a large value of N in order to have a stable estimates of the mean. However, if the process is fluctuating, small values of N provide faster indications of response.

For most estimating problems, some type of moving average is desired that reflects both historical and current trends. A smoothing function is defined as

$$S_t(x) = \alpha x_t + (1 - \alpha)S_{t-1}(x) \tag{5.51}$$

where $S_t(x)$ = smoothed value of the estimated quantity, cost, price, etc.

α = smoothing constant, $0 \le \alpha \le 1$

The α is similar but not exactly equal to the fraction $1/N$ in the moving-average method. Whenever this operation is performed on a sequence of observations, it is called exponential smoothing. The new smoothed value is a linear combination of all past observations. Statistically speaking, the expectation of this function is equal to the expectation of the data, which is its average.

When the smoothing constant α is small, the function $S_t(x)$ behaves as if the function provided the average of past data. When the smoothing constant is large, $S_t(x)$ responds rapidly to changes in trend. Though precise statements cannot be made regarding this smoothing, what follows generally describes the effect of smoothing constant on time-series data:

Variations in α Values

Drift in actual data	Small, $\alpha = 0$	Little, $\alpha = 0.5$	Large, $\alpha = 1$
None	None	None	None
Moderate	Very small	Small	Moderate
Large	Small	Moderate	Large

Consider an example of the exponential smoothing for $\alpha = 0.2$.

Date, t	Data, $\$$ x_i	Smoothed Data, $\$$ $S_t(x) = 0.2x_i + 0.8S_{t-1(x)}$
...	...	...
10	...	67.38
11	63.2	66.54
12	68.3	66.90
13	65.7	66.66
14	78.4	69.00

Initial conditions are established for exponential smoothing, typically $S_{t-1} = M_{t-1}$. If there are no past data to average, then smoothing starts with the first observation, and a prediction of the average is required. The prediction may be what the construction activity intended to do. Those predictions can also be based on similarity with other construction activities that have been observed for some time. If there is a great deal of confidence in the prediction of initial conditions, then a small value of the smoothing constant, $\alpha \rightarrow 0$, is satisfactory. On the other hand, if there is very little confidence in the initial prediction, then it is appropriate to have α as a larger value, $\alpha \rightarrow 1$, so that the initial conditions are quickly discounted. This is the counter to the argument about flexibility of response to a change of the construction. If we believe that the real construction is like the prediction, then there is little reason to have a change. On the other hand, the contrary viewpoint would have a quick response between the prediction and the real construction.

Any practical forecasting system needs to accommodate human judgment. Material escalation, for example, is influenced by complicating factors, and these mathematical models sometimes require the intervention of experienced opinion.

5.4   COST INDEXES

Construction estimates frequently use the popular time-series cost indexes. Because of their importance in escalation and preliminary estimates, we give them special attention. From time to time, escalation adjustment for the contract bid is an urgent need. Once extrapolation through time-series analysis of cost indexes is finished for future periods, this procedure becomes possible. Indexes are important for construction cost analysis.

Index numbers have been used for those purposes for a long time. An Italian, G. R. Carli, devised the index procedure about 1750. Using indexes, he investigated the effects of the discovery of America on the purchasing power of money in Europe.

A cost index is an amalgam of labor, material, and services as it compares cost or price changes between periods for a fixed quantity of goods or services. A cost index is merely a dimensionless number for a given year showing the cost at that time relative to a certain base year. Then we forecast the cost of a similar design from the past to the present or future period without going through detail costing. If discretion is used in choosing the proper index, then a reasonable approximation of cost may result.

Index numbers are useful in other respects. With time for estimating usually being scarce, we make immediate use of previous designs and costs, which are based on outdated conditions. Because costs vary with time due to changes in demand, economic conditions, and prices, indexes convert costs applicable at a past date to equivalent costs now or in the future. If a design cost at a previous period is known, then present cost is determined by multiplying the original cost by the ratio of the present index value to the index value applicable when the original cost was obtained. This may be stated formally as

$$C_c = C_r\left(\frac{I_c}{I_r}\right) \tag{5.52}$$

where C_c = desired cost, present, future, or past, dollars
        C_r = reference cost, dollars
        I_r = index chosen to correspond to time period of C_r, number
        I_c = index chosen to correspond to time period of C_c, number

Consider the following example. Construction of a 700,000-ft^2 warehouse is planned for a future period. Several years ago a similar warehouse was constructed for a unit estimate of \$162.50 when the index was 118. The index for the construction period is forecast as 143, and construction cost per ft^2 will be

$$C_c = 162.50\left(\frac{143}{118}\right) = \$196.93/\text{ft}^2$$

Though general-purpose construction indexes are openly published and have become accepted, their making, alteration, and application are worthy subjects because it may be better for an owner or contractor to develop its own index. Arithmetic development of indexes falls into several types: (1) adding costs and dividing by their number, (2) adding the cost reciprocals and dividing by their number, (3) multiplying the costs and extracting the root indicated by their number, (4) ranking the costs and selecting the median value, (5) selecting the mode cost, and (6) adding actual costs of each year and taking the ratio of those sums. The weighted arithmetic method is the most popular. Though formulas are straightforward, determination of indexes bears little resemblance to formulas because of the variety and complications involved. For most cost-estimating situations, a tabular approach is the best method.

A cost index is a dimensionless number representing the change in cost of material or labor or both over a period of time. Prices, which are the input of an index, must relate to specific material or labor. An index for lumber is based on the price of a specific quantity and type of lumber, such as 1,000 board feet of 2 × 8 in. utility grade pine. Quantity and quality must remain constant over the periods so that price movements represent a true price change rather than a change in quality or quantity. This is difficult for indexes that are charted over many periods.

A cost index expresses a change in price level between two points in time. A cost index for lumber in year 2010 is meaningless alone. An index for material A has no relationship to the index for material B. Similarly, the cost indexes for material A in two geographical areas may not be directly comparable.

To compute a price index for a single material, a series of prices are sampled at an epoch of time, for a specific quantity and quality of the material. Index numbers are usually computed on a periodic basis. The prices gathered for the material may be average for the period (month, quarter, half-year, or year) or they may be a single observed value, as found from invoice records for one purchase. There are numerous sampling biases that need to be understood in the gathering of the data, but we need not discuss them here.

Assume that the following prices have been collected for a standardized unit of a laser glass material. The first data are gathered at period 0, which for this case is the origin. Subsequently other prices are found for all periods.

Period	0	1	2	3*	4	5
Price	$43.75	$44.25	$45.00	$46.10	$47.15	$49.25
Index	94.9	96.0	97.6	100.0	102.3	106.8

* Benchmark period

Index numbers are computed by relating each period to one of the prices selected as the denominator. If period 3 is arbitrarily chosen as the benchmark period, then period 2 price divided by period 3 price = 45.00/46.10 = 0.976. When period 3 index is expressed as 100.0, period 2 index is 97.6. The index can be expressed on the basis of 1 or 100 without any loss of generality. The benchmark period is defined as that period that serves as the denominator in the index calculation and has the index value = 100.

Movements of indexes from one period to another are expressed as percent changes rather than changes in index points:

Index	Value, number
Current, period 5	106.8
Less previous index	102.3
Index point change	+4.5
Divide by previous index	102.3
Equals	+0.044 = +4.4%

The average periodic change resulting from these indexes can be found by using

$$r = \left[\left(\frac{I_c}{I_b} \right)^{1/n} - 1 \right] \times 100 \qquad (5.53)$$

where r = average rate per period, decimal

I_e = index at end of period, number

I_b = index at beginning of period, number

n = number of periods

The time period can vary in length, but monthly to yearly intervals are common. For an index beginning with 94.9 and ending with 106.8 over a 5-year period, the average rate is +2.39%.

If the average index rate is expected to persist, then Eq. (5.53) is reformed to give

$$I_e = I_b\left(1 + \frac{r}{100}\right)^n \tag{5.54}$$

This will give an approximate future value. Suppose that we want an index for period 7, or $I_7 = 94.9(1 + 0.0239)^7 = 112.0$.

If cost $C_2 = \$370,000$ is known from records and for a design that is similar to the index, use of Eq. (5.52) will give for $n = 7$

$$C_7 = 370,000\left(\frac{112.0}{97.6}\right) = \$424,600$$

But construction indexes are more complex than any single material index. A composite index is often required, say, for adjustment of "quote-or-price-in- effect" type of inflation/purchase contracts. Equally important is the updating of estimates of complicated buildings, assemblies, and plants.

An assembly called "10-cm disk aperture laser amplifier" is selected for a composite index. The 10-cm disk amplifier was constructed only during period 0 but cost tracking of selected items has continued. To worry about all amplifier components is too involved, so major items are picked for individual tracking. These prices are gathered over 4 years. Table 5.4 is an example of a tabular development of an index.

The "material" column identifies those items that are significant cost contributors to the laser. Usually, a representative set is selected if the design is complex. If the design is simple, however, all materials may be chosen. The "quantity" column is proportional to the requirements of the design. The "quality specification" column identifies the technical nature of the material.

Cost finding begins once those three columns are determined. Year 0 is the first period of the cost facts, and the index 100.0 is determined once we divide the total by itself. Cost facts are collected for each subsequent period, and each total is divided by the benchmark total to obtain the index. Thus, the indexes for Table 5.4 are 100.0, 94.1, 89.6, and 87.6. A general decline in prices is suggested by those indexes. Apparently, there is technological improvement which reduces prices.

TABLE 5.4   Calculation of composite index.

Material	Quantity	Quality Specification	0*	1	2	3
Laser glass	3–10 cm disks	Silicate	$26,117	$24,027	$22,345	$21,228
Steel tubing	18 kg	AISI 304	1913	2008	2129	2,278
Aluminum extrusion	4 kg	AISI 3004	418	426	439	456
Fittings	3 kg	Mil Std 713	637	643	656	657
Harness cable	4 braid, 4 m	Mil Std 503	2103	2124	2134	2305
Glass tubing	12 m	Tempered	4317	4187	4103	4185
Total			$35,505	$33,415	$31,806	$31,109
Index, %			100.0	94.1	89.6	87.6

*Benchmark period.

One may argue that materials, quantities, and qualities are not consistent, especially over time. Indeed, if technology is active, then a reduction in the cost and index is possible. This could be counter to increases in inflation. A reminder, often forgotten in recent decades,[1] is that deflation is also a possibility. Indexes should reflect basic price movements alone. Index creep results from changes in quality, quantity, and the mix of materials or labor. This phenomenon may be unseen in the collection of long-term information for indexes.

The effects of changes in mix, quantity, and quality on the index scheme are called technology creep. Indexes become unsuitable for high-technology products and active construction progress. Construction indexes can be sustained for longer periods by noting the perturbations when they occur, inputting changes for previous data, and backcasting the previous year's indexes. Every so often it may be necessary to reset the benchmark year whenever delicate effects are influencing the index and are not being removed.

There are several kinds of indexes:

- Material
- Labor
- Material and labor
- Regional effects or composite mixes for material, labor
- Design
- Quality

Virtually any amalgam of materials, labor, services, products, and projects can be evaluated for an index. An interesting contrast to a price type of index is a quality index. Instead of noting price changes, the purpose of a quality index is to remove price effects and show quality changes between the periods.

In the development of the index there is a choice in the selection of the information. Wholesale prices or retail prices, wages or type of construction, and proportion of labor to materials are typical alternative choices. Indexes apply to a place and time, that is, period covered or region considered, base year, and the interval between successive indexes, yearly or monthly. Additionally, indexes are varied as to the compiler and sources used for data. A variety of objectives create diversity for the many indexes.

Every industrialized nation regularly collects, analyzes, and divulges indexes. A government index listing is given by the Statistical Abstract of the United States, a yearly publication that includes material, labor, and construction. A yearly publication of the U.S. Department of Labor is the Indexes of Output Per Man Hour for Selected Industries. This volume contains updated indexes such as output per man hour, output per employee, and unit labor requirements for the industries included in the U.S. government's productivity-measurement program. Each index represents only the change in output per man hour for the designated industry or combination of industries. The indexes of output per man hour are computed by dividing an output index by an index of aggregate man hours. For an industry the index measures changes in the relationship among output, employment, and man hours. The Bureau of Labor Statistics publishes monthly producer prices and price indexes and covers some 4000 product groupings.

---

[1] The last time general wage rates declined in the United States was in 1931, a depression year.

TABLE 5.5 Sample listing of indexes.

Index Name	Cost Measure	Application	Compiler	Description
Bid Price	Construction	Federal highways	U.S. Dept. of Transportation	Derived from average unit bid prices for excavation, surfacing, and structures
Boeckh Building Cost	Building	Construction, insurance, appraisal	E.H. Boeckh Co.	11 building types, 202 U.S. cities, 115 components, 19 trades, 89 building materials
Bureau of Reclamation	Construction	General	Bureau of Reclamation	34 types of dams and water projects
Chemical Engineering Plant Cost	Construction	Chemical plants	*Chemical Engineering*	Equipment, machinery, labor, building materials, engineering and supervision costs
Consumer Price Index (CPI)	Goods	Consumer	Bureau of Labor Statistics	Prices for about 4000 goods and services
Dodge Building	Construction	General	Marshall & Swift	22 trades, material, equipment, crew sizes for 1000 regions
Engineering News Record Building Cost [2]	Building	General	*Engineering News Record*	Hypothetical block of construction in 22 cities requiring skilled labor, steel, lumber and cement.
Means	Construction	General	R. S. Means Co.	16 components in 305 cities.
Producer Price Index	Goods	Producer	Bureau of Labor Statistics	Over 4000 Commodities. Since 1890.

In addition to the federal government, states, contractors, consultants, stock analysts, equipment builders, news magazines, and others regularly collect information and develop indexes. A few indexes are given by Table 5.5

SUMMARY

As is now evident, forecasting is analysis of costs. Estimating, when coupled with the forecasting process, takes the imperfect information and adds the ingredients of judgment and knowledge about engineering designs to provide the setting for decisions. In forecasting

---

[2] www.ENR.com/cost/cost2.asp

processes it should be remembered that data provide the projected information that pretends to be the situation in the future. Thus the forecasting methods of statistics, time series, and indexes uncover the relationships that exist for construction costs, prices, sales, and technology.

The application of time series, single and multiple linear regression, correlation, and graphical analysis are the principal statistical techniques used in forecasting for the future. Moving averages involve a consideration of time and fluctuations that occur during those periods.

Cost indexes are useful for adjustments of estimating information over lengthy periods.

QUESTIONS

1. Define the following terms:

Relative frequency	Regression
Best curve	Transformed units
Moving average	Trend line
Histogram	Least-squares criterion
Confidence limits	Power function
Time series	Indexes
Mean	Intercept
Prediction limits	Correlation
Smoothing constant	Backcasting
Standard deviation	Slope
t table	Multiple linear regression
Cycles	Technology creep

2. "Statistics never lie, yet liars use statistics" is a common statement. Discuss.

3. Why are graphical plots preferred initially over mathematical analysis of data?

4. Is construction-cost estimating more concerned with empirical evidence or with theoretical data? Illustrate both.

5. Cite instances when a cumulative curve would be necessary.

6. Discuss what regression analysis is. What are its underlying assumptions?

7. What is minimized in a least-squares approach?

8. What is meant by correlation analysis? What does $r = 0$ or 1 imply?

9. Distinguish between correlation and causation. Can you have causation without correlation?

10. What is the purpose of a moving average? How does smoothing relate to a moving average?

11. If the construction estimator is confident of past data, then would the smoothing constant be large or small?

12. What are the differences between cycles and trend cycles?

13. Define a cost index.

14. Survey a meta website for statistics. List the topics that are prominent to the compilers of the site.

15. Three moving-average cost trend lines are determined for a material and are shown as Fig. Q5.15. Write a paragraph that evaluates the proposed moving average for the three cases.

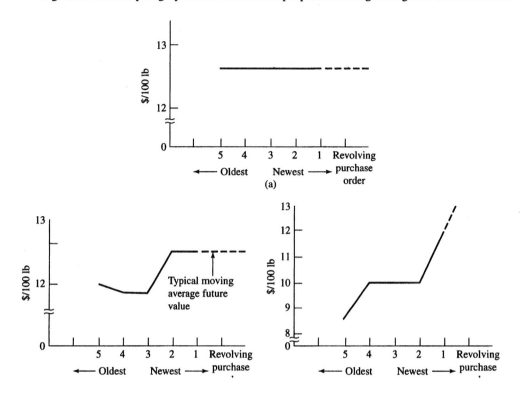

Figure Q5.15

PROBLEMS

5.1 A labor survey is conducted for area contractors, and the following wage groupings and their frequency of occurrence are found:

Wage, $/hr	Number
< 12	5
12 < 15	11
15 < 18	28
18 < 21	36
21 < 24	52
24 < 27	15
27-	13

Sketch the frequency distribution, relative-frequency curve, and cumulative-frequency curve. (Hints: Notice that the wage interval end numbers are open. Use engineering computation paper for the graphs. A sketch is freehand, rendered with the aid of a straight edge and with a 0.5-mm pencil.)

5.2 On a construction site a time-lapse video is made of carpenter-crew work on a concrete-form layup for a large building. The video is observed and tallied, and observations of the time for this completed task are as follows:

Form Layup, Man Days	No. of Observations
man days ≤ 0.04	2
0.04 < man days ≤ 0.08	12
0.08 < man days ≤ 0.12	14
0.12 < man days ≤ 0.16	15
0.16 < man days ≤ 0.20	14
0.20 < man days ≤ 0.24	13
0.24 < man days ≤ 0.28	12
0.28 < man days	6

Plot the frequency-distribution table, relative-frequency curve, and cumulative-frequency curve. Find the mean value of the man days. What is the modal value? (Hints: Use engineering paper with a 0.5 mm pencil, or use a spreadsheet, such as Excel.)

5.3 Find the mean, median, mode, range, standard deviation, and variance for a following set, as selected by your instructor:

(a) 3, 5, 2, 6, 5, 9, 5, 1, 7, 6

(b) 41.6, 38.7, 40.3, 39.5, 38.9

(c) 2, −1, 0, 4, 6, 6, 8, 3, 2

5.4 A regional market survey has determined the value of the residential construction industry over a 10-year period.

Year:	1	2	3	4	5	6	7	8	9	10
10^6:	1467	1216	1360	1400	1518	1678	1818	2160	2290	2460

(a) Graph the data on engineering computation paper by using the "one-half" rule. Find the graphical equation $y = a + bx$ of the line. Forecast the value in year 11 by using the curve and equation.

(b) Find the least-squares equation and forecast the value for year 11. How do the two equations compare? (Hints: Attempt a freehand graph and make notice of the change in slope. This change is a result of market forces. If encouraged by your instructor, execute the graph on a spreadsheet.)

5.5 The index for the union wage rate for welders has followed this pattern:

Year:	0	1	2	3	4	5	6	7	8	9
Index:	100.0	104.4	107.8	112.3	117.2	122.5	132.8	145.6	159.3	171.6

(a) Plot the straight line of the data on arithmetic coordinates by using the "one-half" rule. Find the graphical equation $y = a + bx$ of the line. Forecast the value in year 10 by using the curve and equation.

(b) Visually determine the year for which there is a change in slope, and construct two new straight lines for the data, finding their graphical equations. Also forecast the value in year 10. Does there appear to a change in the data to warrant a dropping of earlier data?

(c) Find the regression equation for the best set of data.

5.6 Aerospace companies estimate the cost of the airplanes, major assemblies, and vendor materials by an important technique known as "learning." This function, which was discovered as an empirical "truth" in the 1930s, states that as production continues, the time or cost to produce one more unit declines at a constant rate between doubled units. The function that models this relationship is $T = KN^s$, where T = time or cost of a specific unit, and N = unit number. The intercept K and slope s are empirical quantities determined by regression calculations. Assume that a contractor determines data ($y = T, x = N$) such as $(450, 15), (325, 30)$, and $(200, 45)$. Determine the regression equation in original and transformed units. Find the estimate for the 60th unit by both methods.

5.7 (a) A model of the form $y = ax^b$ is to be fitted to (x, y) data: $(1, 100), (10, 10)$, and $(100, 0.1)$. Find the regression equations in transformed and original units. What are the values for $x = 150$? (Hints: For this simple set of data, use your calculator. The data are in powers of 10, so use logarithms to the base 10 rather than the Naperian system for any calculation. "Transformed" implies logarithm conversion of the data.)

(b) Calculations for the data, which are not given, are as follows:

$$n = 4$$

$\Sigma y = 14$	$\Sigma x = 10$
$\Sigma y^2 = 14$	$\Sigma x^2 = 30$
$\Sigma \ln y = 4.787$	$\Sigma \ln x = 3.178$
$\Sigma (\ln y)^2 = 6.200$	$\Sigma (\ln x)^2 = 3.129$
$\Sigma xy = 40$	$\Sigma \ln x \ln y = 4.516$
$\Sigma x \ln y = 13.487$	$\Sigma y \ln x = 13.405$

The regression model is believed to be $y = ab^x$. Find the regression parameters in transformed and original units. What is the estimated quantity for $x = 10$ in both original and transformed units?

(Hints: Note that quantities are given in arithmetic and natural logarithms. Be selective in your choice of which quantities to substitute into the standard equations. Also, there is no distinction for the regression equations for either base-10 or natural-base logarithm quantities.)

5.8 (a) Find t_α if $\alpha = 0.10$ and $\nu = 10$ degrees of freedom.

(b) Find $t_\alpha = 0.05$ if $n = 20$. (Hint: Assume that the t_α is necessary for regression of two variables.)

(c) Compare t_α and Z (from Appendixes 1 and 2) for $\alpha = 0.05$ and $n = 120$.

(d) If $t_\alpha = 2.8$, then find α and n for $\nu = 16$.

5.9 (a) Find the mean and individual dependent value for Table 5.1 for year 16.

(b) Determine the variance of the mean value y and the individual y for $x = 16$ from Table 5.1.

(c) Determine the confidence and prediction limit for parts (a) and (b) for 95%.

(d) Repeat the above for $x = 17$.

5.10 An index for the electrician wireman gross hourly wage is given for 6 years:

Year:	0	1	2	3	4	5	6
Index number	100.0	106.0	111.1	117.2	121.3	125.3	128.0

(a) Graph the data on engineering computation paper by using the "one-half rule."

(b) Compute linear trend values, find a least-squares line fitting the data, and construct its graph and compare to the manually fitted line.

(c) Predict the price index for year 7 and compare with the true value. 132.6. What is the range for the individual year 8 by using a 95% prediction interval? If the union contract value for year 8 is $27.15, what is your expected labor cost per hour, and its likely range? (Hint: Use the most appropriate technology for the arithmetic.)

5.11 Determine the values of parameters a and b of the linear regression equation using the method of least squares and then find the correlation coefficient.

Year	Index, y
0	95.1
1	97.7
2	98.4
3	100.0
4	101.1
5	102.2
6	103.5
7	104.9
8	106.6
9	109.7

Find the 90% prediction interval for the year-10 value. Estimate the 95% confidence interval for the slope.

5.12 Construction time studies are conducted on three operations. (Hint: Your instructor will assign a subset. Consider the application of spreadsheet software for the work.) Which are the independent and dependent variables? Plot the raw data as a linear line on engineering computation paper. Find the regression equation and overplot the equation on the eyeball plot. How good was your personal eyeball plot? Find the correlation coefficient. Find the 95% confidence interval for the slope and intercept. Calculate the regressed value and 95 % confidence interval for

(a) 10.5 ft

Depth of shoveling, ft	Cu yd of sand shoveled in 10 hr per man
3	21.4
5	14.1
8	10.8
10	8.2

(b) 8 ft

Depth of shoveling, ft	Cu yd of wet clay shoveled in 10 hr per man
3	5.6
5	4.4
8	4.5
10	3.8
15	2.6

(c) 2 cu yd.

Dipper capacity, cu yd	Power shovels and tractor loaders handling loose materials, cu yd per hr
$\frac{1}{2}$	43
$\frac{5}{8}$	67
$\frac{3}{4}$	119
1	158
$1\frac{1}{4}$	190
$1\frac{1}{2}$	205
$1\frac{3}{4}$	210
2	270
$2\frac{1}{2}$	310
3	340
4	425
5	525

5.13 The theory of learning (which uses the function $T_u = KN^s$, where T_u = unit time or cost for Nth unit, K = man hours or cost estimate for unit 1, and s = the slope of the improvement rate) is applied to construction work and major projects for preliminary estimates if successive jobs are similar in content. Construction of residential houses and natural-gas power plants (where there were different contractors for the gas plants) are real examples for this application. Transform a data set into a log relationship, and determine the log regression line. Also determine the equation in original units. Estimate the (a) man hours or (b) cost for $N = 10$ in both log and original units.

(Hints: For example, house number 2 required 8125 actual hours from job tickets. No design information is given other than the house number and the hours. Nor are the megawatts and other details of the gas plant provided. It is assumed that the similar units are built in successive order.)

(a)

N, No. of house	Contractor direct labor work force, T, total man-hours for house N
2	8125
3	7926
5	7721
8	7603

(b)

N, No. of high-plains natural-gas fired power plant	Owner cost, 10^6
3	716
4	693
6	713
9	684

5.14 Historical data of direct labor for wiring diesel generator sets in remote electrical substation locations are plotted on arithmetic graph paper. Refer to Figure P5.14. The engineer/designer has aggressively designed features into these generator packages to ease their installation, thus allowing the reduction of direct labor hours for the wireman.

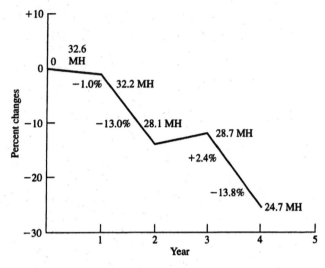

Figure P5.14

(a) Replot these data on semilog paper. (Hint: Have the y axis as the logarithm. The x axis needs a zero location, and thus is requiring arithmetic scale.)

(b) Find the semilog regression equation. Use the theoretical equation $T = Ks^x$, where T = man hours, x = year, K = intercept, and s = slope.

(c) What is the expected value for year 5 from your log plot and equation?

5.15 Technical catalogs are useful in many ways. For example, a catalog provides cost and horsepower information for enclosed capacitor motors, fan-cooled, 1725-rpm, having a 5/8-in. keyway shaft. The listing gives costs on 1/4, 1/3, 3/4, and 1 hp of $67.50, $75.00, $88.50, and $105.00. Determine the cost equation by using

(a) Semilog equation

(b) Power equation

(c) Which equation, using the test of $\Sigma(y - \hat{y})^2$, is best? What is the cost of $1\frac{1}{4}$-hp motor?

5.16 A large contractor has records of its employment and the value of earned contract moneys. Is there a correlation between the number of employees and contract award dollars? What is your assessment?

Contract Award, 10^6	No. of Work Force Employees
573	1573
606	1550
648	1530
720	1550
765	1540
798	1550
848	1560

5.17 Analyze the following data by using multiple linear regression: The data are two dimensional, where weight and length are the independent variables. The response dependent variable is cost/1000 units.

Cost/1000 Units

Weight	Length		
	1.50	1.75	2.00
60.5	5070	4770	4540
52.6	4540	4215	3930
42.3	3660	3660	3120
33.3	2390	2390	2100

5.18 Tunnels are bored for applications such as water power, irrigation, rail and automotive transport, and mining excavation. Data of an equipment delay rate versus length of tunnel are tabulated. (Hints: Understand that the unit estimating rate is influenced by the length of the bore. The data are gathered from the history of tunnel boring and then compared against the unit rate for that length. The data are tabulated against the nearest 500-meter length of tunnel).

(a) Determine an exponential and a second-degree polynomial equation.

Length of tunnel, meters	1000	2000	3000	3500	4000	4500	5000
Increase in tunneling rate, %	5.2	6.5	6.8	8.1	10.2	10.3	13.0

(b) Which model gives the smallest error sum of squares? Which model is preferred for future estimating?

5.19 Material costs are collected from historical purchase orders, and the information is arranged by period and quantity. The quantity represents the volume of units purchased during the period that is specified. The dependent variable is listed as dollars per unit. (Hint: For example, during period –3, or three periods ago, and for the quantity of 2 units, the cost per unit is $8.00.)

Period	Quantity		
	1	2	3
–3	10.00	8.00	7.00
–2	10.70	8.60	7.50
–1	11.10	8.95	7.85
0	11.25	9.00	7.90

(a) Use multiple linear regression to find the unit cost equation y, where y = cost/unit, x_1 = period and x_2 = quantity.

(b) Find the unit cost for period 1 and quantity 2. Discuss the implications of inflation and quantity purchasing upon costs. (Hints: Notice that the vertical column, period, has increasing cost in any column which we assume is due to inflation, while the row, quantity, has declining unit costs as the quantity of the order increases. Find the regression equation using spreadsheet software.)

5.20 For the information given below, determine a three-period total and moving average Also, using an exponential smoothing function, calculate the smoothed data for $\alpha = 0.33$.

Period	Unit Price	Period	Unit Price
0	$60.0	6	$65.2
1	63.2	7	65.8
2	64.2	8	66.4
3	64.3	9	67.5
4	64.4	10	70.0
5	65.0		

5.21 **(a)** Plot the following time-series data:

Period	Value	Period	Value
1	$7.5	11	$4
2	8	12	2.5
3	9.5	13	1.8
4	9.7	14	0.8
5	10	15	0.2
6	9.9	16	0.1
7	9.8	17	0.3
8	9	18	0.9
9	7.6	19	2
10	6	20	4

(b) By using a smoothing function with $\alpha = 0.1$ determine smoothed data and overplot the raw information. Discuss business cycles with the familiar function that the information resembles.

5.22 The standardized cost of Washington State construction-quality Douglas fir used in residential construction is given as:

Years Ago	10	9	8	7	6	5	4	3	2	1	Now
Cost (cents/BF)	23.2	24.1	26.3	25.7	26.8	27.2	28.0	27.8	28.0	28.5	28.3

Analyzed these data using a time-series model. Make a table for a 3-year moving total and average price. Plot the average price versus time and describe the movement. Discuss the practical factors that influence construction-grade lumber price. What is your recommendation for N? Can you see any indication from the data?

5.23 A warehouse construction job is planned in 2 years. A similar layup steel-wall warehouse was constructed for a unit price of $105 per square foot of wall when the index was 107. The index now is 143, but in 2 years it is forecast to be 147. What is the unit estimate for the construction? What is the bid cost of a warehouse having 700,000 square feet of wall?

5.24 A construction man-hour index, based on $1987 = 100$, is given as 1969, 41.9; 1979, 48.0; 1989, 107.6; and 1999, 205.8. Convert the index to a 1999 basis $= 100$. (Hints: Assume the indexes are linear between years. An index given as 100 is also 1.00. This problem demonstrates that indexes can be adjusted to other years, if the composition and value of the index are acceptable. This is a subjective evaluation of the quality of the index.)

5.25 A regional construction index for Los Angeles is given as

Year	Index
5	1200
15	3108

A structure built in year 10 costs $10 million. How much will the same structure cost in year 20, should construction be delayed until then? (Hint: It is first necessary to find the average rate of increase.)

5.26 Indexes for buildings in Denver and New York are as follows:

Year	Denver	New York
1	400	500
5	600	750

(a) If the building cost $400,000 in year 1, then what is its cost in year 6? (Hints: Consider cities separately. The index growth in years is 4, or $5 - 1$.)

(b) If the building costs $400,000 in year 2, find its cost in year 6.

5.27 Determine the cost for a pressure vessel. This vessel is fabricated for a chemical refining plant and is a part of the cost of the construction estimate. The base cost estimates the vessel fabricated in carbon steel to resist internal pressure of 50 psi with average nozzles, manways, supports, and design size. Design: diameter, 8 ft; height, 15 ft; shell material, stainless steel specification 316 plate; and operating pressure, 100 psi. Estimating data: carbon steel material costs = $240,000; factor for noncarbon steel material = 3.67; factor for nonstandard pressure = 1.06; and construction 3 years hence with a 5% material escalation per year.

(Hints: A pressure vessel is a cylindrical shell capped by two elliptical heads. There are various "factors" or indexes that adjust for a nonstandard material pressure, and for the increase in material cost a "multiplier" type of index.)

5.28 A construction firm uses indexes to estimate the cost of concrete forms. Material quantities, labor man-hour data and material unit costs have been gathered for three periods, and are given as

Item	Quantity	Rate for period		
		0	1	2
1. Materials				
Sides, 2 × 12 in.	200 BF	0.06	0.057	0.053
Stakes and braces	75 BF	0.055	0.053	0.052
2. Install forms				
Carpenter	3.5 hr	24.60	27.52	31.02
Laborer	1.75 hr	21.85	23.66	26.12
3. Stripping forms,				
Laborer	1.5 hr	21.85	23.66	26.12

(a) Find the indexes for the three period and determine the index for period 4. (Hint: cost for period is Qty × Rate for period.)

(b) An actual job cost is $175,000 during period 1. A new job that is similar to the job during period 1 is to be estimated for period 4. Calculate the future cost for period 4. (Hint: The benchmark period is 0.)

MORE DIFFICULT PROBLEMS

5.29 Raw data of national unemployment for steelworker journeymen and apprentices are gathered for two separate years where "underwork" is for four weeks or more:

Age Range	March, First Year	March, Second Year	Age Range	March, First Year	March, Second Year
			34–35	38	47
16–17	87	93	36–37	53	42
18–19	636	709	38–39	36	85
20–21	206	191	40–44	89	30
22–23	202	50	45–49	101	97
24–25	81	229	50–54	86	107
26–27	15	37	55–59	111	67
28–29	13	29	60–64	117	173
30–31	19	73	65–69	144	180
32–33	25	83	70+	101	102

(a) Construct the frequency-distribution table, relative-frequency curve, and cumulative-frequency curve for both years. Describe their appearance.

(b) Repeat part (a), using intervals 16–19, 20–24, 25–34, 35–44, 45–54, 55–64, and 65+. Describe the appearance of the curve. Comment about nonuniform-sized classes and any deliberate or unintentional statistical distortion. How would you present the data?

5.30 A study of past records of installed insulation cost (including labor, material, and equipment to install) for central steam-electric plant equipment (including turbine, boiler, pipes) revealed the following data:

Equipment Cost, 10^6	Insulation Cost, 10^5	Equipment Cost, 10^6	Insulation Cost, 10^5
5.5	3.5	14.1	9.2
10.7	5.5	14.8	9.3
34	28	15.5	14.1
2	1.4	15.3	13.8
6	6.4	21.3	15.0
1.5	2.1	34.0	15.8
8.1	7.2	24.1	9.8
10.1	6.4	26.0	15.8

(a) Plot a chart of this information on arithmetic coordinates; on log-log coordinates. Which is more suitable for cost-analysis purposes? Major equipment costs have been estimated as $3 million for a new project. What is the estimate for the installed insulation cost?

(b) Find the arithmetic regression equation and correlation. What is the insulation cost value for $3 million of equipment?

(c) Find the logarithmic equation and the value of insulation cost for $3 million of installed equipment.

5.31 Requirements for a construction material are consolidated, and purchase requests are issued for a quantity each period. The following table is a summary of the purchased cost history of a single material; where 3–represents three periods ago and is the oldest recorded history. Period 0 is the "now" time of a purchase request. The next period is the integer 1. The units of purchase are of small lots, 1, 2, and 3 units. For example, in period–2, we purchased 3 units, and the price was $7.50 per unit. There is no information for values not shown.

Period	Quantity		
	1	2	3
−3	10.00		
−2			7.50
−1		8.95	
0			7.90

(a) Use multiple linear regression and find the equation where x_1 = period, x_2 = quantity, and y = material cost per unit.

(b) Find the unit cost for period 1 and a quantity of 2.

5.32 A contractor uses extensive amounts of cast iron pipe for its principal activity of laying water mains. Purchased pipe constitutes the major cost for its bids. Past data of a standardized amount of this cast iron pipe are as follows:

(a) Determine a four-period total and moving average. Plot the unit price. What typical characteristic curve does it resemble?

(b) Use an exponential smoothing function and determine the smoothed data for $\alpha = 0.25$. Find the price for period 25.

Period	Unit Price	Period	Unit Price
0	60.0	11	72.8
1	63.2	12	73.9
2	64.2	13	74.8
3	64.3	14	74.9
4	64.4	15	75.0
5	65.0	16	75.4
6	65.2	17	75.8
7	65.8	18	76.2
8	66.4	19	77.2
9	67.5	20	80.0
10	70.0		

5.33 A contractor is comparing two producers for a standardized quantity of reinforcing-steel cost for a road and bridge project outside the United States. The construction site is equally spaced between the suppliers in the United States and France. However, the efficiency index is 1.3

French man hours = 1.0 U.S. man hours for identical work. Indirect cost percentages are 120% France = 75% U.S. The direct unit cost (labor only) is $99.65 in France while it is $153.53 U.S. Which of the two suppliers should this contractor buy from? Find the cost of the reinforcing steel for the two countries.

(*Hints:* Since the construction site is equally distant from both producers, transportation costs of the material to the construction site are equivalent. There is no conversion of French euros to U.S. dollars, as the cost of labor in France already has this conversion considered. Ignore any basic material costs, as that is assumed to be equal between the United States and France. Use the model of total unit cost = (direct cost of labor + indirect cost of labor) × efficiency.)

5.34 Indexes, wage rates and the index proportion of construction trades are given for a large contractor for three periods:

Trade	Index Weight, %	Period 0	Period 1	Period 2
Carpenter	31.4	100.0	104.4	107.8
		20.40	21.30	22.00
Electrician	13.6	100.0	102.2	104.3
		23.00	23.50	24.00
Laborer	10.1	100.0	106.1	110.9
		14.70	15.60	16.30
Plumber	15.3	100.0	107.6	110.3
		22.30	24.00	24.60
Painter	11.5	100.0	105.7	108.5
		17.60	18.60	19.10
Others	18.1	100.1	105.0	107.9
		19.80	20.80	21.30
Weighted average Index		100.0	105.0	107.9

(a) Forecast the period 3 index for carpenters; for all trades.

(b) A new job is to be estimated, but it will not have any painting component. Forecast a revised index free of the painting trade and determine the next period index. (Hints: The benchmark period, or the start of the index, is time 0. The wage for carpenters in period 0 is $20.40, and the index for this period is 1. In period 1 the wage increases to $21.30, and the index becomes 104.4 (= 21.30/20.40). Forecast to period 3 using the average-rate-of-growth formula, regression methods, or graphical procedures.)

(c) Discuss: While the rates relate to the amount paid the tradesman for 1 hour, the indexes do not consider the efficiency with which that hour is utilized in period 3.

PRACTICAL APPLICATION

This chapter concentrates on forecasting, and you have learned that indexes are vital information in construction-cost analysis. For this practical application, you are to develop an index for residential construction. Consider the objectives. What geographical area is germane to your index? What are the expected limitations? Is there any distinction in the residential market that you need to be sensitive to? Do you have first-hand knowledge of the proportion of labor, material, equipment, and subcontractor's contribution?

(*Hints:* This index is for the benchmark period only, which for practical purposes is "now", or zero time. It is not necessary to consider either early or later periods. Identify the contents of the index, their proportion, and be careful with the specification of the items. Plan your survey carefully and be professional in your requesting of information from the various sources. Your instructor will suggest the teaming and the separation of duties in the work. Prepare the work in report form using word-processing software.)

CASE STUDY: MARKET-BASKET INDEX FOR GROCERIES

Analyze the "market-basket" index of these several food items. The benchmark year is 0. The prices of these items were collected under similar circumstances over five sampling epochs.

	Period				
Item	*0*	*1*	*2*	*3*	*4*
1. Grade *A* pasteurized homogenized vitamins *A* & *D* skim milk, $\frac{1}{2}$ gal.	1.32	1.63	1.69	1.64	1.72
2. Premium light ice cream, vanilla, $\frac{1}{2}$ gal.	3.49	3.99	4.99	4.99	5.02
3. 100% frozen concentrated orange juice, 12 oz (Minute Maid)	1.99	1.90	1.85	2.02	2.10
4. 37% vegetable oil spread, 1-lb plastic tub (I Can't Believe It's Not Butter)	1.69	1.89	1.69	1.79	1.89
5. 100% natural peanut butter, 1-lb jar	2.35	2.40	2.45	2.49	2.45
6. Decaffeinated instant coffee, 2-oz jar (Folgers)	1.09	1.99	2.49	2.69	2.49
7. Flour, all purpose, 5-lb paper bag (Pillsbury)	1.69	1.75	1.79	1.81	1.83
8. Ground beef, 9% fat, 1-lb package	2.49	2.45	2.59	2.65	2.67
9. Potatoes, U.S. one 5-lb bag	2.09	1.89	1.99	1.91	1.94

(a) Determine yearly indexes. (Hint: Have the base year as 0.)

(b) What is the average percentage rise over the 5-year period?

(c) What are the major and minor items contributing to this increase?

(d) Eyeball a time-series plot of the index. (Hint: Anchor the line at 1 for Period 0, and then use the "one-half" rule for the other points. Begin with a straight line.)

(e) Determine a least-squares equation fit and an exponential smoothing table for this time series. What is the expected index for time period 7?

(f) If those items constitute 3% of the grocery bill for a hypothetical time period and family, what are the gross dollars lost to inflation for time period 4 from the benchmark period 0?

(g) By using inquiry methods at your local grocery store, determine current prices for those items and compare to year 0.

(h) Discuss the hypothetical difficulties of indexes, even with well-known national commodities.

CHAPTER 6

Estimating Methods

Earlier chapters provided fundamental concepts for construction cost analysis. Even though construction designs differ greatly, estimating methods are remarkably similar. That is seen in this chapter, which introduces general estimating methods. These procedures are discussed and their advantages and shortcomings are noted. Methods rely on a range of resources from experience and judgment to mathematics.

6.1 DESIGN AND THE ENTERPRISE

Estimating is a contributor of information to the enterprise and the design. Most projects begin with the recognition of a need for new construction. The range of the projects is indeed broad, be it residential, commercial building, heavy and public works, or industrial. Estimating and economic analysis are connected intimately with the life cycle of the projects, from beginning to the end.

Long before designers begin CAD, and certainly before field construction can commence, broad-scale planning is undertaken. This planning inevitably gives thought to the cost and financing that is required. The owner conducts concept and feasibility studies and asks many questions, and this study gradually converges into engineering and design.

Preliminary engineering and design stress technical process alternatives, size and capacity decisions, comparative economic studies, and architectural concepts. These steps evolve from the concept and feasibility stage. Once the preliminary engineering and design are essentially complete, there is an extensive scrub-down in both technical and business terms before detailed work is allowed to proceed. In private enterprise, the review focuses on seeking approval from executive management, the board of directors, and sources of external financing. Further, the review could be external to the firm with regulatory bodies, and it will look for compliance with codes, licensing procedures, safety, and environmental impact, etc. In public enterprise, the voters may have to give approval, or the funding side of the government, legislature, etc., is involved. Traditionally, early-stage analysis is conducted by the owner alone or with professional consultants.

In the early stages of its development a design is without conformity and order. For instance, we may have progressed through the design steps of problem definition, concepts, engineering models, and evaluation, with the final CAD remaining. An estimate is requested during the initial evaluation period.

Detailed engineering and design is a large undertaking involving many professionals, such as the engineering disciplines, including chemical, civil, construction, electrical, mechanical,

190

and other types as needed, architects, interior designers, and landscape architects. Many studies are necessitated and, throughout this phase, cost analysis continues.

To some extent engineer/architectural consultants, design-constructors, professional construction managers, or program managers become involved in these stages. But along the trail from the concept to the final stage, cost, profit, and economics are important to the enterprise. As the project broadens in scope, other entities become involved. General contractors, subcontractors, and the owner with a program manager are players in the enterprise of construction. Design and cost estimates continue to be very important. It is helpful to the student to realize that a variety of estimating methods are necessary for the enterprise continuum.

6.2 MUCH ADO ABOUT NOTHING[1]

Estimates are called many things. Much lip service is given to a particular description of the estimate. The *noun estimate* will have an *adjective* modifier, such as preliminary, order-of-magnitude, contractor, owner, final, definitive—and the list is seemingly endless.

Consider the many aspects of estimating a task. For instance, a hail storm severely damages residential roofs. The "estimator" conducts an on-site visit, verifies the area, shape, accessibility, and materials, performs some simple calculations (if any), and provides the quotation for the customer. An array of factors are considered, even in this simple example.

At the other extreme, in response to the request for an estimate-quotation, the owner's engineering-architectural firm will provide detailed working drawings, and the team will conduct a one-of-a-kind cost evaluation. In this more complex situation, as in the simple one, a panoply of factors are evaluated.

The nature of the estimate is defined in the following terms:

- Purpose
- Accuracy
- Timeliness
- Effort
- Accountability

If, as often happens, we identify estimates on the basis of *purpose*, we find that they are made for the verification of a contractor's bid, or for a budget, or for project funding, etc. Estimates are used as evidence for bid preparation or are advisory. The most common purpose for the estimate is to supply the best and most factual information to the bidding process. There is no better information for bidding than the estimate.

Some firms require that the tolerance of the estimate be ±5%, or ±10%, or +25% and −10%. At the other extreme, some companies assign their estimates accuracy limits to a ROM (rough order of magnitude). They reason that *accuracy* is crucial to business practice. Other contractors think of the estimate as a not-to-exceed maximum cost. Some managers view the estimate as a target to measure whether the job is good or bad.

Whether the particular accuracy is ±5% or ±50% is not the point here. However, we assert that methods that are generally more accurate have an increase in cost of prepara-

---

[1] The title of a play written by William Shakespeare (1564–1612) during the period 1598–1600.

tion. Thus, data are purified, design has increased detail, and in actual estimating situations management stipulates that the estimate be within an interval bracketing the future actual value.

An estimate is an attempt to forecast the *actual cost*. It is a goal of this author that the accuracy of the estimate, when compared to actual cost, will be such that in half of the cases it will exceed the actual cost, and in the other half fall short of it. Some would reason that there needs to be a factor of safety in estimates, as there is in design, for example. The factor of safety would increase the costs to assure that none are understated. This author categorically rules against "factor of safety" estimates. These thoughts are further amplified in Chapter 9.

Some companies use a *timeliness* concept to describe the estimate. For instance, the estimate may be called preliminary or final. Other terms used instead include conceptual, battery limit, schematic, and green field. Their purpose is to screen and eliminate unsound proposals without extensive engineering cost.

A preliminary estimate is an estimate made in the formative stages of design. Overlooked in this definition are the accuracy, type of design, nature of the organization, dollar amount, and the purpose for the estimate. A precondition of accuracy for preliminary estimates cannot be imposed, because special designs or objectives create a unique set of requirements. An estimate involving, say, pennies is no less a challenge than one involving millions of dollars, because many of the same methods are used at both ends of the dollar scale.

If early estimates lead to a continuation rather than a dismissal decision, then additional methods are required. Attention turns to methods that are more thorough in preparation and more accurate, as well as more costly in design and estimating. At this time the designer/architect has filled out the details, and the engineer constructs an estimate on enlarged quantities of verified information. In some cases the detail estimate is a reestimate of the preliminary one, because only limited updating needs to be done. Naturally, the effort for these estimates varies.

Detail estimating methods are more quantitative. Arbitrary and excessive judgmental factors are suppressed. While fudge factors are never eliminated, emphasis for the detail estimate shifts to comprehensiveness. Whether the detail estimating model is computer software, recapitulation columnar sheet, computational code, or whatever, the intent is the same: We want to use formal rules in the detail estimate to find cost.

Estimates are prepared by the owner, contractor, subcontractor, etc. This accountability requires modifications to the process, and estimates for the same task will differ. Differences arise because of the source of preparation.

The adjective modifiers that are used for the noun *estimate* are much ado about nothing. Because construction tasks vary from the simple to complex, and with time and resources always limited, it is important to have knowledge about a number of estimating methods. For the sake of versatility, students need to learn these methods.

6.3 OPINION

Personal opinion is inescapable in estimating. It is easy to be critical of opinion, but in the absence of data and with a shortage of time, there may be no other way to evaluate designs but to use opinion. The key to opinion estimating is humanware.

The individual is selected for the job because of his or her expert experience, common sense, and knowledge about the designs. The mettle of the domain expert is tested in judging the economic worth of a design. People respected as "truth tellers" are sustained in this activity. Others are not.

Time, cost, or quantities about minor or major bid items are estimated with this inner experience. That the individual be objective in attempting to measure all future factors that affect the out-of-pocket cost is understood. Opinion estimating is also done collectively in conferences.

6.4   CONFERENCE

The conference method is a nonquantitative technique of estimation and provides a single-value estimate. The procedure, although having many rituals, involves representatives from various functions or skills conferring with engineering in a round-table fashion. Various levels of cost are estimated, ranging from construction-work items to total cost for minor projects. Sometimes, labor and material are isolated and estimated, with overhead and profit being added later.

The conference method is managed in several ways, depending on the available knowledge. A conference moderator provides the questions such as, "What is the labor and material cost for this design?" Various gimmicks sharpen judgment. A "hidden-card gambit" has each of the committee experts reveal a personal value to a question. This could provide a consensus. If agreement is not initially reached, discussion and persuasion are permitted as influencing factors. Sometimes this is called estimate-talk-estimate (or E-T-E). The hidden-card idea prevents a brainstorming session, which generally gives optimistic estimates.

A ranking scheme along the "good-better-best" approach can be applied in a cost sense. We rank two or more designs by giving their single-cost value. Ranking seems to help, provided that the number of choices is not large. Sometimes this scheme is called *ordinal* ranking.

The lack of analysis and a trail of verifiable facts leading from the estimate to the governing situation are major drawbacks to the conference method. Although the procedure lacks rigor and there is little faith in its accuracy, these factors seldom deter its use.

6.5   COMPARISON

The *comparison* method attaches a formal logic to estimating. If we are confronted with an unsolvable or excessively difficult design and estimating problem, we designate it as design *a* and construct a simpler design problem for which an estimate can be found. The simpler problem is called design *b*. This simpler design might arise from a clever manipulation of the original design or a relaxation of the technical constraints on *a*. Too, by branching to *b* we gain information, as various facts may already be known about *b*. Indeed, the estimate may be in final form, or portions may exist and there need be only a minor restructuring of information to allow comparison.

The alternative design problem *b* must be selected to bound the original problem *a* in the following way:

PICTURE LESSON
Damn Big Dam

The Six Companies president, William Wattis, after learning that his company had won the right to build Hoover Dam, said, "Now this dam is just a dam, but it's a damn big dam." The largest single contract ever let by the U.S. government was to a consortium of six companies and rugged individualists. From the president to the superintendent and to the jack-hammer man, a can-do spirit prevailed, but there were many dissenters saying that because of its huge dimensions the dam could not be built. Others said that earthquake would break the dam or cause the reservoir to overflow and catastrophically flood cities downstream.

National events of the Depression accelerated the construction. Starting with the barren landscape of the Arizona and Nevada desert, the area erupted with a shanty town of thousands of desperate people looking for work, many of them without construction experience. The human problem, as epitomized by Ragtown, would have to be solved, just as surely as the engineering ones.

On both sides of Black Canyon, rough walls of red and purple rock soared more than twelve hundred feet straight up. Along the walls, white blazes highlight the dam abutments. On these walls the first phases of construction would begin. The building of a railroad down to link the railhead at Government Camp One with the dam site was one of the early steps. Above the walls in the desert, with a searing heat of 117 °F and higher, heat prostration and exhaustion were feared in the summer time. Within the walls and in the tunnels, the heat was even more oppressive. It was vital that many men be put to work quickly, but first Black Canyon had to be opened, access roads constructed, water and power provided, transportation of work crews and equipment mobilized, and a new city built. Experienced miners, powdermen, machinery operators, electricians, pipe fitters, and carpenters were needed before the unemployed and waiting workers in Las Vegas could go on the payroll. These were the imperatives for 1931.

The March 1934 picture shows the concrete columns of the dam rising up. A jumble of form boxes is winched upward, with each block having cooling tubes to lower the heat generated by the concrete. Bureau of Reclamation engineers calculated that the mass would rise 40 degrees and require 125 years to cool, and that the stresses caused by the setting process could fracture the body of the dam, rendering it useless; thus cooling was necessary. The blocks were no more than five feet thick and were honeycombed with pipes for chilled water. After cooling was completed, the pipes were filled with grout, making each unit solid.

Pouring of the concrete was critical to the success of the Six Companies' gamble. Having quoted at $2.70 per cubic yard, which was 35% below the next bidder, they had to be efficient with the thousands of workers. An efficient network of aerial cable ways was the linchpin to the success of pouring large buckets of concrete, moving material, equipment, and workers over the mile-long construction site.

$$C_a(D_a) \leq C_b(D_b) \tag{6.1}$$

where $C_{a,b}$ = value of the estimate for design a or b, dollars
          $D_{a,b}$ = design a or b

Further, D_b must approach D_a as nearly as possible. We adopt the dollar value of our estimate C_a as something under C_b. The sense of the inequality in Eq. (6.1) is for a conservative stance. It may be management policy to estimate cost slightly higher at first, and once the detail estimate is completed with D_a thoroughly explored, we comfortably find that $C_a(D_a)$ is less than the original comparison estimate.

An additional lower bound is possible. Assume a similar circumstance for a known or nearly known design c, and a logic can be expanded to have

$$C_c(D_c) \leq C_a(D_a) \leq C_b(D_b) \tag{6.2}$$

Assume that designs b and c generally satisfy the technical requirements (but not the economic estimate) as nearly as possible.

Comparison principles apply to any complexity of design, but consider a simple example. Three exterior wall systems for residences are given as shown in Figure 6.1. The

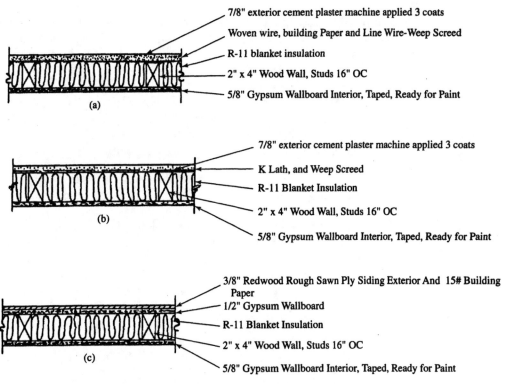

Figure 6.1 Comparison method. (a) Unknown design cost,
(b) higher-cost design, and (c) lower-cost design.

design that has the unknown cost is (a), while (b) and (c) have been explored for costs and standards are available, which indicate that for a 10-ft-long section $C_b = \$5.16/\text{ft}^2$ and $C_c = \$4.93/\text{ft}^2$.

The inescapable judgment for this method initially requires that the unknown design be squeezed between two designs that are above and below the unknown cost. Does the reader believe that design (a) satisfies this requirement? If this initial premise is accepted, then the cost is between these two numbers.

Other common-sense hints are helpful. The tighter the tolerance, the better the comparison. Typically, the contractor will avoid estimating from a zero base of information, and a system of *adds and deducts* begins with a selection of the boundary estimates. The lower or upper bound is the reference to either add or deduct costs for elements that are known.

The comparison logic is simple and useful. Indeed, this author believes that all estimating is basically a comparison of the unknown to the fathomed. For the estimating process does not start without knowledge of design and costs. The comparison method is sometimes called *similarity* or *analogy*.

6.6   UNIT

The previous methods were qualitative and depended very little on facts. The unit method, on the other hand, uses historical and quantitative evidence and leads to a cost driver easily understood. The unit method is the most popular of all estimating methods. Other titles describe the same thing—average, order of magnitude, lump sum, function, parameter, module estimating—and involve various refinements. Extensions of this method lead to the factor estimating method, discussed later. Examples of unit estimates are found in many business and economic activities:

- Cost of house construction per square foot
- Cost of electrical transmission per mile
- Construction cost per hospital bed
- Chemical plant cost per barrel of oil capacity

Though typically vague in those contexts, the strongest assumption necessary is that the design to be estimated is like the composition of the parameter used to determine the estimate. Notice that the estimate is "per" something.

Perhaps the most common unit cost estimate for buildings is the *cost-per-square-foot*. Area is perceived to have a powerful effect upon costs, and thus its popularity. A similar one is the *cost-per-cubic-foot* estimate. Another type of the unit-cost estimate is the cost-per-unit or trade-unit estimate. The building is broken down into basic trades, and cost and percentage factors are given. Various cost encyclopedias provide extensive information for unit-cost estimates, and that discussion is deferred to Chapter 7.

The unit method is used extensively, such as average material prices, man hours, and labor costs. The unit estimate is defined as the mean, where the divisor is the principal cost driver, or

$$C_a = \frac{\Sigma C_i}{\Sigma n_i} \tag{6.3}$$

where C_a = average dollar cost per unit of design
C_i = value of design i, dollars
n_i = design i unit (lb, in., ft, ft^3, yd^3, count, BF, etc.)

These costs are nothing more than the *average*. An average like Eq. (5.1) where $\bar{x}$ is found from a set of data is one simple way to find the statistic.

Consider the simple example of field-installed gate valves with carbon steel piping systems. There are three historical observations of total material and labor cost:

Design Weight, lb	C_i, $
2	$ 20
3	30
4	60
Σ = 9	Σ = $ 110

The cost per pound is $12.22 (= 110/9). The cost for a different gate valve is found by using its weight and multiplying by $12.22.

Effectively, the unit method is nothing more than the general slope b, where $a = 0$, or $y = 0 + bx$. For in the estimating of C_a = $12.22 to a design involving the weight parameter, the cost is zero for no weight, because 12.22 is a linear multiplier.

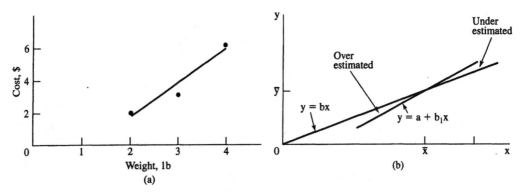

Figure 6.2 (a) Plot of three observations and (b) unit model compared to $y = a + bx$.

The unit estimating method is improved when the function $y = a + bx$ is statistically fitted with the data. When using $C_a = 12.22$, the unit method either overestimates or underestimates the design parameter of weight when compared to $y = a + b_1 x$. Notice Fig. 6.2(b), which shows the two simple methods compared. The only location of weight where there is no estimating error between the two methods is at $\bar{x}$ and $\bar{y}$ of the data, or average weight = 3 lb and $C_a = \$36.67$. Note that for these data the value of the intercept a is a negative fixed cost, an unlikely situation. Whenever $a < 0$, we have a signal for faulty data, or the equation model is suspect.

Certainly the reliability of the value of $12.22/lb$ is improved by employing more observations, but the method suffers from a more fundamental fault. If the data are plotted, and values of $y = a + bx$ are found by regression methods of Chapter 5, then Cost = $-23.30 + 20.0$ (lb). Notice Fig. 6.2(a), where the three observations are plotted with the line.

The unit method fails to apply the principle of economy of scale. For example, a 2-lb gate valve costs $24.44 each, while a 4-lb gate value costs $48.88. At least a linear regression improves on the estimated value as $16.70 and $56.70, which is not simple doubling.

The simple unit method, while an enhancement over opinion methods, is refined by other methods of estimating, which we now consider.

6.7 UNIT QUANTITY

Development of the unit method and man-hour data into a consolidated unit-quantity method is natural. This method is important for detail estimating for assemblies.

The method starts with the take-off of the material quantity requirements for the design. We add labor, equipment, and subcontractor requirements, along with the material and quantity for the design. The unit-quantity model is given by

$$C_o = \sum_{i=1}^{m} \sum_{j=1}^{n} n_i \left(R_i + H_{ij} R_{ij} \right) \qquad (6.4)$$

where C_o = cost for construction cost code or detail or assembly, dollars

n_i = quantity for design material i, in dimensional units compatible to design

R_i = unit cost rate for material i

R_{ij} = unit cost rate for requirement i, j

H_{ij} = unit man or equipment hours for requirement i, j associated with material n_i

i = material $1, 2, \ldots, m$ from take-off associated with construction work

j = requirement $1, 2, \ldots, n$ to satisfy installation of material i

Overhead and profit are not included in the cost at this point. Those matters are deferred to Chapter 8.

The n_i quantity is the important material take-off. The H_{ij} unit man hours are determined from productivity tables and were discussed in Chapter 2. Unit-cost rates are found from forecast costs for the material, labor, and equipment, and they correspond to the skills, time, and equipment that are necessary to install the materials.

Consider the cross-section view of the foundation wall shown by Fig. 6.3. Corners, offsets, or openings are not a part of this 100-ft wall design. Total cost of concrete, materials, and labor is required. A spreadsheet of the analysis is given by Table 6.1.

Different materials are required for this design. But forms and excavation are direct materials that are not evident from the foundation wall design, and they are required because of the method of doing this work. The symbol n_i is the quantity required for the several components for the design. It includes excavation, forms, reinforcing steel, concrete, and backfill. Observe that 59.3 yd^3 of excavation is necessary for a $4 \times 4 \times 100$ ft trench. R_i is the unit cost of material and R_{ij} are rates for labor and equipment. Obviously, excavation does not need material, and $R_i = 0$, but it does need labor and equipment.

Forms, on the other hand, have a material and labor component, but equipment is unnecessary, so $n_2 H_{23} R_{23} = 0$.

H_{ij} is the man hours per unit of n_i quantity. For example, $n_1 H_{12} = 1.60 \, (= 0.027 \times 59.3)$. R_{12} is the unit labor cost, which may be for one or more workers, including crew work. An n_i quantity does not always require material or equipment. Concrete material has been identi-

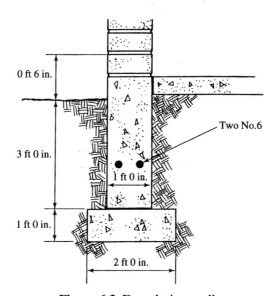

Figure 6.3 Foundation wall.

TABLE 6.1 Spreadsheet of the unit-quantity method for the foundation wall, Fig. 6.3.

Work and Material Take-off	Quantity, n_i	j = 1 Materials		j = 2 Labor				j = 3 Equipment				Row Total
		Unit Cost, R_i	Sub-total, n_iR_i	Unit Man-Hours, H_{i2}	Total Man-Hours, n_iH_{i2}	Unit Cost, R_{i2}	Sub-total, $n_iH_{i2}R_{i2}$	Unit Hours, H_{i3}	Total Hours, n_iH_{i3}	Unit Cost, R_{i3}	Sub-total, $n_iH_{i3}R_{i3}$	
Excavation, 4 × 4 × 100 ft	59.3 yd³	0	0	0.027	1.6	70	$112	0.027	1.6	$90	$144	$256
Forms, 2 at 4 × 6; 100 ft + 10%	990 ft²	1.60	$1584	0.035	35	65	2252	0	0	0	0	3836
Rebar, 2 No. 6; 100 ft + 10%	330 lb	1.18	389	0.023	8	42	319	0	0	0	0	708
Concrete, 2 × 1 × 100; 1 × 3 - 6 × 100	20.4 yd³	63.5	1295	2.0	41	40	1632	0	0	0	0	2927
Backfill, compact	41 yd³	0	0	0.02	1	25	21	1	41	94	3854	3875
Column total			$3268				$4336				$3998	$11,602
Cost per lin ft			$32.68				$43.36				$39.98	$116.02
Cost per yd³			$160.19				$212.54				$195.98	$568.73

fied as requiring contractor labor in Table 6.1, but the delivery service for concrete is included by the concrete supplier in the cost of concrete.

The subtotal labor amount of $112 is required for excavation. It is not necessary that this match the equipment requirement, as $H_{12} \neq H_{13}$. The capital consumption cost or rental cost for the equipment is identified as R_{13}.

Row and column totals are extended, and 100 ft of a wall will require a total of $11,602. A cost per lineal or cubic foot is found and may be used for subsequent estimating. Obtaining the cost of a parameter, such as $/ft^2 or $/yd^3, is a significant advantage of the unit-quantity method. If a different design is to be evaluated because of business costs or engineering changes, then we will choose between adjusting the table or accepting it as "close enough." The method is discussed again in Chapter 7.

6.8 FACTOR

The *factor* method is an important one for plant and industrial construction estimates. Other terms, such as *ratio* and *percentage,* describe the same thing. To estimate a plant with all of the materials, systems, and minutia, using detail approaches is too time consuming and, perhaps, inaccurate. The factor method eliminates these drawn-out details. Essentially, it determines the construction estimate by summing the product of several quantities, or

$$C = \left(C_e + \sum_i f_i C_e \right)(f_I + 1) \tag{6.5}$$

where C = cost of design, dollars

C_e = cost of selected major equipment, dollars

f_i = factors for estimating major items, building, structural steel, instrumentation, etc., that correlate to major equipment, number

f_I = factor for estimating indirect expenses, such as engineering, overhead, etc., number

$i = 1, \ldots, n$ number

The factors f_i are uncovered by several techniques. An owner's statistical analysis of the firm's operations, industry at large, business associations, contractors, news journals, or state and federal governments are sources.

A natural simplification of Equation 6.5 leads to the preliminary unit estimating model $C = fC_e$, where one factor is used to find the composite cost. There are variants, such as $C = \Sigma_i f_i D$, where D is the design parameter (for example, area of construction of building, or capital investment) and f_i is the factor in dimensional units compatible to the design.

The unit-cost estimating method discussed earlier is limited to a single factor for calculating overall costs. The factor method achieves improved accuracy by adopting separate factors for different cost items. For example, the cost of an office building can be estimated by multiplying the area by an appropriate unit estimate such as the dollars-per-square-foot factor. As an improvement, individual cost-per-unit-area figures can be used for heating, lighting, painting, and the like, and their value C can be summed for the separate factors and designs.

Plant construction cost can be estimated by the factor method. Consider the following case study. A hydrobromination flow chart, given by Fig. 6.4, describes a microprocess

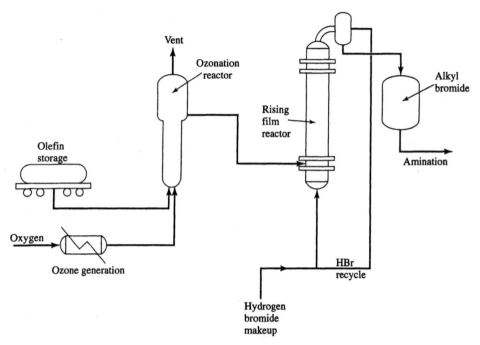

Figure 6.4 Flow chart for hydrobromination process.

plant, and we want to estimate the construction cost of this plant. A hydrobromination process is one step in the sequence of making a liquid soap formulation.

The flow chart and the specification sheet are primary input data to the engineer. The basic item (or items) of the process is identified using the flow chart. This *basic item(s)* is a major item for a building, such as the structural shell, or tons of concrete for a highway, or process equipment in a chemical plant. First, the cost of the basic item is determined. The next step is to find the cost relationship of other components as a percentage, ratio, total cost, or factor of the basic item.

For a building, the structural shell may be selected as the basic item, and the regression factors would be brick, concrete, masonry, reinforced concrete, carpentry, finish trim, electrical work, plumbing, and so forth. For a chemical process, the major item could be high-pressure high-temperature reactors, and equipment erection, piping and direct materials, insulation, instrumentation, and engineering are factors correlated to process cost. Regression is statistically found by using the methods in Chapter 5.

In some instances there are costs that are independent of the variation in the cost of the basic item. Those independent components (for example, roads, railroad siding, and site development for a new chemical plant are not related to the equipment) must be estimated separately by other methods.

There are common-sense considerations for the factor method. For chemical plant construction, variations are due to the size of the basic equipment selected, materials of fabrication/construction, operating pressures, temperatures, technology (such as fluids processing, fluids-solids processing, or solids processing), location of plant site, and timing of construction.

Practically speaking, if the physical size of the basic item becomes larger and therefore more costly, the factor relating, say, the engineering design cost is relatively smaller.

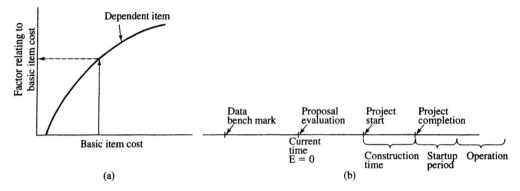

(a)   (b)

Figure 6.5 (a) Factor relationship to basic item cost, and (b) time line for project evaluation.

If the basic item is constructed with more expensive materials, such as stainless steel or glass-lined materials instead of plain carbon steel, then the factors become nonproportional to the item being estimated.

Factor behavior is described by Fig. 6.5(a). The basic item cost is the x-axis entry. A vertical intersection with the line from the x-axis factor value leads to a factor, as read from the y axis. If the x-entry value is doubled, then we assume that the factor is not doubled, thus assuring the principle of economy of scale. This is apparent because the factor lines in Fig. 6.6 show a declining nonlinear scale.

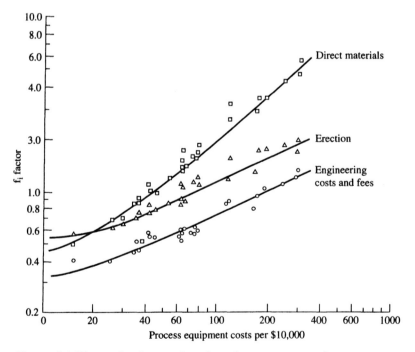

Figure 6.6 Illustrative factor chart based on process equipment costs for benchmark year.

The factor data are typically benchmarked to an index. An owner or industrywide or government index with a specified benchmark year is used, and all data are referenced to this benchmark year. Indexes are important in the factor method. Indexes are discussed in Chapter 5. Notice Fig. 6.5(b).

The first step in the factor method estimates the important *basic item* costs. For the hydrobromination plant, the rising film and ozonation reactors are chosen as the important basic items from the flow chart. Because this process is high pressure and high temperature, we expect the piping, supports, instrumentation, foundation, and so on to correlate with these critical reactors. The basic item costs recognize the distinction between carbon steel and stainless steel, and the factor reflects the distinction.

Notice Fig 6.6, which shows the factors of engineering costs and fees, erection, and direct material costs for the construction of the hydrobromination plant. The data points are shown along with the lines. The entry *x-axis* variable is process equipment costs. Notice that the scales are logarithmic and the lines are nonlinear. All data are left as points on the graphs, an important practice in cost analysis.

The two reactors from Fig 6.4 are estimated for current time costs, which correspond to the time of the estimate at $E = 0$, as shown by Fig.6.5(b). The reactor costs could be estimated internally by the plant engineering staff, or fabricators can be asked for bids. In either case, once the basic items are estimated, we have

Major Process Design Equipment	Cost, $
Rising film reactor	2,200,000
Ozonation reactor	700,000
	C_e = $2,900,000

The estimate of the basic items is in *current-time dollars*. In doing factor estimating an owner maintains indexes allowing it to *backcast* the C_e value to the benchmark year of the cost data. Once the basic item cost is backcast to the period of the factor curve, the factors are read off the curve, or determined from the data. At this point the indexed base-item cost value is multiplied by f_i. Next, the benchmark cost is indexed forward to that point in time when the cost is expected as out-of-pocket, project start, completion, or when progress payment periods are stipulated by the contract between the contractor and the owner.

We revise Eq. (5.52) to show

$$C_r = C_c\left(\frac{I_r}{I_c}\right) \tag{6.6}$$

where C_r = cost at benchmark time, dollars

C_r = cost of the equipment at current time, $2,900,000

I_r = index of 100 for benchmark year

I_c = index of 114.1 for current time for equipment

then $C_r = \$2,900,000\left(\dfrac{100}{114.1}\right) = \$2,541,630$

With this adjusted major component cost as the entry, we find the factors that correlate to the equipment. Figure 6.6 yields engineering costs and fees, 1.1; erection, 1.7; and direct materials, 4.1.

Factor data for plants are often regionally located in a construction area where significant activity is expected. As an example, chemical plants are frequently benchmarked to the Gulf Coast of the United States. Construction at other sites requires regional indexing to adjust for differences in labor and material cost between the plant site and the benchmark area.

Notice Table 6.2, where the factors become a multiplier for the benchmark cost, or $4,320,770 (= 1.7 \times 2,541,630)$ for erection. Next, this cost is inflated to the project start point or the progress payment time stipulated by the contract clauses. This would, for instance, be $4,320,770 \times \left(\dfrac{189}{100}\right) = \$8,166,250$.

The data in Fig. 6.6 are Gulf Coast information, although our hydrobromination plant is constructed in the Midwest where direct materials are 8% more expensive, erection costs are 13.2% higher, and engineering costs and fees are identical. Further, those regional indexes account for the mix of the components. The regional factors are applied to current values. The Gulf Coast cost becomes $8,166,250 \times 1.132 = \$9,244,200$, adjusting for geographical distinctions of cost.

Finally, the owner's capital investment, $34,630,210, is provided as the raw construction estimate. A factor for indirect costs, contingency, and profit is applied later to convert raw capital investment into the full cost for construction.

The example deals with plant construction and equipment estimating, but the factor method is used more broadly to estimate labor, materials, utilities, and indirect costs as a

TABLE 6.2 Spreadsheet of the factor method for the flow chart of the hydrobromination plant, Fig. 6.4.

Item	Current time cost, $	Current index	Major Item Bench Mark Cost, $	f_i	Bench Mark Cost, $	Project Start Index	Gulf Coast Project Start Cost, $	Midwest Regional Index	Project Start Plant Site Cost, $
Equipment	2,900,000	114.1	2,541,630	1	2,541,630	118.0	2,999,120	-	2,999,120
Major process item Erection				1.7	4,320,770	189.0	8,166,250	1.132	9,244,200
Direct materials				1.7	10,420,680	127.0	13,234,260	1.08	14,293,000
Engineering costs and fees				1.1	2,790,000	145.0	4,053,890	1.0	4,053,890
Building site development									1,250,000
Site development									
Process building									2,100,000
Railroad spur									40,000
Utilities									650,000
Total									$34,630,210

multiple of some other estimated or known quantity. Operating costs as a percentage of plant investment or as a percentage of the product selling price are popular.

6.9 COST- AND TIME-ESTIMATING RELATIONSHIPS

Cost-estimating relationships (CERs) and time-estimating relationships (TERs) are mathematical models or graphs that estimate cost or time. Simply, CERs and TERs are statistical models that characterize the cost of a project or task as a function of one or more independent variables. These principles were discussed in Chapter 5, but forecasting deals with time-serial and minor analytical problems. However, CERs and TERs are formulated to estimate various end items that are significant. Rules of thumb and the unit-quantity method are not recognized as CERs.

The word *parameter* is often used in the CER context. Although it usually refers to the coefficients found in an equation [e.g., $C_a = -23.30 + 20.000$ (lb)], it has different meaning for the CER. After the development of a CER, and in the estimating phase, we substitute the new value for the design, and that value is loosely called the "parameter." Then we calculate a cost. For example, the parameter of a gate valve weighing 5 lb, which we substitute in the equation, gives a cost of $76.70. Sometimes CERs are called *parametric* equations.

These approaches are not new. We use output design variables (i.e., physical or performance parameters such as lb or yd^3) to predict cost, since the preliminary design parameters are available early. Sometimes those parameters are called *cost drivers*. CER estimates are considered preliminary.

As in all functional estimating models there must be a logical relationship of the variable to cost, a statistical significance of the variables' contribution, and independence of the variables in the explanation of cost. This is sometimes referred to as *causality*, where the design factor contributes significantly to the cost.

CERs are developed by using a variety of steps. We suggest this approach: obtain actual costs, interview experts who have knowledge of cost and design, find cost-time drivers, plot data roughly and understand anomalies, replot and conclude regression analysis, review for accuracy and communication, publish, and distribute the CERs to engineers and cost analysts. Those steps constitute the basis for much of cost analysis, of course.

We study two renowned CERs. They are known as *learning* and the *power law and sizing* models. Other statistical equations can be labeled as CERs, but the student will need to refer to other books for more information.

6.9.1 Learning

It is recognized that repetition with similar construction projects reduces the time or cost invested in the project. This phenomenon can be modeled with ordinary estimating techniques. The improved performance is called *learning*.

The first applications of learning were in airplane manufacture, which found that the average number of man hours spent in building an airplane declined at a constant rate over a wide range of production. Though the initial airplane application stressed only time, practice has extended the concept to cost. Learning has been convincingly found in construction, as well.

This learning practice suggests that time or cost is lowered with increasing quantity of construction or experience. Knowing how much the time or cost can be lowered, and at what point learning is applied in estimating procedures, are reasons for studying learning. The learning model rests on the following assumptions:

- The time or cost amount required to complete a unit of construction is less each time the project is undertaken.
- Unit time or cost decreases at a decreasing rate.
- Reduction in unit time or cost follows a specific estimating model, such $y = ax^b$

To state the underlying hypothesis, the man hours or cost necessary to complete a unit of construction will decrease by a constant percentage each time the construction quantity is doubled. A frequent stated rate of improvement is 20% between doubled quantities. This establishes an 80% learning, which means that the man hours to build the second unit will be the product of 0.80 times that required for the first. The fourth unit (doubling 2) will require 0.80 times the man hours for the second, the eighth unit (doubling 4) will require 0.80 times the fourth, and so forth. The rate of improvement (20% in this case) is constant with regard to doubled construction quantities, but the absolute reduction between amounts is less. Note Figure 6.7, which generalizes for the doubling concept, where the known information is an 80% learning experience and actual value, called a.

The notion of constant reduction of time or cost between doubled quantities is defined by a unit formula:

$$T_u = KS^s \tag{6.7}$$

where T_u = project man hours or cost per unit of construction required to construct the Nth unit

N = unit number

K = constant, or estimate, for unit 1 in dimensions compatible with T

s = slope or function of the improvement rate

Learning slope s is negative because time or cost decreases with increasing units of construction. T_u plots as a curved line on arithmetic coordinates, or as a straight line on logarithmic coordinates. Notice Fig. 6.8.

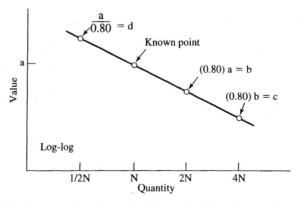

Figure 6.7 Constructing learning line on basis of known data for an 80% curve.

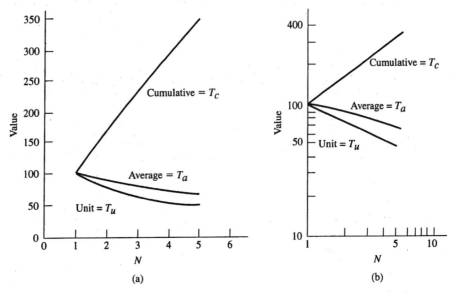

Figure 6.8 Learning function: (a) arithmetic plot and (b) logarithmic plot.

Reviewing algebra, we convert both sides of Eq.(6.7) with a logarithm, and we have

$$\log T_u = \log K + s \log N \tag{6.8}$$

which is of the form $y = a + bx$, the equation of a straight line. Recalling the ideas of transforming the variables, a topic discussed in Chapter 5, then $y = \log T_u, a = \log K, b = s$, and $x = \log N$. From this point, we are able to statistically determine the parameters of the equation, K and s. The student is encouraged to review Table 5.2.

To understand the presentation of learning on logarithmic graph paper, first compare the characteristics of arithmetic and logarithmic graph paper. On arithmetic graph paper, equal numerical differences are represented by equal distances. For example, the linear distance between 1 and 3 will be the same as from 8 to 10. On logarithmic graph paper the linear distance between any two quantities is dependent on the ratio of those two quantities. Two pairs of quantities having the same ratio will be equally spaced along the same axis. For example, the distance from 2 to 4 will be the same as from 30 to 60 or from 1000 to 2000.

Learning is usually plotted on double logarithmic (log log) paper, meaning that both the abscissa and the ordinate use a logarithmic scale. The exponential function $T_u = KN^s$ plots in a straight line on log log paper. This function can be plotted from either two points or one point and the slope (e.g., unit number 1 and the percentage improvement). Also, by using log log paper, the values for a large quantity of units can be presented on one graph, and values can be read relatively easily using the graph. Arithmetic graph paper, on the other hand, requires many values to sketch in the function. Refer to Fig. 6.8(b).

Define Φ as the decimal ratio of time or cost per unit required for doubled construction, and then

$$\log \phi = s \log 2 \tag{6.9}$$

$$\text{and } s = \frac{\log \phi}{\log 2} \tag{6.10}$$

This relationship is tabulated in Table 6.3.

TABLE 6.3   Table of decimal learning slopes for ratio Φ.

Φ	1.0 (no learning)	0.95	0.90	0.85	0.80
Exponent, s	0	−0.074	−0.152	−0.234	−0.322

There is a need to extend the unit formulation to other relations. The cumulative time or cost that a project will require is found using:

$$T_c = T_1 + T_2 + \cdots + T_N = \sum_{u=1}^{N} T_u \tag{6.11}$$

where T_c = cumulative total cost or time from unit 1 to N

Refer to Fig. 6.8, where T_c is shown.

Additionally, the average time or cost, T_a, for a unit N can be found by consecutively dividing the cumulative value by the quantity and is found as

$$T_a = \frac{\sum_{u=1}^{N} T_u}{N} = \frac{T_c}{N} \tag{6.12}$$

where T_a = average unit time or cost from unit 1 to N

Notice Fig. 6.8, where T_a is positioned below the unit line. Observe that T_u, T_c, and T_a join together at $N = 1$.

An approximation to Eq. (6.12) is given by the following:

$$T_a \cong \frac{1}{(1 + s)} K N^s \tag{6.13}$$

where the approximation is improved after 20 or so units. The T_u and T_a lines become parallel at that point. Prior to $N < 20$, T_a is joined to T_u at $N = 1$ and it remains above T_u.

If only two actual project experiences are known, then it becomes possible to find the slope using

$$s = \frac{\log T_i - \log T_j}{\log N_i - \log N_j} \tag{6.14}$$

where s = slope of learning function, decimal

T = time or cost, units of time or dollars

N = unit number, $N_j > N_i$

Imagine a commercial development, where the units are similar. Data of a past development are analyzed using regression statistics. The business development demonstrates learning, and the experience is found to have a slope parameter of 95%. A planned new development will have similar units, and the construction pattern of the units is in a serial fashion. The first unit is estimated to require $180,000. The learning equation is identified as $T_u = 180,000N^{-0.152}$. As $T_u = 180,000(2)^{-0.152} = \$162,000$, the unit cost value for the second is entered in Table 6.4. Other rows are similarly calculated. Then the other columns are completed. These values, say for $N = 1, 2, 3, \ldots$, become estimates that are useful in various ways.

It is important to stress that other units have been previously analyzed and form the basis for the new estimates. The previous development may have determined the slope s, and the new commercial development will have a new intercept because of design and specifications. Then the value of K at $N = 1$ is found, but the slope allows extension to other units. This is a common practice in the application of learning.

Learning theory has been successfully uncovered in the construction of gas-fired power plants, nuclear power plants, and other engineering construction. Its best application arena is in the estimating of large-scale projects. Small jobs, with little or no engineering design improvement and follow-on and planning, such as residences are unsuitable.

How does learning happen? Even if the actual value for the first several units increase as compared to the estimates, the owner/contractor needs to insist that the learning philosophy prevail. This is so, as there is a perception that management persuasion assists the learning phenomena. The process of construction learning does not happen without engineering effort and construction planning. There is no serendipity in the learning process, and *wishing so* does not make it happen. The construction enterprise needs to understand that learning will happen if there are engineering activities that make it happen.

The phenomena of learning results from the consequences of better planning, improved productivity, and competition, and these effects are driven by engineering design and construction management. We cannot separate these driving factors for any day-to-day estimating. Learning is best practiced as an overall preliminary method.

Practitioners will have general knowledge of slopes s and learning rates ϕ. In some cases, this information is openly published and available. In the estimating process, we estimate the first few units, and then the remaining units are calculated using the learning percentage. The slope and learning-rate factors are transferred from actual situations. The first unit K is identified by any of several estimating techniques.

Learning is more suitable for projects having certain characteristics. The learning phenomenon is practiced with low project quantity, as few as two project units. It is improbable that the project units are identical, but that they be *similar* is important. The degree of

TABLE 6.4   Calculation of costs for commercial development.

House No., N	$T_u = KN^s$, Unit Cost, $	T_c, Cumulative Cost, $	T_a, Average Cost, $
1	180,000	180,000	180,000
2	162,000	342,000	171,000
3	152,316	494,316	164,772
4	145,800	640,116	160,029
5	140,940	781,056	156,211

similarity is subject to the interpretation of experienced managers. Further, the slope percentage ϕ is crucial to the sanity of estimating with the learning equations. The slope percentage is important, i.e., while 75% learning is desirable as compared to 95%, it may be impossible or uneconomic. Projects that are routine may be unsuitable, especially if engineering design and construction management are not consciously driven to "improve."

6.9.2 Power Law and Sizing

The *power-law-and-sizing* CER is used for estimating process and plant equipment. This CER models designs that vary in size but are similar in type. The unknown costs of a 200-gallon kettle can be estimated from data for a 100-gallon kettle, provided that they are of similar design. We would not expect the 200-gallon kettle to be twice as costly as the smaller one. The *law of economy of scale* assures this. In general, costs do not rise in strict proportion to size.

The power-law-and-sizing CER is given as

$$C = C_r\left(\frac{Q_c}{Q_r}\right)^m \tag{6.15}$$

where C = cost for new design size Q_c, dollars
  C_r = known cost for reference design size Q_r, dollars
  Q_c = design size of equipment expressed in engineering units
  Q_r = reference design sized in consistent engineering units
  m = correlating exponent, $0 < m \leq 1$

If $m = 1$, then we have a strictly linear relationship and deny the law of economy of scale. For chemical processing equipment m is frequently near 0.6, and for this reason the model is sometimes called the *sixth-tenths*[2] model. The units on Q are required to be consistent, as it enters only as a ratio.

Further, the model considers changes in cost due to inflation or deflation and effects independent of size, or

$$C = C_r\left(\frac{Q_c}{Q_r}\right)^m \frac{I_c}{I_r} + C_i \tag{6.16}$$

where I_r = index time - associated with reference equipment, dimensionless number
  I_c = index time - associated with new design, dimensionless number
  C_i = cost items independent of design, dollars

Assume that 6 years ago an 80-kW naturally aspirated diesel electric set and its plant cost $1,600,000. The engineering staff is considering a 120-kW unit of the same general design. The value of $m = 0.6$, and the price index for this class of equipment 6 years ago was 187 and now is 194. Differing from the previous design is a precompressor, which when

---

[2] But in a study by D. S. Remer and L. H. Chai, "Design Cost Factor for Scaling-Up Engineering Equipment," *Chem. Eng.* 86(8), 77-82, 1990, the average value of m was found to be closer to 0.7.

isolated and estimated separately costs $180,000. The future cost, adjusted for size of the design and for inflation, is

$$C = 1,600,000\left(\frac{120}{80}\right)^{0.6}\left(\frac{194}{187}\right) + 180,000 = \$2.3 \text{ milion}$$

The measurement of m is important to the success of this model, and statistical methods given in Chapter 5 are useful in finding those values. If the statistical analysis assumes constant dollars, then the index ratio I_c/I_r is used for increases or decreases for inflation or deflation effects.

The model usually does not cover those situations where the estimated design Q_c is greater or less than Q_r by a factor of 10. Values of m are published, and a few exponents for equipment cost versus capacity are given by Table 6.5.

The value of m is important in several ways. If $m > 1$, then we deny the economy-of-scale law, as shown with the centrifugal compressor. In fact, when $m > 1$, we have *diseconomy of scale*. If $m = 0$, then we can double the size without affecting cost, an unlikely event.

An equation expressing average cost C/Q_c can be found, or

$$C\frac{Q_r}{Q_c} = C_r\frac{Q_r}{Q_c}\left(\frac{Q_c}{Q_r}\right)^m = C_r\left(\frac{Q_c}{Q_r}\right)^{-1}\left(\frac{Q_c}{Q_r}\right)^m$$

$$\frac{C}{Q_c} = \frac{C_r}{Q_r}\left(\frac{Q_c}{Q_r}\right)^{m-1} \tag{6.17}$$

Total cost varies as the mth power of capacity in Eq. (6.15), but average cost C/Q_c will vary as the $(m - 1)$st power of the capacity ratio.

6.9.3   Other CERs

CERs are popular because of their association to historical records and the ease of their statistical calculation. Sometimes parameter values that are tried and true exist, making CERs effective. Relationships, in addition to those mentioned, exist, such as

$$C = KQ^m \tag{6.18}$$

where K = empirical constant for plant, equipment, or system, dollars

           Q = capacity expressed as a design dimension

TABLE 6.5   Illustrative values of the exponent m for the power-law-and-sizing CER.

Equipment	Size Range	Exponent, m
Blower, centrifugal (with motor)	1–3 hp	0.16
Blower, centrifugal (with motor)	$7\frac{1}{2}$–350 hp	0.96
Compressor, centrifugal (motor drive, air service)	20–70 hp drive	1.22
Compressor, reciprocating (motor drive, air service)	5–300 hp drive	0.90
Dryer drum (including auxiliaries, atmospheric)	20–60 ft^2	0.36

Difficulties exist in using Eq. (6.18) because it is necessary to know the capital cost of an identical plant. Nor is the scale factor m constant for all sizes of the design. Generally, scaling up or scaling down by more than a factor of 5 should be avoided.

The ineffectiveness of Eq. (6.18) can be overcome somewhat by separating the direct capital costs, which are subject to the economy-of-scale-rule, and the indirect or fixed-cost element, which is not. Another relationship is expressed as

$$C = C_v \left(\frac{Q_c}{Q_r}\right)^m + C_f \tag{6.19}$$

where C_v = variable element of capital cost, dollars
           C_f = fixed element of capital cost, dollars

A multivariable CER is also possible. For instance,

$$C = KQ^n N^s \tag{6.20}$$

where the symbols have been previously defined. Coefficients are determined by regression methods.

Equation (6.20) is a relationship dealing with the cost principles of *economy of scale* and *economy of quantity*. These are two important principles, and this relationship attempts to tie them together.

It is possible to have multivariable CERs with more than two variables. Indeed, these equations can have many variables, but the knowledge gained from the statistical parameters becomes murky. In construction, we desire a more hands-on understanding of the variables, where the connection to the variable gives a cause-and-effect relationship that is apparent. Another specialty field is *econometrics*, which is the association of social economics and statistics. Econometrics pays attention to gross national income, expense, and political programs, and measures those effects. These equations can have many variables[3] and the source of the raw data could be national census and federal and state governmental information.

While CERs are equations, they also can be plotted as a graph, which is an effective way to describe cost behavior. All the data are shown, together with the statistically fitted line, whenever a graph is plotted. The original data remain on the graph, since erasing the points destroys the sense of variation. Visual plots give a gut feeling for the data. Plots are limited to two variables.

Notice Fig. 6.9, which is an example of the plot of pressure vessels, where the x axis is "gallons per minute" multiplied by "feet head." The line is nonlinear on logarithms, and the standard deviation, shown as $s = 8.78\%$, is used instead of the absolute value of the standard deviation. A percentage standard deviation is constant for all values of the variables. The absolute standard deviation varies over the range of the variables and it is a minimum at the means, $\bar{x}$ and $\bar{y}$.

6.10   PROBABILITY APPROACHES (OPTIONAL)

As usually prepared, the estimate represents an "average" concept. Nor does the estimate reveal anything about the probability of these expected values. It uses information that is called certain, or *deterministic*.

---

[3] One equation has 150 active variables.

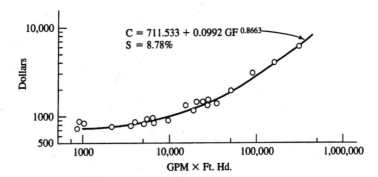

Figure 6.9 Illustrative CER plot and equation derived by using physical variables.

6.10.1 Expected Value

For this discussion we assume that the engineer is able to give a *probability point estimate* to each element of uncertainty as represented by the economics of the design. This assignment associates nonnegative numerical weights with possible events. For example, if an event is certain, then its associated probability equals 1. The cost of a 2×4 in. stud is $8.79, and this is 1, which implies a probability of 0 to any other stud cost.

Sometimes two events A and B are *mutually exclusive*, meaning the probability of the event "either A or B" equals the sum of the probability for each of the events.

Probabilities are a numerical judgment of the likelihood of future events. The techniques for deriving probabilities include:

- Analysis of historical data to give a relative-frequency interpretation, such as the histogram
- Convenient approximations like the normal distribution function
- Expert introspection, or what is called *opinion probability*

The first two choices are known as *objective probability*. Seldom is there disagreement about how values are found using objective probability methods.

Opinion probabilities call for judgmental expertise and a pinch of luck. Better success is assured when past data are analyzed. On the other hand, data may be unavailable, and it should be remembered that while data are past, probabilities should be indicators of the future. Sometimes both past data and a reshuffling of probabilities are undertaken. This discrimination is not new to professional cost-analysis practice.

For the most part, we prefer to deal with the simplest case: certainty. When we say "material cost is $8.79," we imply, but leave unstated, that the probability is 1, a *certain event*. Any other material cost has the probability of 0. Despite this assertion of certainty, it seldom exists.

The category involving *risk* is appropriate whenever it is possible to estimate the likelihood of occurrence for each condition of the design. These estimates describe the true likelihood that the predicted event will occur. Formally, the method incorporates the effect of risk on potential outcomes by using a weighted average. Each outcome of an alternative is multiplied by the probability that the outcome will occur. This sum of products for each alternative is called an *expected value*. Mathematically, for the discrete case,

$$C(i) = \sum_{j}^{n} p_i x_{ij} \qquad (6.21)$$

where C = expected value of the estimate for event i, dollars
  p_j = probability that x takes on value x_j, $0 \le p_j(x_j) \le 1$
  x_{ij} = design event

The p_j represents the independent probabilities that their associative x_{ij} will occur with $\sum_{j=1}^{n} p_j = 1$. The expected-value method exposes the *degree of risk* when reporting information in the estimating process.

   Consider the following example: An electronics manufacturing firm is evaluating a portable television that has special design features of wireless email and real-time computer functionality. Market research indicates a substantial market available for a small lightweight set if priced at $850 retail. This implies that the set will have to be sold to wholesalers for approximately $650. Several questions need to be answered before the decision can be made to enter the market. Three important ones are: What will be the first year's sales volume in units? How much will the sets cost to produce? What will be the profit? To answer those questions, marketing furnishes an estimate of the first year's sales in units.

Annual Sales Volume	Probability of Event Occurring
15,000	0.2
20,000	0.2
*25,000	0.6

*Most frequent case

The total cost per unit is estimated:

Cost per Unit, $	Probability of Event Occurring
*450	0.7
500	0.2
550	0.1

*Most frequent case

   The illustration can be divided into

- Risk not apparent
- Risk apparent.

Assume that marketing and estimating use the most frequent case number from their studies and do not report any risk. Profit is calculated as

$$\text{Profit} = (650 - \text{cost}) \times \text{volume}$$
$$\text{Most frequent cost} = \$450$$
$$\text{Most frequent volume} = 25{,}000 \text{ units}$$
$$\text{Profit} = (650 - 450)25{,}000 = \$5 \text{ million}$$

TABLE 6.6 Calculation of joint probability for cost and volume of product.

(650 − Cost)Volume, $	Joint Probability of Occurrence	Expected Value Profit, $
(650 − 450)15,000 = 3,000,000	0.14	420,000
(650 − 450)20,000 = 4,000,000	0.14	560,000
(650 − 450)25,000 = 5,000,000	0.42	2,100,000
(650 − 500)15,000 = 2,250,000	0.04	90,000
(650 − 500)20,000 = 3,000,000	0.04	120,000
(650 − 500)25,000 = 3,750,000	0.12	450,000
(650 − 550)15,000 = 1,500,000	0.02	30,000
(650 − 550)20,000 = 2,000,000	0.02	40,000
(650 − 550)25,000 = 2,500,000	0.06	150,000
Total	1.00	Profit = $3,960,000

Conversely, assume that the organization encourages a policy of reporting risk in estimates. We use opinion probabilities to have three values for cost and three for volume. We calculate nine profit possibilities (Table 6.6).

Another approach is possible. Both marketing and engineering could have computed an expected value as 470 (= 450 × 0.7 + 500 × 0.2 + 550 × 0.1) and 22,000 (= 15,000 × 0.2 + 20,000 × 0.2 + 25,000 × 0.6). The expected profit becomes 22,000 (650 − 470) = $3.96 million, which is the same as in the table. In the case of *risk not apparent* the total profit is overstated.

It is important to point out that there is a distribution or histogram that can be found from these data. Technical people relate to curves, distributions, and so on, and individual cost values are sometimes incomprehensible by themselves, particularly when dealing with dollar values that have little or no relevance to personal experience.

The calculations for this example require that the cost per unit be *independent* of volume and that cost be inversely related to volume, which is generally not the situation. One of the difficulties in applying the expected-value method is the inability of experts to guess opinion probabilities. Seldom are there personal experiences that allow for a development of this ability.

6.10.2 Range

Knowing the weaknesses in information and techniques, we recognize that there are probable errors in the estimate and its procedure. The recognition that cost is a random variable opens up the topic of *range estimating*. A *random variable*, in statistical parlance, is a numerically valued function for the outcomes of a sample of data. Finding the mean or standard deviation, for example, using equations with sample information gives a random variable. This notion introduces an important probability improvement to single-valued estimating.

The range method involves making three estimates for each major cost element. We bracket a *most frequent* or *modal* estimate value for each cost element. This forms the basis for range estimating.

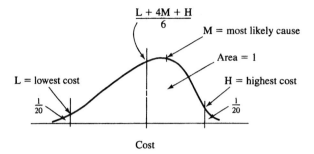

Figure 6.10 Location of estimates for Range method.

The following procedure is based on a method developed for PERT (program evaluation and review technique) and involves making a most likely cost estimate, an optimistic estimate (lowest cost), and a pessimistic estimate (highest cost).

Those estimates are assumed to correspond to the unimodal *beta distribution* in Fig. 6.10. The beta distribution has rich properties that are useful for cost estimating. For example, only positive costs are allowed. Remember in the normal or *t*-distribution, negative costs are theoretically allowable, as the range of the variable can be negative. This particular beta distribution is skewed left. Symmetric and skewed-right distributions are also possible. The total area under the probability distribution is 1. The mathematics of the beta probability density are complicated, but their expected values lead to simplified equations.

Almost any method in making the individual estimate is acceptable, from preliminary to detail. For discussion of this range method, we assume that the estimates are already made. A mean and variance for the single cost element are calculated as

$$E(C_i) = \frac{L + 4M + H}{6} \tag{6.22}$$

$$\text{var}(C_i) = \left(\frac{H - L}{6}\right)^2 \tag{6.23}$$

where $E(C_i)$ = expected cost of distribution i, $i = 1, 2, \ldots, n$, dollars
         L = lowest cost, or best-case estimate of cost distribution i, dollars
         M = modal value, or most likely estimate of cost distribution i, dollars
         H = highest cost, or worst-case estimate of cost distribution i, dollars
    $\text{var}(C_i)$ = variance of cost distribution i, $i = 1, 2, \ldots, n$, dollars^2

If a dozen or more elements are estimated this way, and the elements are assumed to be independent of each other and are added, then the new distribution of the total cost is approximately normal. This follows from the *central limit theorem*. The mean of the sum is the sum of the individual means, and the variance is the sum of the variances. The distribution of the sum is normal, despite the reasoning that says that the individual elements were beta shaped. A normal shaped distribution is shown by Fig. 6.11.

$$E(C_T) = E(C_1) + E(C_2) + \cdots + E(C_n) \tag{6.24}$$

$$\text{var}(C_T) = \text{var}(C_1) + \text{var}(C_2) + \cdots + \text{var}(C_n) \tag{6.25}$$

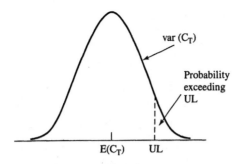

Figure 6.11 Normal curve summed for individual probabilities.

var (C_T)

Probability exceeding UL

$E(C_T)$ UL

where $E(C_T)$ = expected total cost of independent subdistributions i, dollars
var(C_T) = variance of total cost of independent subdistributions i, dollars^2

It is seen that the total expected cost and the variance of the total cost are merely sums of component expected costs and variances, respectively. Also, the distribution of total expected cost is nearly normal regardless of the shape of the distribution of the estimated cost elements.

Various statements regarding the interval for a future estimate value can be made with a normal distribution. What is the probability that a certain cost will be exceeded, or Pr[cost $> E(C_T)$]? What are the probability boundaries for a bid? With the normal distribution assumed for C_T, it is possible to examine this distribution and answer those questions. This probability is found by using

$$Z = \frac{UL - E(C_T)}{\left[\text{var}(C_T)\right]^{1/2}} \tag{6.26}$$

where Z = value of the standard normal distribution, Appendix 1
UL = upper limit of cost, arbitrarily selected, dollars

The square root of var(C_T) is the *standard deviation*. The value of Z is used to enter Appendix 1, and the probability of exceeding the value is found by using $0.5 - \text{Pr}(Z)$. The normal distribution is symmetrically centered, and the area of 1 is divided into parts with 50% above and 50% below the value of $Z = 0$. Refer to Appendix 1 and observe the characteristics of the numeric values. Manipulation of the normal distribution is a straightforward calculation.

Table 6.7 shows an application of this approach. The expected cost sum of the individual cost elements is $10,600, and the variance representing probable error is 22,300 dollars^2. We have assumed that there are a number n of individual estimates, and they are independent of one another. In other words, the flash-lamp cost is unrelated to the cost outcomes of a data grid. The total sum and variance are assumed to be normally distributed, even though that the individual cost distributions were something else.

The application of Eq. (6.26) allows the finding of risk for a project. For example, UL can be the bid value or ceiling price for the project, and management wants to know the chance of exceeding the bid price with project cost. The location of UL is greater than $E(C_T)$, which is what is usually anticipated. Assume that management is interested in knowing the risk of an actual cost exceeding $10,885. Now we use Eq. (6.26):

$$Z = \frac{10,850 - 10,602}{(22,301)^{1/2}} = 1.66$$

then $P(\text{Cost} > 10,850) = 0.05 = 5\%$

TABLE 6.7   Calculation of expected cost and variance for the range-estimating method.

Design Element	Lowest Cost, L, $	Most Likely Cost, M, $	Highest Cost, H, $	Expected Cost, $E(C_i)$, $	Variance, $\text{var}(C_i)$, $\$^2$
1. Flash lamp	370	390	430	393.33	100.00
2. Data grid	910	940	1,030	950.00	400.00
3. Computer	200	210	270	218.33	136.11
4. Optical isolator	170	180	190	180.00	11.11
5. Power supply	260	290	350	295.00	225.00
6. Switching	171	172	176	172.50	0.69
7. Capacitor	875	925	975	925.00	277.78
...	...	...	...	...	...
n. Frame	2,000	2,100	2,600	2,166.67	10,000
				$E(C_T) = \$10,610$	$\text{var}(C_T) = \$^2 22,301$

The value of Z is entered in Appendix 1, and an *upper-tail probability* can be determined. Thus, the range method is able to say something about the risk of exceeding a particular value, an important assertion by the cost estimate.

According to PERT practices, the optimistic L and pessimistic H costs would be wrong only once in 20 times, if the activity were to be performed repeatedly under the same conditions. The most likely cost, M, is the cost that would occur most often if the activity were repeated many times, or the cost that would be given most often if qualified people were asked to estimate its value. When M is eventually compared to the actual cost, we should find that with a number of estimates, M falls above and below the actual cost equally.

Construction work is estimated only once, so those precise requirements are never met. We suggest that these costs be viewed as *best, most likely, or worst* cost. Further, it is unlikely that a distribution of major actual cost elements can ever be known. Finally, costs, if they are internal to a firm, are seldom independent of each other, although for a project estimate of various subcontractors' costs, the assumption of independence is more valid.

The expected-value method discussed earlier required selection of opinion probabilities, which we believe is more tenuous. In practice, estimating opinion probabilities is difficult, and experts often give distorted values. However, the range method requires three values of cost; and if rational methods of selection are employed, then the range method is acceptable as a preliminary technique.

There are advantages to this method. We are able to identify elements of cost that have great uncertainty. Management is then able to gather more information, or more bids, or improved engineering, and so forth, to reduce the variation. The range method is discussed again in Chapter 8, where it is helpful for contingency analysis.

6.10.3   Monte Carlo Simulation

Models are descriptions of systems. In the domain of physical sciences, models are based on theoretical laws and principles. The modeling of complex construction systems may be as difficult as modeling a physical system, because few fundamental laws are available, many procedural elements are untractable, and policy inputs are hard to quantify. Surveys indicate that simulation is a tool that is sufficiently robust to handle construction systems. Business decisions under uncertainty have been modeled this way since the early 1960s.

Simulation is divided into the following classes:

- Real versus abstract
- Machine versus man-machine
- Deterministic versus probabilistic
- Possibilistic-probabilistic for fuzzy logic

Testing of a prototype airplane or a pilot plant is *real* simulation, while mathematical and logical statements employing a synthetic model are *abstract* simulation.

Machine simulation attempts to program all eventualities for computer running. On the other hand, *man-machine* simulation allows program interdiction by the human being at strategic points where the skill and experience of the human are considered superior to the computer. This division of labor allows the computer to do what it can do well, and encourages the ability of people to interpret qualitatively rather than quantitatively.

While construction business is a probabilistic undertaking, these situations are usually characterized by constant-value models. In a *deterministic* situation, we assume that the *constants* substituted for the business parameters are an ideal approximation for that design. In a *probabilistic* situation, however, it is necessary to introduce random events such that the operating parameters of the modeled system are affected. Sometimes probabilistic models are titled with the interesting term *Monte Carlo*.

Although simulation uses mathematics, it is not mathematics *per se*. In simulation you *run* problems, not *solve* them as you do in mathematics. The intent is the collection of pertinent data from the experiment as one runs and watches the outcome of many simulation trials. On the one hand, the actual experiment of a design provides realism. On the other, orthodox mathematical solution to business and construction problems remains an abstraction. "So far as the laws of mathematics refer to reality, they are not certain. And so far as they are certain, they do not refer to reality."[4] Figure 6.12 illustrates the contrast of simulation to closed-form single-value solutions.

A simulation model is formulated to look like the actual construction system. Other chapters give many examples of systems. Simulation procedures, then, use the common estimating elements of labor, material, overhead, etc., which are coupled to the governing business condition. This is the pattern for estimating.

Assume a simple case where direct cost is expressed as $A = x + y$, where x and y are probability distributions. The x and y can represent direct labor and direct material, for instance. Further, we have field data for the distribution of $f(x)$, given as

Cost of x, 10^6	Relative Frequency	Monte Carlo Numbers
1.1	0.05	0.00–0.05
1.2	0.10	0.06–0.15
1.3	0.15	0.16–0.30
1.4	0.20	0.31–0.50
1.5	0.20	0.51–0.70
1.6	0.15	0.71–0.85
1.7	0.10	0.86–0.95
1.8	0.05	0.96–1.00

[4] Albert Einstein, Theoretical physicist and Nobel laureate, "Geometrie und Erfahrung," Lecture to Prussian Academy, 1921.

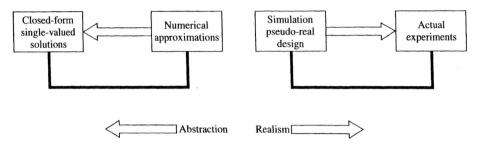

Figure 6.12 Comparison of abstraction and realism for the solution of construction-business problems.

Monte Carlo numbers are found similarly to the development of the cumulative frequency interval for empirical distributions, as shown in Sec. 5.1. Thus, $1.1 million has a Monte Carlo interval of 0.00 to 0.05. In simulation we assume that random numbers are the probability of the functional cost item. If a random number of 0.18 is drawn from a hat containing a well-mixed collection of 100 separately numbered decimals or from a random-number table, both having the number range between 0 and 1, we say that 0.18 corresponds to $1.3 million after entry in the Monte Carlo table, because it falls within that range. Tables of random numbers are found in other texts. We provide a small sample when required for a problem.

The collection of field data and determining of Monte Carlo numbers is one approach. A second method uses theoretical distributions fitted with empirical or trusted coefficients. There are many distributions, both continuous and discrete, and the student is referred to a text on probability and statistics for more information. In this brief summary, we do the following:

A *probability density function* for the continuous random variable Y is defined as

$$f(y) \geq 0$$

$$\int_{-\infty}^{+\infty} f(y)\, dy = 1 \tag{6.27}$$

$$P(a \leq y \leq b) = \int_{a}^{b} f(y)\, dy$$

As an immediate consequence, the *cumulative distribution* $F(x)$ of the random variable is given by

$$F(y) = P(Y \leq y) = \int_{-\infty}^{y} f(y)\, dy \tag{6.28}$$

In Monte Carlo simulation we have various random numbers, 0 to 1, substitute for $F(y)$ and then we solve for the upper value of integration, y. The student will recognize this as the inverse method of calculus.

We continue the illustration of the direct cost A as a sum of x and y. For example, we may assume that the exponential function represents a type of cost. Now assume that $f(y)$ is given by an exponential density with a functional form of

$$f(y) = \frac{1}{8} e^{-y/8}, \quad y \geq 0$$

To find the cumulative probability distribution, we use

$$F(y) = \int_0^{C_y} f(y)\, dy = \int_0^{C_y} \frac{1}{8} e^{-y/8}\, dy = 1 - e^{-C_y/8}$$

then $C_y = -8 \ln \left[1 - F(y) \right]$

When random numbers, 0 to 1, are individually selected and substituted, we find the Monte Carlo value of C_y.

Our example for the direct cost A requires that the random variable x be added to the random variable y. If a random number is found from a random-number table or computer file, it is set equal to the cumulative probability distribution. A random number of 0.73 is drawn for design x, and a corresponding value is $1,600,000. In a similar way, we set a random number of 0.18 to the functional model for y, and then $C_y = -8 \ln (1 - 0.18) = 1.587$ or $1,587,000. Our solution is cost A $= 1,600,000 + 1,587,000 = \$3,187,000$, which is accepted as a single trial of our cost A.

This procedure is repeated many times, and finally a distribution of direct cost A is found. The shape of this distribution varies from a jerky histogram to a smooth normal distribution. This depends on the number of trials. See Fig. 6.13. For this example the expected mean is $9,450,000 while the standard deviation is 8002. For a typical construction-business system, the distribution is tested for its cost-estimate average, standard deviation, confidence interval, and other statistical properties.

In using simulation to estimate construction-business cost, it is usually necessary that input relationships be independent of each other. And typically we apply the *central limit theory* to say that the concluding distribution is a normal distribution.

There are numerous offspring of statistical-probabilistic and numerical methods that are not discussed in this chapter. Decision trees, Bayesian, neural nets, artificial intelligence, etc. are interesting, yet space limits our choices to the more popular methods. For additional information on these approaches, the student is directed to the References.

6.11 SINGLE-VALUE OR PROBABILITY-DISTRIBUTION COMPARISONS? (OPTIONAL)

Suppose that the statistical distributions of two designs, A and B, are to be compared and that minimum cost is the criterion for selection. A hypothetical question is asked: Which design is the preferred choice if all else is considered equal with the exception of cost?

If single value estimates are adopted, and it is necessary to compare between competing choices, say A or B, it is straightforward to select one based on the criterion of minimum cost. But if probability cost distributions are available for designs A and B, the choice is not always clear cut. Notice Fig. 6.14 for this discussion.

For Fig. 6.14(a) all probable costs of A are lower than those for B and there is no difficulty in choosing A. Notice that the *expected costs* are identified by C_a and C_b.

The situation for (b) has a probability that the actual cost of A is higher than that for B, or $C_a > C_b$ even though the mean cost of A is less than the mean cost of design B. Given that it is a small chance that $C_a > C_b$, we may comfortably select A. But as the amount of

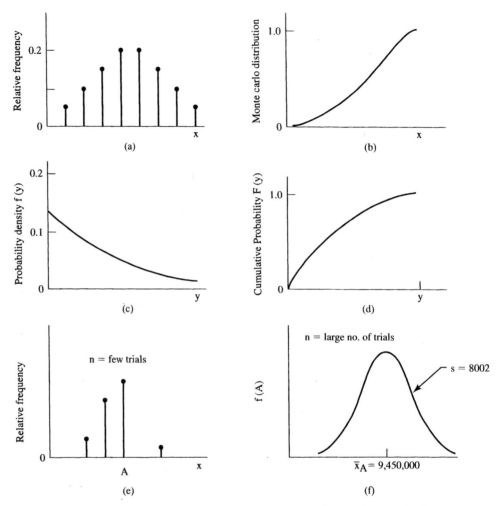

Figure 6.13 Progression of individual effects of the estimate via simulation: (a) discrete histogram obtained from field studies or opinion; (b) Monte Carlo or cumulative distribution of frequency distribution; (c) probability density idealized to an economic factor; (d) cumulative probability distribution of density; (e) histogram of several simulation trials; and (f) distribution of simulation after large number of trials.

overlap between distribution A and B increases, the expected-cost estimate may not be given a clear mandate.

For (c) the average cost estimates are equal, although their distributions are not. Certainly, the cost distribution of B is greater and there is a chance of having a lower cost than A. On the other hand, A is less variable. The appraisal of minimum cost and risk will serve as the guide here. How would you make the choice?

In (d) the expected-cost estimate of B is lower but less certain than that of A. If only the expected-value estimate is used in this case, then we are likely to choose the more desirable alternative B.

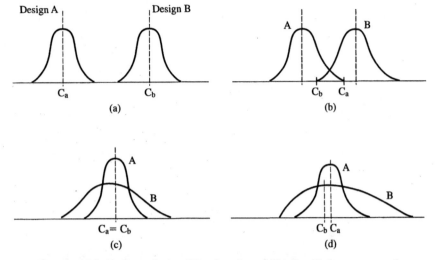

Figure 6.14 Cost analysis of Design A and Design B shown as probability distributions.

Which is preferred: single-value or probability-distribution estimating? The answer is not always obvious. We see that statistical/probability science may not make the choice of either design A or B any more certain.

It is tantamount to estimating practice that at the beginning of the 21st century single-value, or deterministic, estimating is dominant. But the future holds newer and bolder ideas, which may increase the reliability of the estimate. A new idea, interestingly called *fuzzy logic*, introduces a procedure for estimating, which we discuss next.

6.12   FUZZY LOGIC CONCEPTS[5] (OPTIONAL)

Construction is an ideal business for firms who enjoy taking chances. Too, owners and contractors are exposed to risk situations where the traditional methods of this book may be unlikely to give much advice on what to do. This perplexity is due to a lack of historical data, shifts in technology, or the magnitude of the job and its risk. There may be bidding opportunities where the input variables are nonprobabilistic. But the owner or contractor is able to give linguistic assertions that are useful for cost estimates. An application called fuzzy set or fuzzy logic is appropriate for special and narrow risk situations.

Fuzziness represents situations where membership in sets cannot be defined on a yes-or-no basis because of the vagueness in boundaries. In classical set theory, a sharp distinction exists for members of the set, those that belong and those that do not. If the *A* element is a member, then the value is 1 or yes, and if it is not a member, the value is 0 or no. Notice Fig. 6.15(a). The Venn diagram of Fig. 6.15(b) shows elements *B*, *C*, *D*, and *E*. *B* and *C* represent areas, while *D* and *E* are point values that are not included in the sets *B* and *C*. Sometimes it is difficult to make a sharp distinction between members of set, and

---

[5]"Fuzzy logic" is a bona fide title and is respected in academe. While there are specialized software and understood steps, fuzzy logic is more of a research technique than a day-to-day construction estimating tool. The beginning of the era of fuzzy logic is attributable to Lotfi Zadeh, "Fuzzy Sets," *Inf. Control*, vol. 8, p 338–353, 1965.

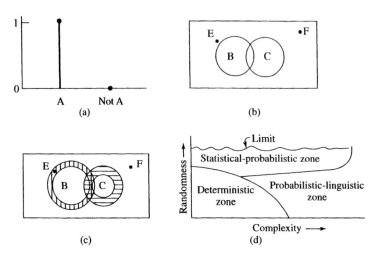

Figure 6.15 (a) Classical set theory; (b) Venn diagram; (c) gradual fuzziness to unclear boundaries; and (d) hypothetical definition of zones for cost analysis.

the boundaries are fuzzy, or not crisp, as shown in Fig. 6.15(c). The transition from members of a set to nonmembers appears gradual rather than abrupt, and if the element is a member of a fuzzy set to some degree, then the defined value of its membership function ranges between 0 and 1.

Notice Figure 6.15(d), where the world of construction business is sliced into three hypothetical areas. Axes are called Complexity and Randomness, and they need to be weighed in the larger sense of the complications that those terms suggest. The deterministic zone is where we are able to comfortably make assertions, find information, and estimate. For example, the cost of a field-installed gate valve with carbon steel piping is given as cost $= -23.30 + 20.0$ (lb). This crisp type of description is the basis of much of estimating for the construction business today. Most of this book deals with deterministic estimating.

The statistical/probabilistic zone is one that is familiar, and we have studied some approaches for the estimating needs for construction. In probabilistic analysis, statistical measures (mean and standard deviation) of input variables are used to estimate the mean and standard deviation of the results. This method is precluded if the input variables have a nonprobabilistic nature.

The *possibility/linguistic zone* deals with what is called fuzzy logic. We are unable to find the probabilistic parameters, or write their equations, and we have incomplete knowledge of the events, which must precede the computation for the probabilities. It may be possible to express linguistic or lexical statements about the events. Roughly, this description is provided for the possibilistic/linguistic zone, Fig. 6.15(d).

As the complexity of the construction-business regime increases, our ability to make precise and yet relevant statements about its behavior diminishes. There is a threshold beyond which we make practical and linguistic statements about its events. Consider an example: When a statement about a high-temperature reactor is made that "it is costly," the speaker and listener may have a general agreement as to its cost value because of their mutual circumstances. Conversely, when the statement is that it is "very likely that the reactor is extremely cheap," the nuance of the statement is shaded differently, and the speaker and the interpreter will have a different understanding.

TABLE 6.8   Linguistic components of fuzzy logic.

Fuzzy Component	*Linguistic Terminology*
Predicates	Big, small, heavy, expensive, cheap
Predicate modifiers (hedges)	Very, quite, more or less, extremely
Truth values	True, very true, false, fairly false
Quantifiers	Most, much, few, almost, usually
Possibilities	Likely, very likely, highly unlikely

But fuzzy logic attempts to be more than a philosophic analysis of linguistic terms, as it attempts to quantify the hazy terms into numerical standards that are assuage by human powers of reasoning. A fuzzy system has variables which range over states that are fuzzy numbers or fuzzy intervals representing concepts described in natural language. The natural language is also known as linguistic variables.

Notice Table 6.8 for the linguistic components of fuzzy logic. Throughout this book, these linguistic components are applied. This is a characteristic of our rich language. For example, the cost of a reactor is *not likely* to *increase substantially* in the *near future*. The underlined words are the linguistic terminology that we want to numerically understand.

A fuzzy set is a set whose boundary is not sharp. A central concept of fuzzy set theory is the membership function, which represents numerically the degree to which an element belongs to the set. The value of this membership is between 0 and 1, which is unlike classical theory because fuzzy membership can have values in between values of 0 and 1. Notice Fig. 6.16(a). A fuzzy set is characterized by a membership grade function in the following way:

$$\mu : X \rightarrow [0,1] \tag{6.29}$$

where X = crisp universal set

   A = label of the fuzzy set defined by μ

$\mu_A(x)$ = grade of membership of x in A

   = degree of compatibility of x with the concept represented by A

$A_\alpha = \{x \in X \,|\, \mu_A(x) \geq \alpha\}$

where $\alpha \in [0,1]$ is α - cut of fuzzy set A and α - cuts form a nested family of subsets

   $\alpha_2 > \alpha_1 \Rightarrow A_{\alpha_2} \subset A_{\alpha_1}$

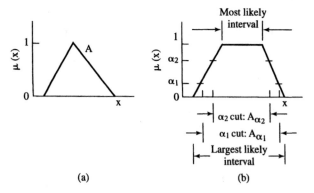

(a)

(b)

Figure 6.16 Membership functions: (a) triangular, and (b) trapezoidal.

This membership assignment is shown in Fig. 6.16(b). There are more ways to assign membership values to fuzzy variables than there are ways to assign probability density functions to random variables. This assignment process can be intuitive or it can be based on algorithmic or logical operations. Geometrical reasoning of triangles, for instance, permits membership assignments.

The fuzzy expert rule makes an assertion of a relationship between cost driver and costs using variables stated in linguistic terms. This is an advantage if we are unable to give statements in more precise terms. The premises and conclusions are formulated by fuzzy sets. The fuzzy sets are defined by membership functions on a universe of discourse which covers all possible values for the cost driver and the cost. It is not a probability, as axioms of probability are not met, but a statement of our relative belief. It is a measure of *possibility* rather than probability. These possibilities are obtained using expert opinion, or they may be supported by some historical data.

An expert method can be developed in the following way: We develop rules to capture the knowledge in the form of "if … then …" statements. For example, if the maximum temperature of a reactor is between 3,774 and 3,819° F, then the direct labor and direct material cost are $86.5 million. These rules have two parts: antecedent clauses and a consequent clause. The *antecedent* clauses deal with the cost driver, and the *consequent* clause is the cost. The antecedent clauses may be composed of several conditions. If the antecedent clauses are true, then the consequent clause is also true. If any one is false, then the consequent portion of the rule is false. While the consequent cost as discussed above is a dollar value, a cost formula is possible too. This approach treats the antecedent and consequent response on a true or false basis similar to a yes or no condition. A project analysis may have many such if-then clauses.

But it is possible that the response is neither 0 nor 1, as it may be somewhere in between. A fuzzy logic expert approach to the making of rules with the premise and the conclusion linguistically stated is as follows: "If the maximum temperature of a reactor is hot, then the cost is very high." Another one is "If the maximum reactor temperature is moderate, then the cost is high." Also, "If the reactor temperature is cool, then the cost is very low." Values for the outcomes in between 0 and 1 are possible for these situations. These inferencing procedures produce a fuzzy set of responses, not a crisp set. Too, values of the response can overlap, as shown in Fig. 6.17. Notice that moderate and cool overlap, but hot is separate.

In this illustration, the cost of the refinery system for North Sea light petroleum crude is composed of two major cost drivers: the reactor and design complexity. The universe of discourse for design complexity is normal or difficult. Once again, the domain engineer uses opinion and other rituals to assess complexity, as shown in Fig. 6.17(b). The universe of discourse for reactor system ranges between $62.1 and $95.3 million, as shown in Fig. 6.17(c).

With 3 fuzzy sets for temperature and 2 for design complexity, there are 6 possible rules, and we have the 3 × 2 table, or

| | Design Complexity | |
Reactor Temperature	Normal	Difficult
Cool	Low	Moderate
Moderate	Moderate	High
Hot	Moderate	High

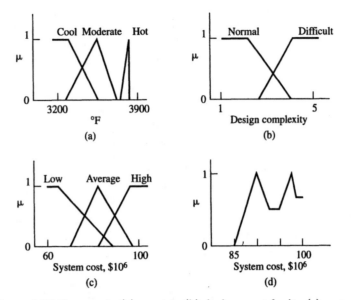

Figure 6.17 Fuzzy sets. (a) reactor, (b) design complexity, (c) system cost, and (d) aggregate system cost adjusted for risk.

The body of the table is cost. If the reactor temperature is cool and design complexity is normal, then the direct labor and direct material cost are low. Or if the temperature is hot and design is difficult, then the experts believe that the cost is high.

Suppose the expert can say that the temperature will be 3809° F, and after entry in Fig. 6.17(a) the membership value is 0.79 from the fuzzy set hot. There are no other intersections, so the other values are 0. This implies that the membership value is 0.79 for any rule which states that the temperature is hot. If the temperature is 3410° F, then the membership value is 0.45 for cool and 0.64 for moderate.

Procedures continue and one criterion, which is called min-max, leads to a fuzzy set of final values for the reactor system cost, shown as Fig. 6.17(d). An array of fuzzy set values is possible, but a crisp value needs to be determined. One possible method is to determine the centroid of Fig. 6.17(d), giving a $94.725 million system cost from area interpolation. This final value is the aggregate conclusion. The set of Fig. 6.17(d) suggests the magnitude of uncertainty.

A project analysis may have many such if-then clauses, and specialized software shells are available that allow the structuring of these fuzzy logic approaches. The student will want to visit the references for additional information on fuzzy logic.

SUMMARY

Estimates are assembled on available information. If there is no information, then there can be no estimate. Conversely, an estimate is unnecessary if actual costs are available. The engineer operates within those limits. In this in-between region we separate estimating methods into preliminary and detail. Preliminary methods are less numeric than detail methods. Accuracy of the estimate is improved by attention to detail, but offsetting this are the speed and cost of preparation that favor preliminary methods. The methods are preliminary because they correspond to a design not well formulated. Frequently, a preliminary estimate leads to the management decision of further consideration for the design.

Before detail methods of estimating are attempted, information that has been deliberately collected, structured, and verified to suit the method of estimating must be available. Most construction-work estimates are shaped by using detail methods. Large-scale projects are less dependent on detail, and they lean more on information specially developed by the owner or contractor.

In the next chapter the methods of detail estimating are examined further. Our study ranges from simple cost factors to encyclopedias of productivity and extensive tables of material unit costs to estimate construction work.

Here we return to the question: How do we select the right method for a particular application? Regrettably, no textbook can supply an infallible set of rules. You will have to rely on learning and experience.

QUESTIONS

1. Define the following terms:

Preliminary estimating	Unit-quantity method
Detail estimating	Power law
Conference method	Assembly method
Hidden card	Economy of scale
Unit method	Opinion probability
Cost driver	Economy of quantity
CERs and TERs	Range method
Parameter	Central limit theorem
Learning	Modal cost
Cumulative learning	Benchmark costs
Sixth-tenths model	Monte Carlo simulation
Fuzzy logic	If-then statements

2. Discuss the timing of preliminary estimates. Is the timing a precise point in a well-ordered organization?

3. Use the conference method to estimate the following: (a) The price for a clean-air car using a gasoline fuel-cell electric wheel drive in 2010. (b) The ticket price for a 5000-mile one-way air trip. (c) The cost of a year's college education in the year 2020. (d) The time to dig a trench $2 \times 4 \times 10$ ft in soft clay. (e) The improvement percentage in the efficiency of handling mail of a central post office near you after converting to automated methods.

4. What advantages can you cite for the conference method? Disadvantages?

5. Assume that design A can be redesigned into designs B and C with known costs. What safeguards can you suggest to assure that A is properly estimated?

6. Discuss some substitute unit measures for "cost per square foot of residential construction." What are your local values?

7. Give the steps for the development of data for the factor method.

8. Point up the human frailties in determining opinion probabilities. Would you think that one would underestimate or overestimate those point probabilities?

9. Describe the kinds of applications suitable for the factor method. What kinds are suitable for the unit-quantity method?

10. What is meant by the law of economy of scale? Apply it to the factor method. Does the unit method follow this law?

PROBLEMS

6.1 Raw data for flooring are collected for factory-built houses. The cost data are converted to a consistent and common square-foot basis for estimating and comparison:

Material	Size	Material Cost, $/ft²	Labor Cost, $/ft²
Laminated oak blocks	$\frac{1}{2} \times 9 \times 9$ in.	6.60	3.70
Parquet	5/16 × 12 × 12 in.	6.20	5.30
Strip flooring, oak	25/32 × $2\frac{1}{4}$ in.	7.20	4.50
Resilient tile	9 × 9 × 1/8 in.		
A grade		3.20	3.40
B grade		3.70	3.40
C grade		3.80	3.40
Vinyl sheet	9 × 9 × 1/16 in.	4.10	3.40
Softwood, C grade	S4S, 1 × 6 in.	7.30	3.60
Linoleum	1/8 in., plain	5.00	3.50
Slate, irregular flags	Irregular flags	14.00	10.00
Terrazzo	$\frac{1}{4}$ in. thick	11.50	9.90

Notice that cork tile flooring is not in the table. Using only your opinion, where would you cost cork tile? Where would you cost carpet? These prebuilt houses are delivered to the site by trucks. What are the nonfeasible alternatives for flooring under this circumstance? Conduct an analysis of the data, as well as you can, to help sharpen your value after you give your opinion.

6.2 An inventor has fashioned a new mailbox design with an interesting feature. When the mail deliverer closes the hinged cover, a spring-loaded plastic flag pops up at the back of the steel box, telling the owner that she has received mail. The inventor surveys popular catalogs and finds the following data:

Size, in.	Weight	Material	Cost, $	Features
14 × 7 × x4	4 lb, 4 oz	Steel	47.20	Holds magazines, wall mount
13 × 7 × 3	3 lb, 6 oz	Steel	31.20	Hold magazine, wall mount
14 × 6 × 4	4 lb, 15 oz	Aluminum	71.92	Holds papers, magazines, wall mount
14 × 7 × 4	4 lb, 12 oz	Steel, aluminum	55.20	Holds magazines, wall mount
6 × 2 × 10	4 lb, 8 oz	Some forged iron	64.80	Letter size
6 × 2 × 10	2 lb, 11 oz	Steel	35.20	Letter size
5 × 2 × 11	2 lb, 11 oz	Steel	20.40	Liberty Bell emblem
10 × 3	1 lb, 8 oz	Brass	47.20	Mail slot
10 × 3	1 lb	Plastic	14.16	Mail slot

(a) In using comparison estimating, where do you place the inventor's design in the table?

(b) Construct a rough unit estimate from the data to help in the evaluation. (*Hint*: Do not expect too much from an analysis of this sort, but it may give insight.)

(c) Discuss reasons that will support a different analysis. How much "forgiveness" is necessary for a survey of this type? How would you sharpen the approach?

6.3 The cost of a residential home with 2000 ft^2 livable space and a basement and a two-car garage is $146,000. This cost is exclusive of any lot costs. The house dimensions are such that 1200 ft^2 is on the first floor. The volume, including house, garage, and basement, is 30,200 ft^3. Determine unit estimates for future homes having the same configuration.

6.4 A residential builder constructs a 2000-ft^2 two-story home on sandy loam soil for $124,000. This cost is exclusive of land, taxes, and utility hookup costs.

(a) What is the unit estimate for the next home?

Item	%	Item	%
Rough lumber	13.0	Earthwork	2.6
Rough carpentry	9.2	Flooring	6.4
Plumbing	13.6	Hardware	2.1
Finish lumber	1.2	Heating	1.5
Finish carpentry	2.8	Insulation	1.5
Cabinets	6.4	Lighting	0.5
Concrete	5.8	Painting	7.1
Wallboard	6.0	Roofing	2.2
Electric wiring	5.5	All other	11.1

(b) An additional breakout of costs revealed the following percentages for this same job: A 3200-ft^2 two-level home is to be built. Estimate the total and item costs. Which elements would you adjust if this were a single-story home? Up or down?

6.5 A contractor is asked to bid on a concrete retaining wall. The wall, on the average, is 2 ft high, and it will rest upon a 2 × 1 ft foundation. Wall length is 75 ft. The wall is similar to Fig 6.3, except for the height. Find the cost that is given by the unit-quantity method. What is the cost per yd^3? (*Hint*: Adjust Table 6.1 to accommodate to the new wall height. Excavation is 4 × 3 ft high × 75 ft.)

6.6 An electrical contractor receives a request for a cost quote for the sketch given as Fig. P6.6. The material is rigid galvanized steel conduit. It does not include the wire. Her craft hours-per-ft standards, which do not perfectly conform to the job requirement, are given below: Gross hourly labor costs are $34 per wireman journeyman-hour.

Dia, in.	Hr/ft Standard					Cost, $/100 ft
	Exposed	Slab Deck Wood Frame	Trench	Furred Ceiling	Notched Joists	
$\frac{3}{4}$	0.046	0.033	0.023	0.030	0.039	$190
$1\frac{1}{2}$	0.080	0.062	0.040	0.054	0.063	443

Develop a unit-quantity table for 100 ft of a 2-in. conduit from the available information. Then state what her costs will be for 7000 linear ft. (*Hints*: Use some algebraic method to estimate labor standards and material costs before setting up the unit-quantity table. Consider the material cost to be "all in" including screws, hangers, etc.)

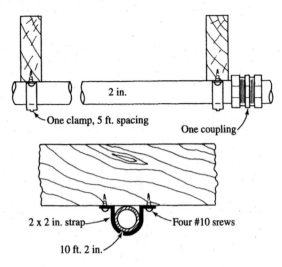

Figure P6.6

6.7 An interior wood-framed partition wall is 8 ft high × 10 ft long and is illustrated by Fig. P6.7 below. Raw data are provided by the following information:

Description	Quantity	Material Unit Price	Labor Unit Man Hours	Labor Rate	Subcontract Unit Price
2 × 3 in. wall framing, double wall	160 ft^2	$0.35	0.020	$30.00	
R-11 blanket insulation	80 ft^2				$0.45
5/8-in. gypsum wallboard, 2 sides	160 ft^2				$1.20

Develop a unit-quantity table, and find the total cost and cost per linear ft and ft^2.

6.8 An equipment bid for a rising-film and ozonation reactor is received for $1,750,000. The current index for this class of equipment is 121.5. Construction cost for a chemical plant is required.

    (a) Find the benchmark cost for the equipment. What are the factors for engineering, erection, and direct materials?

    (b) An overall index of 123 is forecast for the costs during the construction period. The regional factor for the plant site is 7% more than for the Gulf Coast. Items independent of the factor-estimating process are found to cost $2,000,000. What is the total factor estimate for the chemical plant?

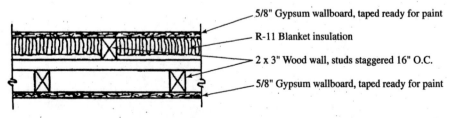

Figure P6.7

6.9 Imagine that a set of four similar construction projects, though not identical in design, are built in serial order. Project management believes that the data conform to the learning model.

Project No.	1	2	3	4
Unit-Value Cost, 10^3	7815	7424	7205	7053

Plot the unit, cumulative, and average values on both arithmetic and logarithm coordinates. Find the future unit, cumulative, and average values for the fifth and sixth units. (*Hints:* The unit values are determined after the conclusion of the project. Each project is separated by a start and a conclusion from its neighboring project, and there is no intermingling of values. It is the unit value that is considered *actual*, and the cumulative and average values are derived. Use engineering computation paper for the arithmetic plotting.)

6.10 (a) Find the unit time of a project for the fourth and sixth units when unit 1 has a learning rate of $\phi = 87\%$ and a unit time of 18,000 hours. (*Hints:* Use the notion of doubling to find the answer for unit 4. Use Eq. (6.7) to find the answer for the sixth unit.)

(b) Find the value for unit 6 when the value for unit 3 is 7961 hours at 75% learning.

(c) Find the intercept value K when the tenth unit is 6054 hours with 95% learning. What is the expected unit time for the eleventh unit?

(d) If unit 1 is $2 million, then find the dollars for units 2 and 4 with learning rates of 92% and 100%. What observation can you make for 100% learning?

(e) Units 1–50 have accumulated 10,000 hours. Calculate the time for the twenty-fifth unit if least square slope for the learning rate is 92%. What is the time for the fiftieth unit? (*Hint:* Use an approximation.)

(f) If the time at unit 10 is $100,000, then find the unit, cumulative and average cost at unit 11 for a learning rate of 92%.

(g) A contractor has collected actual data for units 3 and 4 as 7729 and 7225 hours. Find the future unit, cumulative, and average hours for the fifth and sixth units.

6.11 A project has an actual first unit cost of $72.8 million. Four additional and similar units are to be financed by bank loans and constructed in sequential fashion. The bankers are anxious that a learning model be adopted for budgeting and financial management. Based on other experiences, the bankers insist that costs follow the 93.7% learning model.

Determine a table of unit, cumulative, and average costs for the next four projects.

6.12 There are two competing construction designs, called A and B. Additional construction of later units is to be sequential. Detail estimates are found for the rival designs, and learning rates are chosen based on comparison to similar projects.

Estimate	Design A	Design B
Initial unit	$600,000	$710,000
Learning rate	95%	85%
Final no. of units	3	3

If the designs are equal in function and intrinsic value, which design is the preferred economical choice?

6.13 An 80-kW diesel electric set, naturally aspirated, cost $1.6 million 8 years ago. A similar design, but 140 kW, is planned for an isolated site. The exponent $m = 0.6$ and the index $I = 187$. Now the index is 207. A precompressor is estimated separately at $190,000.

(a) Find the estimated equipment cost.

(b) Repeat for $m = 0.7$.

6.14 A cost estimate is desired for a 34-cm-dia. disk laser amplifier assembly. A previous design of 22 cm dia. was estimated for a cost of $16,459,000. The annual Producer Price Index of 7.5% for this type of equipment will be used for a 3-year escalation. A value of $m = 0.7$ is adopted. Find the cost of the new equipment. (*Hint:* Consider the PPI to be a compound effect.)

6.15 The cost of turbine equipment is suspected to follow the power-law CER. Information on turbine cost is gathered as follows:

kW	Cost, 10^6
3500	18
5000	25
7000	32
8000	35

A similar design, except that it is larger at 8750 kW, is to be estimated. Determine the slope m and estimate the cost of the new unit. (*Hints:* Use $\log C = \log C_r + m \log \dfrac{Q}{Q_r}$ and let $Q_r = 3500$ kW. Plot the relationship on logarithmic graph paper. Have Q/Q_r as the x-axis. You can graphically find the slope by using the rise/run triangle.)

6.16 **(a)** A project is estimated to cost $50,000. The contractor believes that a bid of $56,000 has a probability of 0.3 of being low, and a bid of $53,000 has a probability of 0.8 of being low. Which one has the best expected profit?

(b) A salesperson makes 15 calls without a sale and 5 calls with an average sale of $200. What is the expected sale per call?

(c) A construction reseller takes old construction equipment as *trade-in* for new models and sells the returned equipment through a secondary party outlet. Analysis shows that the markup is $2500 on 70% of the equipment and $4000 on the rest. What is the expected markup?

(d) An insurance company charges $20 for an additional $50 increment of insurance (from $100 to $50 deductible). What is their assessment of the risk for the increment of insurance?

(e) A company sells two different designs of one item. A study discloses that 65% of its customers buy the cheapest design for $75. The remaining 35% pay $110 for the expensive model. What is the expected purchase price?

6.17 A student is interested in selling her car instead of trading it in at some point in the future. Her estimating model, she reasons, is the sale price of a new car − (depreciation + major maintenance cost). Other driving costs are the same regardless of whether she drives a new car or not. The original sticker price is $26,274 and a major maintenance cost is $4000. The depreciation of cars is known, and she tabulates that value in a table. Her *opinion probability* for a major maintenance cost is given as follows:

Life	Cumulative Probability of Major Maintenance	Cumulative Decline in Depreciation
2	0.2	0.49
3	0.4	0.64
4	0.7	0.75
5	1.0	0.83
6	0.1	0.89
7	0.2	0.93
8	0.4	0.96
9	0.7	0.98
10	1.0	0.99

When should she sell her car? Initially assume that the price of her next car, whenever she buys it, will be equal to that of her first car. Next, assume that the new car's price increases by

3% compounded per year. (*Hints*: Disqualify any first year for trading-in the car. The price of a new car, if increasing, is given by price $\times 1.03^n$.)

6.18 Revisit the joint-probability example given in the chapter. Find the *expected profit* for the following changes in the probabilities:

Annual Sales Volume	Probability of Sales Volume	Cost per Unit, $	Probability of Cost per Unit
15,000	0.1	450	0.6
20,000	0.3	500	0.3
25,000	0.6	550	0.1

6.19 **(a)** Find the expected total mean cost and variance.

Cost Item	Optimistic Cost, 10^6	Most Likely Cost, 10^6	Pessimistic Cost, 10^6
Direct labor	79	95	95
Direct material	60	66	67
Indirect expenses	93	93	96
Fixed expenses	69	76	82

(b) What is the probability that the future project mean cost will exceed $325 million?

6.20 A five-element project is estimated as follows:

Cost Item	Optimistic Cost, 10^6	Most Likely Cost, 10^6	Pessimistic Cost, 10^6
1	4.0	4.5	6.0
2	10.0	12.0	16.0
3	1.0	1.0	1.5
4	4.0	8.0	12.0
5	2.0	2.5	4.0

(a) Determine the elemental mean costs, total cost, and elemental and total variance.

(b) Find the probability that cost will exceed $26 million.

6.21 Refer to the illustration given in Sec. 6.10.3 and determine a new value of the Direct Cost A using these random numbers:

Random Number, Direct Labor	Random Number, Direct Material
0.24	0.64
0.82	0.98
0.83	0.25
0.18	0.94
0.66	0.03
0.76	0.23
0.07	0.96
0.62	0.80
0.61	0.64
0.96	0.99

Sketch a histogram of the random variable Direct Cost A. Find the mean and standard deviation of direct cost, assuming that each of the ten random pairs is a trial.

(*Hints*: It is necessary to make a wide-interval selection of direct cost to allow more than one value in each histogram interval. Perhaps, select only 4 or 5 intervals. Have the lower and upper intervals as open ended. The vertical axis of the histogram is *count*.)

6.22 Direct cost is equal to the sum of direct labor and direct material. Let direct labor be distributed exponentially with the mean of 3. Direct material is distributed continuously with $f(y) = 2/9y$, where $0 \leq y \leq 3$. Find the mean of direct cost. (*Hint*: For convenience use the 10 random-number sets from Prob. 6.21.)

6.23 A simulation model is defined as $C = x/y$, where cost x is given by

$$f(x) = \begin{cases} \dfrac{1}{b-a}, & 2 \leq x \leq 8 \\ 0 & \text{elsewhere} \end{cases}$$

and the cost variable y is given by the frequency:

Cost of y, $	Occurrence
1	0.15
2	0.25
3	0.40
4	0.15
5	0.05

Find the mean C after five simulation trials. (*Hint*: Use the first five random-number sets of Problem 6.21.)

MORE DIFFICULT PROBLEMS

6.24 A wood-beam and wood-joist floor is designed for a residence The outside size is 40×28 ft. Two cross sections are shown from opposite views. Notice Figure P6.24 A and B. The wood material is No. 1 Douglas fir. Concrete foundations are not included in this estimate.

Description	Quantity	Material Unit Price	Labor Unit Man Hours	Labor Rate
Treated pier pads, $2 \times 6 \times 6$ in.	14 ea	$1.80	0.08	$30
Pier posts, $4 \times 4 \times 12$ in.	14 ea	2.16	0.05	30
Treated sill, 2×6 in.	136 ft	1.80	0.08	30
Floor joists, 2×8 in. $\times$ 10 ft	117 ea	17.36	0.22	30
Rim joist, 2×8 in.	136 ft	1.76	0.028	30
Blocking, 2×8 in.	80 ea	1.76	0.07	30
Beams, 4×6 in. $\times$ 10 ft	8 ea	32.36	0.34	30
T & G Underlayment, 5/8 C.D.X.	1120 ft^2	1.81	0.013	30

Develop a unit-quantity table and find the total cost and cost per ft^2. Also find the total material and total labor cost, and cost per ft^2.

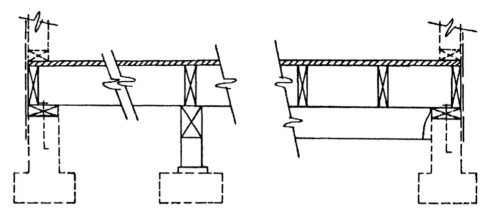

Figure P6.24

6.25 Construction of a small food factory is planned near an area of truck farms. A design-build firm selects standard equipment for the plant. The following equipment prices are found from specialty catalogs:

Process	Cost, $	Process	Cost, $
Roller grader	10,000	Extractor	36,000
Peeling and coring	18,000	Two cookers	7,000
Pulping	30,000	Can seaming	54,000
Instrumentation	57,000	Packaging	22,000
Conveying	37,000	Six motors	35,000
Filler press	14,000	Heat exchanger	44,000
Tanks	10,000		Total $374,000

The *design-build* firm has factors to obtain installed cost. These factors include the cost of site development, buildings, electrical installation, carpentry, painting, contractor's fee, foundation, structures, piping, engineering overhead, and supervision. The factors are listed for various equipment as

Physical Process	Factor	Physical Process	Factor
Process centrifuges	2.0	Can machines	0.8
Compressors	2.0	Cutting equipment	7.0
Heat exchangers	4.8	Conveying equipment	2.0
Motors	8.5	Ejectors	2.5
Graders	1.6	Blenders	2.0
Tanks	2.1	Instruments	4.1
Fillers	0.8	Packaging equipment	1.5

Estimate a plant cost based on this information. Discuss other ways to improve on the quality of this estimate.

6.26 The learning phenomenon was confirmed with the construction of similar natural-gas-fired central-steam power plants in the western plateau region of the United States. In this actual case, construction used different designers and general contractors. An owner is interested in capitalizing on this feature for future plants. Actual data are recorded as (N = unit no., T = $ 10^6$, and year no.) or (1, 98.75, 1), (2, 89.86, 2), and (3, 82.85, 3). The data are adjusted for escalation.

Find unit, cumulative, and average costs for the future construction of units 4, 5, and 6. The future index values are (year, index no.), or (4, 1.2), (5, 1.3), and (6, 1.5).

(*Hints*: The actual cost data are deescalated to the benchmark period. Unit 4 will be built in year 4, and so on.)

6.27 A construction supplier sells totally enclosed capacitor motors, fan cooled, 1725-rpm, $\frac{5}{8}$-in.-O.D. keyway shaft, and publishes its costs on $\frac{1}{4}, \frac{1}{3}, \frac{3}{4}$ and 1 hp as $90, $100, $118, and $140 in a catalog. A 3450-rpm, 1-hp motor costs $134. Determine m for the 1725-rpm motor series, and estimate the cost for a $1\frac{1}{2}$-hp motor, which is not published in the catalog. Should the 3450-rpm motor be included in the sample to estimate the 1725-rpm motor?

6.28 Imagine that bid cost is the sum of three major items: estimated cost, contingency, and profit. The estimated cost is a fixed value of $25. Contingency has the form $f(c) = \frac{1}{4}e^{-c/4}$, where $c \geq 0$. A profit frequency, determined from analysis of past bids for projects, is given as

P, $	f	P, $	f
1	4	5	15
2	18	6	2
3	22	7	8
4	30	8	1

Find the values of the mean and the standard deviation, and graph the statistical distribution. Use spreadsheet software to find these values for the following sample sizes. $n = 10, n = 100$, and $n = 10,000$.

(*Hints*: Note that the P field data are bimodal. It is necessary to generate random numbers, and spreadsheet software should be used. For the smaller sample sizes, the statistical numbers will vary among the members of the class, but as the sample size increases, the comparison between simulations becomes closer. Observe the changes in the appearance of the bid cost distribution as n increases.)

PRACTICAL APPLICATION

Use the conference method to estimate the cost of a construction task. Selection of the task and forming of the team will be guided by your instructor. The general description of the task should be familiar to the team, but the exact cost value is a mystery. Candidates for the task, for instance, are the cost of interstate highway construction per mile or the cost per square foot of residential construction. Then check the value of the design by making formal inquiry with experts. Write a report detailing the process.

(*Hints*: As important as determining the cost of the construction task is to the conference method, it is significant to grasp the advantages and foibles of the procedure. The report should evaluate the process. *Evaluation* is where the learning takes place.)

CASE STUDY: INDUSTRIAL PROCESS PLANT

An industrial process plant is to be constructed. Major basic items are separately quoted by fabrication contractors in current-time cost as $2,720,000 during an index period of 136 as compared to a benchmark index of 100. The estimating graph, which is shown as Fig. C6.1, is based on the Gulf Coast

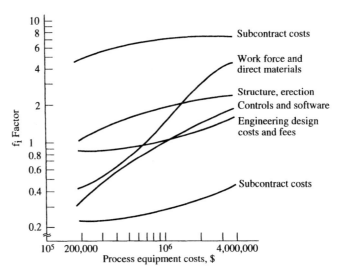

Figure C6.1

region. It is anticipated that the plant will be built in five years, and the index for that period's cash flow is forecast as 172. The plant site is in the northeastern region of the United States, and construction for that area is expected to cost more than the base area by 8.3%. Plant site development is independent of the basic items and is expected to cost $8.2 million.

Find the plant cost. Which element is the most expensive?

(*Hints*: The independent costs are at current time and are estimated in the region of the plant site. The estimate does not ask for indirect costs.)

CHAPTER 7

Work Estimating

This chapter presents the principles for the work estimate. Owners, contractors, and subcontractors, while differing in many significant ways, need a framework for planning this estimate. Students need to know these steps. Labor, material, tools, and equipment costs are the grist for the detail work estimate, and those elements must be determined efficiently. The work estimate is the starting point for the project estimate, which gives the very important bid.

Data warehouses are popular sources of information to estimate construction work. These remarkable data sets are readily obtainable and are helpful for reckoning construction work.

7.1 INITIATION OF THE WORK ESTIMATE

The heart of construction estimating is a talent for dividing a larger task into smaller ones. We are given a design, either formal or informal (imagine a design sketched on the "back of an envelope" or stated in words). We use cost equations, data, and tables, which are subsumed into the computer in various software, that model the business details of the design.

Designs are used for estimating, planning, scheduling, methods improvement, and construction. The designs communicate the description and sequence of construction and are the hub around which, individually and collectively, many decisions, both great and small, revolve.

Outmoded practices of construction estimating take the total cost of the job and divide by a total-job parameter. Residential construction, where the cost per ft^2 is found, is typical of a method of estimating that is suspect. Historical records of similar construction are used without the consideration of alterations for different factors. If these simple methods have a virtue, it is for sanity checking of gross values.

Construction estimates have several purposes. Foremost is the finding of the bid for the construction in order to obtain new work. Other reasons are to compare different design ideas, verify quotations submitted by subcontractors, and help determine the better economic method, process, or material for constructing a design.

Thousands of labor and material construction operations exist. Regrettably, our choice of explanation extends only to a few. Trade books, handbooks, and company sources must be consulted for the data for any real-life estimating. Some practical sources of information are listed in the References.

Before an estimate is started, some authority must formally solicit this activity; that is, we need a request for quote (RFQ) from the owner, a work order, an employee suggestion, or an engineering/architect direction. These requests are the commissioning order to prepare the estimate. The request will include information about the customer, design and working drawings, specifications, site location, period of activity, and quality requirements and will state business assumptions for the work estimate. Many estimates are initiated with an informal request, via a telephone call, and must be carefully considered.

Electronic commerce is changing the business scheme too. Opportunities for new construction business are found on the Web and internet, as RFQs are posted in many locations. One example is the U.S. Commerce Business Daily, a Web location, where bidding instructions are given. Owners, businesses, and government are employing the internet to give wide solicitation of construction opportunities.

7.2   ELEMENTS OF CONSTRUCTION-WORK COST

Estimating has many variations. Each of the various approaches calls for microanalysis, which subdivides a construction design into both physical and economic work elements. We begin by subdividing the design task into large portions of labor and material. Progressively, finer detail is determined until a description of labor and materials is very broad. At this point, dollar extensions of labor and material are made to reflect the cost of the design.

Because detail estimates are costly to prepare, the chance of achieving success must be offset by the cost of preparation. If the chance were not minimal, then the estimates for the construction would not be started. This risk is measured by a *capture percentage*, which we define as (estimates won)/(estimates made) $\times$ 100. If the capture percentage is low, quotations are prepared by using preliminary procedures. If the capture percentage is high, or the estimating cost is not proportionately high compared to the benefits from success, then detail estimating is mandatory.

Though labor, material, equipment, and procedures differ from design to design, techniques for evaluation are based on the same principles and practices. Construction operations are necessary to produce a change in economic value. The value of the constructed end item is altered through workers and activities involving work, tools, and equipment. The measure of the change is cost.

A *construction-cost estimate* consists of direct labor, direct material, tools, equipment, and subcontract materials. Figure 7.1 is a layer chart illustrating these elements. A bid has other cost components, including the work estimate, but these developments are deferred to Chapter 8, "Project Estimating."

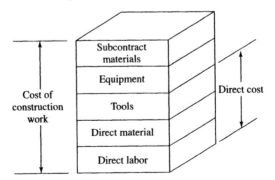

Figure 7.1 Descriptive layer chart of the elements included in a construction-work estimate.

PICTURE LESSON
Golden Gate Bridge

In light of the terrible San Francisco earthquake of 1906, and with a bridge to be constructed in that vicinity, a critical requirement was that, in the event the bridge should fail
and fall, sufficient draft would be available for warships to pass without obstruction. Once
assured of this specification, the War Department agreed to the project. Construction work
confidently commenced in January 1933 and the bridge opened in May 1937.

The design is a suspension type with two cables, 36 inches in diameter. The main span
is 4200 feet between the towers. The deck is 90 feet wide with trusses below it for stiffening. The cables were spun by methods developed on the George Washington Bridge, but

with refinements. By laying looped steel cable wire over the towers one loop at a time, the complete cable avoids hoisting heavy cable, as well as the risk of damaging the tower. From the reel of wire at one end of the bridge a loop is carried over the span by means of a spinning wheel attached to a hauling rope. When the loop of wire reaches the far anchorage, the wire is pulled off the spinning wheel and connected to an anchorage. The weight of the anchorage blocks, constructed of steel and reinforced concrete, holds the cable ends in place. Units of the deck and stiffening truss weighing up to 204 tons were erected on land, floated out, and then hoisted from floating barges, the lifting being done by cranes suspended immediately below and held by the main cables.

The foundations had to be made in the ocean floor, and their construction presented difficulties. Caissons were required to be sunk through deep water in a current of $7\frac{1}{2}$ knots and to a depth beyond that which men can work under compressed air. The bay is tidal and the sea water sweeps in and out. Much of the perilous underwater work on the foundations could be done only during slack water, as during other times the divers would be swept out to sea by the tidal currents. Financing of the bridge was through a toll system, necessary to pay back the bonds that financed the structure initially. This view of the world-renowned bridge is looking north, showing the sea side of the structure. It remains one of the longest suspension bridges in the world. Chief engineer was Joseph P. Strauss. It is an international symbol of civil engineering.

Direct labor is the work that transforms materials from one state to another value-added condition. Carpenters, masons, steel setters, heavy equipment operators, and wiremen electricians are typical titles for direct labor. There is also *indirect labor*, for example, timekeeper, watchman, and secretary. These classifications of indirect labor are costed by overhead calculations, a topic developed in Chapter 4. The wage of direct labor may be a base wage, or it may include all necessary fringes, which is called Gross Hourly Wage. These distinctions are developed in Chapter 2.

Material that is required to satisfy the construction of the design is classified as *direct material*. It appears in the constructed end item or is essential to its building. The material is determined from the quantity takeoff, as developed in Chapter 3. There is a difference between direct material and subcontract material. Direct materials end up in the construction design and are placed by the bidder's work force.

Comprising the third element in a construction work estimate are the *tools* used by direct labor. They range from simple hand tools to powered electric/pneumatic tools, etc. As inconsequential they may be, their cost must be estimated. In principle, it will not do to ignore these costs. But the greatest distinction deals with absorbing the cost of the tool. A hammer, for instance, as vital as it is to a carpenter, is too insignificant to itemize and prorate its cost to a specific job. Regardless of their insignificance, tools need to be considered, and several methods are available. A dollar allowance based on field experience may be provided, or it may be a part of the overhead. In practical terms, tools may be estimated as a percent of the direct labor cost, say 0.1% to 2%.

Most construction work requires *equipment*. The purchase of equipment, or its rental, is a significant cost. The cost of equipment involves ownership, operation, maintenance, moving, setup and teardown. Equipment costs for construction work can be calculated specifically for the job or can become a part of the overhead. In this text, we prefer the finding of the cost of equipment for each job, that being more accurate than general cost recovery by overhead methods.

Subcontract materials involve the principle that subcontract labor is the work force installing the subcontract materials, and it is this dual nature that makes it different from direct materials. The subcontract will include both the material and the subcontract labor. For example, in the construction of a building, subcontracts may be let for plumbing, electric wiring, painting, and landscaping. In the construction of a bridge, a subcontractor may build the foundations and piers, and another may build the approaches on the foundations. The subcontract cost is provided by the subcontractor, after an RFQ given by a general contractor. Subcontractors and general contractors use many of the same estimating practices.

7.3 PRIMITIVE WORK ESTIMATING

Construction work is varied and complex. Consider an example[1] of simple hand labor without equipment where we are to estimate the time required to excavate earth by hand under different conditions.

"For the first operation, the soil is a sandy loam. The maximum depth of the trench will be 4 feet. Climatic conditions are good with the temperature averaging about 70°F. The work requires light loosening with a pick prior to shoveling. The laborer should easily move 5 cubic yards of earth per hour using a long-handle round-pointed shovel; and he should require about 150 loads to move a cubic yard of earth. If a laborer can handle 2.5 shovel loads per minute, he will remove 1 cubic yard per hour. These rates of production were for 2 man-hours to loosen and 1 man-hour to remove a cubic yard of earth of trench or a total of 3 man-hours per cubic yard."

"The second operation the soil is a tough clay, which is difficult to dig and lumps badly. The maximum depth of the trench is 5 feet. The temperature is estimated about 100°F without shade. Under these conditions a laborer may not loosen more than 0.25 cubic yard per hour. Because of the physical condition of the loosened earth, it is estimated to require about 180 shovel loads to move a cubic yard of earth. It is estimated that a laborer can handle 2 shovel loads per minute or 120 shovel loads per hour, which is equivalent to 0.67 cubic yard per man-hour or 1.5 man-hours per cubic yard. Given these rates of production, it is necessary to have 4 man-hours to loosen and 1.5 man-hours to remove a cubic yard, or a total of 5.5 man-hours per cubic yard."

Perceiving varying rates of production for hand operations, we calculate the number of man hours per cubic yard. For a given labor cost per hour, including fringes and other factors, the cost estimator determines the cost of excavation per yard. In this example, a shovel and pick are the simple tools that are used. Those tool costs, which are minor, must be considered in some way to have the full cost of labor with the tools (or equipment in more conventional work).

This discussion is historical, and the estimator makes inferences on work that might be similar. Indeed, topical history, such as this one, comprised the testament lessons by which technical estimating of construction was conducted during the first half of the twentieth century. Magazines of that era regularly featured these case histories, and the construction estimating pioneers of that time published these articles in handbooks.

---

[1] R. L. Peurifoy, *Estimating Construction Costs*, 3d ed., New York: McGraw-Hill, Inc., 1978.

"I have seen 700 two-horse wagons, holding $\frac{3}{4}$ cu yd each, loaded per 10-hour day; and I am informed, that with good management and an easy soil, and a 10-hr day, and with three-horse wagons, the output on the Chicago Drainage Canal was 500 cu yd of top soil."[2]

From these first-hand accounts, the next step of the progress of construction estimating was to interpret those experiences into mathematical constants, variables, tables, and formulas. Then we were able to estimate with the information more reliably. Eventually, these histories and time/method measurements were codified and formed into tables that permitted greater use and understanding. Data warehouses became commercially available. Company information manuals/software and openly published handbooks are now widespread.

7.4 CONTEMPORARY WORK ESTIMATING

Data sets and computer software applications depend on mathematical approaches. Computer software is easily used, and extensive experience with these applications is the norm. Those developments have allowed formulas to be consistently applied. A formula, first presented in Chapter 6, Eq. 6.4, consolidates observed data into a format that becomes more precise than an individual's interpretation. Reconsider the model for estimating based on the unit-quantity method:

$$C_o = \sum_{i=1}^{m} \sum_{j=1}^{n} n_i \left(R_i + H_{ij} R_{ij} \right) \qquad (7.1)$$

where C_o = cost for construction code, assembly, or design detail, dollars
n_i = quantity for material i, in dimensional units compatible to design i
R_i = unit cost rate for material i
H_{ij} = unit man or equipment hours for requirement i, j associated with material n_i
R_{ij} = unit cost rate for requirement i, j
i = material $1, 2, \ldots, m$ from takeoff associated with construction work
j = requirement $1, 2, \ldots, n$ to satisfy installation of material i

The model focuses on the quantity of material, n_i. There are two types of material: direct and indirect. Direct materials are, for example, the concrete for a wall, are found on the working drawings of the design, and are further clarified by the specifications.

Indirect materials are a result of the construction process—for example, the wooden forms necessary for the emplacement of the concrete for a wall. The subscript i considers all the relevant materials, meaning that all direct and indirect materials that are not included in any allowance or overhead rate are evaluated. If the material is also included in the overhead rate and the unit-quantity method, there is double estimating, a serious blunder. Equation (7.1) and its table show the contemporary takeoff method, so popular in construction estimating.

R_i is the unit-cost rate for the material. If the dimensional units for the material are lb, bft, ton, 100 ft^2, etc., the cost rate is in identical units. For example, the cost for lumber

---

[2] Halbert P Gillette, *Handbook of Cost Data*, New York: Myron C. Clark, Publisher, 1905.

PICTURE LESSON
Civilian Conservation Corps

During the Great Depression of the 1930s, unemployment was rampant, with over 25% out of work during 1933 alone. President Franklin Roosevelt proposed the CCC for the relief of unemployment through good public work. The CCC's mission was to promote conservation of natural resources and provide employment and vocational training for youthful citizens. The Department of Labor organized a nationwide recruiting program, and the Army provided conditioning, basic military skills, and transportation of enrollees to the camps. The National Park and Forest Service supervised the work assignments. Conservation projects were diverse—including, for example, construction of primitive bridges, trails and roads, insect control, and reduction of fire hazards. As the picture shows, a Washington camp gathers for a moment of glory in a ribbon-cutting ceremony that dedicates the bridge.

A CCC enrollee received $8 in cash per month, with another $7 being placed in a savings account and $15 mailed to his family. The bulk of the hires were from large urban centers. The enlistees, as they were called, had a period of enrollment before they mustered out. The early camps were tents, and mess halls were framed wooden structures. Subsistence, military clothing, and medical care were provided. Daily life in the camp was similar to Army regimen with reville, inspection, and recreation, in addition to work assignments.

The typical enlistee is described as being 17–18 years old, having completed the ninth grade, and recently out of work; his family was likely to be on relief. The enlistee had a commendable belief that the CCC would teach him how to work, and he liked the idea. He had no feeling that hand labor was a disgrace, and in his view, happiness did not depend on having "lots of money," according to a survey. The CCC involved a total of 2 million enrollees in 3350 camps in the National and State Parks. The CCC is credited with creation of 711 new state parks. Shortly after the start of World War II, the CCC was terminated.

---

may be given as $150 per 100 bft, and it is necessary that the quantity n_i be expressed in similar 100-bft units.

H_{ij} is the required time to install the takeoff material units. R_{ij} is the corresponding cost rate either for labor, or for equipment, or for any of the satisfying requirements. For instance, if the material is lumber, and the units are board feet, and the matching H_{ij} is given in hr/bft, and then R_{ij} is in units of $/hr. We now turn to a contemporary problem illustrating the unit-quantity method.

7.5 UNIT-QUANTITY EXAMPLE

Many thousand construction-work variations can be illustrated with the unit-quantity method. We have selected the work of concrete: forms, reinforcing, mixing or delivery, pouring, and curing.

7.5.1 Concrete-Forms Work

The major cost for a concrete structure is sometimes the cost of the forms. Concrete structures can be shaped into any geometry for which it is possible to build forms. Forms are constructed from lumber, plywood, steel, aluminum, and other composition materials, either separately or in combination. If the form is for a special and unique design, lumber is frequently the choice, especially if this material is to be used only a few times. But if there are multiple occasions for reuse, then steel or aluminum may produce a lower cost than lumber.

The cost of forms includes the cost of materials, such as lumber, nails, bolts, form ties, form oil, etc., and the cost of labor for fabricating, erecting, and removing the forms. Waste, salvage, and reuse or one-use-only influence the allowance that affects the cost of some of these materials, especially the lumber form.

Work for concrete forms includes the use of equipment, which for lumber forms can be simple, such as saws, drills, nailers, and hand tools. Too, the tools can be powered, such as electric/pneumatic, entailing operating costs for their use.

7.5.2 Concrete Wall Design

Estimate the design shown by Fig. 7.2, which shows the materials for forms. The forms require plywood, studs, wales, sills, scab lumber, and braces. For those materials we assume an average price per board foot. This average number is forecast for the period of purchase for the lumber. If we had itemized the different lumber for the quantity takeoff, then

TABLE 7.1 Schedule for lumber requirements for Fig. 7.2

Lumber Material Takeoff	Quantity, n_i
1 × 48 in. × 8 ft plywood	768 bft
Studs, 2 × 4 in. × 10 ft, 34 pc	227 bft
Wales, 2 × 4 in. × 12 ft, 20 pc	160 bft
Wales, 2 × 4 in. × 14 ft, 20 pc	187 bft
Sills, 2 × 4 in. × 12 ft, 2 pc	16 bft
Sills, 2 × 4 in. × 14 ft, 2 pc	18 bft
Scab splices, 2 × 4 in. × 2 ft, 20 pc	27 bft
Braces, 2 × 4 in. × 10 ft, 10 pc	67 bft
Total	1470 bft

it would be necessary to price each size separately on the material takeoff. For the sake of convenience, we gather a schedule for the various sizes. This is seen in Table 7.1.

The tongue-and-groove plyform is 1 × 48 × 96 in. external AC grade that is surface oiled for protection. Studs and wales are 2 × 4 in., and the wales are double. The spacing is required to meet specification of loading, and spacing for studs is 19 in., while spacing for wales is 25 in.

With the height of the wall as 114 in. and length as 304 in., we need 12 plyforms per side. If this is a one-time use only, then the waste is about 35%, a high value.

The number of 10-ft studs per side is 17 ($\approx 304 \div 19 + 1$). The number of wales per side is 5 ($\approx 114 \div 25$). For each wale, use 2 pieces of 2 × 4 in. × 12 ft and 2 pieces of 2 × 4 in. × 14 ft. It is necessary to have 20 pieces of each. Bracing specifications require 5 pieces per side.

Total lumber is 1470 bft for a two-sided surface area of 481 ft^2. This quantity is important to the development of the unit-quantity estimate. This quantity relates to other requirements. Look at Table 7.2, which is the tabular development of Eq. (7.1). Notice that the first entry is lumber and the quantity is 1470 bft. The cost for lumber, as entered in the table, is $1.50/bft. The subtotal extension is $2205, which is carried to the row total.

Nails, while very minor, are an interesting case. This item is cross related to board feet first, and then cost to its quantity. Notice $i = 2$ for a row total cost of $8.

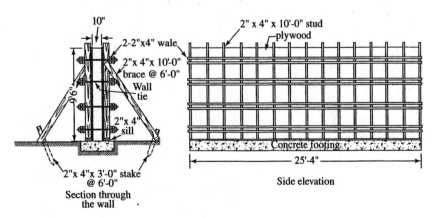

Figure 7.2 Forms for a concrete wall.

TABLE 7.2 Unit-quantity development using Eq. (7.1) for Fig. 7.2.

Work and Material Takeoff i	Quantity, n_i	$j=1$ Materials Unit Cost, R_i	Subtotal, $n_i R_i$	$j=2$ Labor Unit Man Hours, H_{i2}	Total Man Hours, $n_i H_{i2}$	Unit Cost, R_{i2}	Subtotal, $n_i H_{i2} R_{i2}$	$j=3$ Equipment Unit Hours, H_{i3}	Total Hours, $n_i H_{i3}$	Unit Cost, R_{i3}	Subtotal, $n_i H_{i3} R_{i3}$	Row Total
1. Lumber	1470 bft	1.50	$2205									$2209
Surface area	481 ft²			0.05	24	50.50	$1215	0.0025	3.7	1	$4	1215
2. Nails @ 1 lb/100 bft	13 lb	0.50	7									7
3. Oak stakes	10 pc	0.65	7									7
4. Form ties	50 pc	1	50									50
5. No. 4 reinforcing bar	110 lb	0.85	94	0.023	2.5	34.00	86					180
6. No. 5 reinforcing bar	285 lb	0.85	242	0.023	6.6	34.00	223					465
7. Concrete, 4000 psi	7.5 yd³	78.50	589	3.5	26.3	95.00	2494	3.5	26.3	78.50	2061	5144
Column total			$3194				$4018				$2065	$9277
Cost per yd³			$426				$536				$275	$1237
Cost per lin ft			$126				$159				$82	$367
Cost per ft² of form area, two sides			$7				$8				$4	$19

TABLE 7.3   Illustrative relationship of labor productivity to height of form wall.

Height of Wall, ft	Carpenter or Helper: Build, Erect and Remove Forms, hr/100 ft²-side	Carpenter or Helper: Erect and Remove Prefabricated Forms, hr/100 ft²-side
2	2.5	1.5
4	3.5	1.7
8	4.5	2.1
12	5.0	2.3
16	6.5	3.3

The material takeoff specifies the principal cost driver for the labor, which in the case of wooden forms is surface square feet of contact, and a carpenter and a helper are assumed to be the crew. Occasionally, a part-time foreman, who it is assumed is also supervising other jobs, can be a part of the crew. The relationship of wall height to hours per 100 ft² is given by Table 7.3. The value of productivity is shown as H_{12}.

Wages for labor trades vary in the United States. Shown in Table 7.4 are illustrative values of the gross hourly wage rate that are used in this chapter. Details about the gross hourly wage-rate method are given in Chapter 2. For $i = 1$, Table 7.2, a crew is determined to be a journeyman carpenter, helper, and a part-time foreman for 25%, and the unit cost rate from Table 7.4 is $50.50 $(= 25 + 18 + 0.25 \times 30)$, which is entered as R_{12}.

The design specifies that form ties are used. Using the design standards for safety and loading, the spacing of form ties and their assembly units is 31 in. The form ties resist the internal pressure from the concrete and hold the forms in position. The form ties hold the forms apart prior to the pouring of the concrete, and then resist the bursting pressure of the concrete. The assembly consists of a steel tie and two end clamps. The number of form ties per wale is 50 ($\approx 304 \div 31 = 10, 5 \times 10 = 50$). Costs for form ties based on wall thickness are given in Table 7.5.

TABLE 7.4   Illustrative gross hourly wage rates for selected trades.

Trade	Gross Hourly Wage Rate, $/hr
Carpenter	$25
Carpenter-helper	18
Common labor	15
Mason	24
Concrete mixer	17
General foreman	30
Hoist operator	20
Steel setter	19
Truck driver	21

TABLE 7.5   Illustrative costs of form-tie assemblies for concrete wall.

Wall Thickness, in.	Cost of Form Tie and 2 Clamps, $/100 Assembly Units
From 4 to under 8	$85
From 8 to under 12	100
From 12 to under 24	125

Extension of Table 7.2 gives row and column totals. The cost of producing the forms, exclusive of the reinforcing rods and concrete, is $3,488.

7.5.3 Reinforcing Steel Work

Reinforcing for concrete consists of steel bars. The cost of rebars is estimated by the pound, hundred pound, or ton, even though slight variations are found with diameter variation. An additional 5% is added for overlap and waste. Usually bars are fabricated to the required length and shape by shops especially suited to this task, rather than by field forming, which can add to the basic material cost. There are many types, sizes, and material grades for reinforcing bars. Those properties are found in handbooks. Reinforcing bars are identified by numbers, i.e., 2 to 11, and this number represents the diameter in "eighths." Some information for bars is given in Table 7.6. Waste is not included in this information.

Now reconsider Fig. 7.2 for an operation of setting steel reinforcing bars inside the chamber, before the installation of the concrete ties. The horizontal reinforcing steel for the wall consists of No. 4 bars spaced 2 in. in from the wall. There are three pairs of horizontal bars, equally spaced vertically. The horizontal bars extend beyond both wall ends 12 in. Vertical No. 5 reinforcing bars tie to the horizontal bars and are spaced 24 in. on centers. They extend 12 in. above the surface of the wall concrete after the concrete is poured. Bar length is 13 ft 8 in. for No. 4 bars and 10 ft 6 in. for No. 5 bars. The quantity of reinforcing steel for Fig. 7.2 is given in Table 7.7

Labor for placing reinforcing steel bars varies depending on size and length of bar, shape, complexity of the structure, distance and height that the bars must be moved, tolerance of setting, tying, and skill of steel setter. Experience points out that less time is required to place a ton of steel when the bars are long and of large diameter than when they are short and of small diameter. Straight bars are placed more quickly than bars with bends and end hooks. Table 7.8 gives a small sampling of information for steel bar placement. For Fig. 7.2, a steel setter and general laborer are the scheduled crew.

TABLE 7.6 Illustrative information for selected steel reinforcing bars.

Bar No.	Dia., in.	Area, in.^{2}	Weight, lb/ft	Material Cost, $/ft
4	1/2	0.20	0.668	$0.561
5	5/8	0.31	1.043	0.887
6	3/4	0.44	1.502	1.277

TABLE 7.7 Schedule of reinforcing-bar requirements for Fig. 7.2.

Reinforcing Bar Material Takeoff	No. Pc	Weight, Lb/pc	Quantity, n_i, lb
No. 4	12	9.13	110
No. 5	26	11.0	285

TABLE 7.8   Illustrative labor productivity for placing reinforcing steel bars for steel setter and general helper trades.

	Size of bar			
	5/8 in. bars and less		3/4 in. bars	
	Length of bar			
	15–50 ft	Under 15 ft	15–50 ft	Under 15 ft
Bars not tied in place, hr/ton	30	35	25	32
Bars tied in place, hr/ton	40	46	28	34

7.5.4   Concrete Work

Common concrete, that ubiquitous and astonishing building material, is composed of water, mineral aggregate, and Portland cement, in which the separate ingredients coalesce to form a hardened mass. Simple as this definition is, we appreciate the complexity of this material. Discussion of the engineering/architecture, design, placement, specifications and properties, standards and their extensive possibilities is beyond the needs of this text. Students and professionals are referred to other sources.

The cost of the concrete material depends on many factors. Is the job small or large? Is the job a structure, highway, dam, residence, or a small retaining wall? How remote or accessible is delivered ready-to-use concrete? Can concrete be placed using pumping equipment? Will a concrete plant be necessary? What exactly are the design requirements? Placement influences the cost—is the location accessible or distant from the supply? Is a hand buggy, wheelbarrow, crane bucket, or pump-pipe method essential? The better choice among these possibilities depends on the answers provided by the estimates. Our illustrations in this chapter deliberately rule out these many opportunities, favoring instead a basic understanding for simple construction.

Concrete is usually measured net, as placed. Units of measure are cubic yards $\left(yd^3\right)$. Deductions are not made for steel beams, reinforcement, pipes, or pass-through, unless they are greater than some specified amount that is usually recognized by the contractor. Each mix and specification of concrete is separately measured and identified, because different concrete mixes and proportions are costed differently. Concrete recipes are proportioned for strength, durability, and workability. Strength and durability are determined by the water-cement ratio, the workability of the slump, and so on.

Concrete mixing, placing, and curing is sometimes done with low technology, meaning *labor-intensive work*. The raw materials may start with sacks of cement, piles of sand, rock, etc., and water. Next come the mixing, stirring, and hauling with buggies to the point of pouring, followed by the protection for any curing requirements. Table 7.9 gives an example of this highly variable data.

When ready-mixed concrete is used, preparation costs are transferred from the job to a central mixing plant, and these costs decline because of the efficiency of the plant over labor-intensive mixing. Concrete is a *variable cost*, which depends on the size of the job and the number of cubic yards, annual usage by the buyer, location for delivery from the plant, quality, and other special additions such as color, accelerator, retarder, fiberglass, etc., and whether the minimum load charge is met. Table 7.10 is an illustrative example of ready-mixed-concrete costs. Now continue the example of Fig. 7.2. Refer to Table 7.2 where 7.5 yd^3 of 4,000 psi quality concrete is added to the material takeoff. Entry of $78.50/yd^3$ is made for R_7.

TABLE 7.9   Illustrative productivity for labor-intensive work to mix, place, and cure concrete, hr/yd³.

Technology	Application	Concrete Labor	Mixer Labor
Manual, ordinary weather, hand mixing	Small jobs, homes, and patches	3.0	1.9
Manual, cold weather, hand mixing	Small jobs, homes, and patches	3.5	2.1
Manual, ordinary weather, portable-machine mixing	Small jobs, homes, walls, and patches	2.0	0.3

TABLE 7.10   Illustrative delivered-and-ready-to-pour concrete costs.

Quality Concrete, psi	Sacks of Cement/yd³	Amount, yd³	Cost, $/yd³
3000	5.5	< 10	$73.50
	5.5	10- < 20	68.50
	5.5	20- < 100	63.50
	5.5	100-1000	58.50
4000	6.5	< 10	78.50
	6.5	10- < 20	73.50
	6.5	20- < 100	68.50
	6.5	100-1000	63.50

Labor productivity to place ready-mixed concrete will vary with the rate of delivery and the design, location, and accessibility of the structure. If the truck can be driven to the structure, constructed at or below the level of the ground, it may be possible to discharge the concrete directly into the forms using the chute. A prespecified crew will spread, vibrate, and control the pouring.

Concrete for a slab may be discharged into a bucket, hoisted by the crane, and distributed over the slab area with little labor assistance. Many other methods are available to handle concrete. Refer to Table 7.11 for labor hours per yd³ to place concrete. The rate of safe concrete delivery affects these data, and we assume that it is constrained to a maximum of 15 yd³/hr for this application.

Now return to Fig. 7.2, where a crane and bucket are scheduled with a crew of one crane operator, one on the ground handling the bucket, one man guiding the bucket and directing the hopper, one man working with the vibrating equipment, and a foreman. Because of the small size of the concrete requirement for Fig. 7.2, crew time is limited to the placing rate for common labor. The crew of five consumes 3.5 hr/yd³ (=5 × 0.7) and the joint cost of the crew is $95/hr (=20 + 3 × 15 + 30). The concrete flows into the forms at a rate that will not burst the walls. Refer to Table 7.11 for the extension of labor to place the concrete. The truck-mounted jib crane for bucket loading of the concrete has an hourly rate of $78.50. Notice the line $j = 7$ in Table 7.2 for the cost of ready-mix concrete, labor, and equipment to place the 7.5 yd³.

The unit-quantity method allows a final cross check of costs to other jobs. For instance, Table 7.2 shows the total for the job as $9277 and the average cost per cubic yard

TABLE 7.11 Illustrative productivity for placing concrete, hr per yd^3, of ready-mixed concrete.

Design	Discharged Directly from Truck; Method of Handling	Common Labor	Hoist Operator	Foreman
Foundation or wall	Buggies	1.0		0.07
Foundation or wall	Chute	0.5		0.07
Foundation or wall	Crane or hoist, bucket, hopper	0.7	0.07	0.07
Slab on grade	Crane and bucket	0.7	0.07	0.07
Slab above ground level	Crane or hoist, bucket, hopper, and buggies	1.3	0.07	0.07

as $1237. Additional calculation gives cost per lin ft as $367/lin ft and $19/ft^2 of concrete contact surface area. Now if there is another similar design, these cost factors can estimate its costs.

7.6 EQUIPMENT COSTS

If the job exceeds the life of the equipment, then the evaluation of equipment cost for the job is trivial. The full cost of the equipment becomes a part of the direct job cost. Such was the case for the Alaska pipeline from Prudhoe to Valdez, as the equipment and its cost were consumed in the performance of the work. The job work life exceeded the equipment performance life. Salvage value was the only residue of economic worth for the equipment.

Whenever the equipment performance life exceeds the job requirements, it is necessary to estimate that portion of the equipment's economic value which is used up by the job. The pro-rata estimate of equipment economic-value reduction is not trivial. Equipment costs are partitioned into:

- Ownership or rental
- Travel to and from job
- Erection and dismantling costs
- Operating
- Maintenance
- Indirect or overhead

If the contractor owns the equipment, there are no rental costs, and if the contractor rents, there are no ownership costs. Most equipment will have transportation costs to and from the job. Some equipment, like truck-mounted apparatus, will have no or small setup costs. Steel scaffolds with telescoping connections have no operating costs. Tower cranes will have travel, erection and dismantling, and operating costs. Maintenance costs of some equipment can be significant, and sometimes they may equal or exceed the other annual costs of the equipment.

Construction-equipment rental is common, especially because of the availability of rental businesses. These local businesses handle smaller-scale equipment. The rental may be for a half or full day, week, or longer. For large-scale or more specialized equipment, leases can be arranged with commercial banks that specialize in this business. If equipment

PICTURE LESSON
Penstock

This August 1934 picture of the Hoover Dam construction shows the cable way lowering a 30-foot-diameter penstock for placement in the upper Nevada header tunnel. Note the man sitting in the pipe. Two intake towers are seen on the left and right sides, upstream from the dam. The towers are as tall as a 39-story building.

The arch gravity dam uses its weight and the thrust on the canyon walls to resist the hydrostatic pressure of the impounded water. The concrete monolith, 726.4 feet high–as tall as a 60-story building–has its wedge 660 feet thick at bedrock, tapering to 45 feet at top. The top serves as a 1282-foot-long highway connecting Arizona and Nevada. The dam's base sits between transverse fault lines in the canyon floor. The final design allowed for a maximum compressive stress of forty tons per square foot. Trucks, Caterpillar tractors, electric shovels, draglines, and other motorized vehicles of all shapes, sizes, and descriptions were used on the job in unprecedented numbers, making Hoover Dam one of the first major projects in which horses and mules would have virtually no role.

Diversion of the Colorado River required four tunnels. The tunnel work was mechanized as well, having a mammoth motor-driven rig to drill half of the 30-foot tunnel bench at once. The drilling jumbo was fitted with thirty 144-pound compressed-air rock drills. The final diameter of 56 feet would be reduced to a 50-foot concrete-lined tunnel. Four of these tunnels, two on each side, ran for approximately three-fourths of a mile through the rock. Temporary earthen cofferdams, one upstream and a second downstream, walled off water from the construction site.

Two tunnels, first used to separate the water from its ancestral course, would give water to the spillways for irrigation in other states. Two other tunnels would use the penstocks and divert water to adits supplying the water to the powerhouse turbines.

Once the tunnels were completed the diversion of the river would begin in earnest. Just below the entry portals, trucks deposited muck and rock every 15 seconds into the channel from the trestle bridge. With the cofferdams exploded to allow water to enter the tunnels, first as a trickle. Then, as slowly the new rock dam climbed, the water took a turn and flowed into the tunnels. Following the November 1932 diversion of the river, the site was pumped dry ready for dam and tower excavation. In September 1935 President Franklin D. Roosevelt dedicated the dam, then the world's largest. Other dams were to follow throughout the Western United States, but the Hoover Dam remains the icon.

can be purchased, it is said, it can also be leased. The contractor may have a shortage of capital money, and thus he is unable to buy equipment, or she chooses to lease because of attractive terms. The lease may be for a specific job or for long term. Understanding the lease contract is beyond the needs of this text. But it is important to realize that the cost of the lease is passed on to the future work, as designated by the estimate.

If the contractor has equipment that is *capital cost* in nature, it is necessary to determine the cost of the equipment. Ownership costs must recover the cost of the money that was initially spent to buy the equipment. Small incidental equipment is expensed as a part of the overhead. Capital equipment differs from incidental equipment. The distinction between the two depends on the contractor's business size and cost-estimating policy.

Reasoning about *ownership costs* resembles that about depreciation methods, which are discussed in Chapter 4. The student will want to review that material. Remember that depreciation is a *tax-deferred expense*. The P&L Statement and the Balance Sheet were discussed in Chapter 4, and factors involved in the understanding of capital cost are important using that perspective. *Depreciation* is a cost against income that reduces the income taxes of the corporation. The greater this depreciation cost, the smaller the corporation federal or state taxes for an equal-valued income. The greater the depreciation, the greater the positive cash flow to the company.

Alternately, it is not legally mandatory that the equipment moneys be recovered by the depreciation calculations that are used for tax accounting. The Modified Accelerated Recovery System[3] is the most common method for overhead accounting of depreciation.

---

[3] From 1913 to 1954, the U.S. government gave businesses broad leeway for depreciating company assets. In 1954 Congress authorized accelerated methods. A shift in approach occurred in 1962, which made it more difficult for companies to depreciate their assets. In 1971 another modification was made, and in 1986 Congress modified the tax laws concerning depreciation, now known as the Modified Accelerated Recovery System, which specifies the recovery period of construction equipment.

However, it does not give equal yearly amounts of depreciation throughout the life cycle of the equipment, nor does the method consider the equipment's salvage value. And if the equipment is fully depreciated, no additional depreciation charge may be made for it when determining profit and loss for income-tax purposes. Depreciation methods are inappropriate to determine the cost of the equipment for estimating purposes.

For the purpose of equipment cost recovery the consumption rate of capital cost may be expressed as a cost per unit of time, or it may be expressed as a cost per unit of work performed. The calculation of the pro-rata share of the cost of this equipment and its inclusion in estimates is determined in the following way:

$$D_{cc} = \frac{(P - F_s)}{N} \qquad (7.2)$$

where D_{cc} = equal yearly-or-production capital-consumption cost, dollars
$\qquad P$ = total initial investment, dollars
$\qquad F_s$ = future salvage value of investment, dollars
$\qquad N$ = life in years or total units of production for equipment, number

If the equipment is used for more than one job, it is necessary to prorate the cost of the equipment to these job estimates. Formula (7.2) gives a straight-line approximation of the consumption of equipment cost. Formula (7.2) is familiar as an out-of-date depreciation method.

The total investment value P includes the cost of the equipment, its transport to the contractor's work site, any taxes, etc. If the equipment is to have a specified life, and then the contractor will use the residual of the equipment for income, there is a salvage value F_s. Life N can be estimated in years or units of production that the equipment will provide. This annual prorated value D_{cc} of the equipment is divided by the number of either annual working hours or units of production for the equipment.

Typically equipment will have an average cost value, remembering that at the moment that the equipment arrives at the contractor's location, and even before any work, there is a reduction in value from its first cost. But initially the equipment is higher in value, and eventually its actual value declines as time and jobs continue.

$$\bar{P} = \frac{P(N + 1) + F_s(N - 1)}{2N} \qquad (7.3)$$

where $\bar{P}$ = average equipment investment during life of equipment, dollars

There are *annual investment costs* in addition to the first cost of the equipment. These investment costs can be proportional to the average investment cost in the following way:

$$\text{Annual investment cost} = \bar{P}(1 + I + i + t + s) \qquad (7.4)$$

where I = insurance cost, decimal
$\qquad i$ = interest rate on unpaid balance of loan, decimal
$\qquad t$ = tax rate on appraised value of equipment, decimal
$\qquad s$ = storage cost, decimal

The contractor wishes to protect the investment from any loss or damage, and insurance is typically the way financial protection is gained. Insurance cost can be expressed as a percentage of the average investment.

Equipment can be purchased similar to a car, i.e., down payment and monthly reductions in the mortgage. The purchase agreement and the repayment plan will specify an interest rate, i. The equipment will be stored when not in active use. These several costs are often 10%–25% of the first cost of the equipment. The annual investment cost is divided by the number of working hours or the number of production units for the year.

There can be travel-to-and-from-job costs to move equipment job to job. Haulage companies are contracted to move equipment from one site to another. Or the contractor is able to move the equipment with the firm's flatbed truck. The equipment may be truck mounted. These costs need to be determined.

The equipment may require setup and teardown. A concrete pumping truck is an example. Both at the beginning and at the conclusion of the concrete pumping, the plumbing needs to be cleaned for the next job.

Maintenance cost is significant and is calculated as a percentage of the investment, using the contractor's historical evidence. This would include the cost for replacement parts and the labor to keep the equipment in ready-to-use serviceable condition. Some contractors employ specialized maintenance shops instead of their own employees. Maintenance costs vary with the type of equipment and duty cycle. Some historical evidence suggests that the yearly maintenance cost for crawler tractors, for example, is approximately 50–100% of the capital consumption cost. Naturally, the longer the useful life, the more cost expended on maintenance. It is expected that shovels, hoes, draglines, clamshells, and lifting cranes exhibit different cost histories.

Engine operation requires gasoline or diesel fuel and engine oil during operation. A concrete pumper truck requires energy for its ongoing operation. Electrical energy needs are common. These costs are determined for an operating hour and added to the equipment cost.

Equipment has *indirect costs* that are too varied to determine for a specific job. The principles for overhead calculation were developed in Chapter 4. These indirect costs may be fuel, management, supervision, dispatching, licensing, fees, operator training, and the like. These costs can be collected for the year and ratioed against the job costs. When done in this fashion, these indirect costs increase the job cost.

7.6.1 Equipment-Cost Example

A concrete pumping truck having a maximum capacity of 195 yd^3/hr is purchased for $1,200,000 delivered cost. It has a service life of 5 years before the first overhaul. The overhaul will nearly restore the equipment to its initial performance and appearance, and this major maintenance cost is $125,000. At the end of the next 4 years, the unit is salvaged for 15% of initial cost. The climate and business conditions for this equipment allow a 6-day work week and 3000 operation hours per year. Pumping-time utilization of the equipment is 2600 hours per year. Average pumping capacity is 125 yd^3/hr. Investment costs are 15% of the average investment value.

Travel time, including the jockeying of the truck into position for work, is 0.125 hr/mile, as determined from the home base of the pumper. Annual cost for periodic and unscheduled maintenance is 2.5% of the first cost. Ten tires are replaced annually, and a new set together with its installation and downtime costs $10,000.

Setup and teardown time and supply setup for each job are 64 minutes, as determined from a nonrepetitive one-cycle time study. See Chapter 2 to find the productivity of the concrete-pumping operation.

The unit is cleaned with a Portland cement slurry, which is forced through the plumbing after the conclusion of the pumping of the concrete. The cost of the cement is $8/sack. This semifixed cost is related to the number of yd^3 pumped. Historical experience for this cleaning indicates the following:

Job Concrete yd³ Pumped	No. of Sacks for Slurry	Cost for Sacks
15 < 25	7	$56
25 < 60	13	104
60 or more	18	144

The equipment crew is one driver, and gross hourly wage cost is $21/hr. Indirect costs are 11% for this equipment.

A solution is as follows: The capital consumption cost, D_{cc} = $133,333 per year (=1,200,000/9), or $44.44/hour (=133,333/3,000). The average investment is $746,667 $\left(= (1,200,000 \times 10 + 180,000 \times 8)/18 \right)$ and its annual investment cost on a per hour basis is $37.33/hr (=112,000/3,000).

Lifetime maintenance cost is composed of regular, unscheduled and one-time overhaul costs, and is $17.96/hr (=485,000/27,000). A schedule for maintenance costs gives the following:

Maintenance Item	Calculation
One overhaul	$125,000
Annual maintenance excluding tires	$30,000 = 0.025 × 1,200,000
Tires, annual replacement	$10,000
Total maintenance and overhaul for lifetime	$485,000 = (30,000 + 10,000)9 + 125,000
Maintenance cost per hour	$17.96/hr = 485,000/27,000

The gross hourly cost for the operator adds $21/hr to the hourly calculations. The total hourly cost is $120.73 (=44.44 + 37.33 + 17.96 + 21.00). This cost is related to the average pumping rate for the equipment, and is $0.966/$yd^3$ (=120.73/125).

Travel time cost is $15.09/mi (=0.125 × 120.73). Travel cost depends on the total round-trip distance from the home base of the equipment.

Erection and dismantling cost is $128.46 per job (=64/60 × 120.73). This a fixed cost for each job.

The cost for the equipment is composed of job costs, pumping of concrete, distance to the job site, and cleanup. It is seen that the types of costs are *fixed, variable with respect to the quantity* of concrete that is pumped, *variable with respect to the job distance* from the home station of the equipment, and *semifixed with respect to cleanup*. Fixed, semifixed, and variable costs are discussed in Chapter 4; refer to that material for definitions.

It should be noted that overhead can be additionally applied for scheduling and management of the equipment, and the job cost is without a profit markup. These topics are reserved for Chapter 8.

An equation is written that characterizes the equipment cost per job for pump delivery of concrete, or

$$\text{Cost per job} = 128.46 + 0.966 \times \text{No. of yd}^3 + 15.09 \times \text{No. of miles}$$

$$+ \begin{bmatrix} \text{yd}^3 \text{ pumped} & \$ \\ 15 < 25 & 56 \\ 25 < 60 & 104 \\ 60 \text{ or more} & 144 \end{bmatrix} \tag{7.5}$$

For example, a job is 5 miles distant from the equipment's home location and requires 55 yd³. The cost is $436.49 ($=128.46 + 0.966 \times 55 + 15.09 \times 10 + 104$). With an indirect cost percentage of 11%, the final job cost is $484.50 ($=1.11 \times 436.49$). Profit calculations are applied at this point.

For a subcontractor depending exclusively on this equipment business, profit is a markup to these calculations. If the equipment is a part of the general contractor's work equipment, this and other costs are eventually summed for the work estimate. These equipment costs become a part of the unit-quantity calculation.

This analysis for equipment cost is static in time. Other analyses for the recovery cost of equipment are covered in Chapter 10. These more advanced analyses are dynamic and cover time-value-of-money concepts.

7.7 SUBCONTRACTOR WORK

Subcontractor costs are prominent in any construction work estimate. Indeed, subcontractors' cost may be 60%–80% of some bids. A general contractor may solicit one or more subcontractor bids, and the contractor will select the better bid for the total contract bid. The lowest bid is not always selected, thus the term "better." Actually, most general contractors behave as subcontractors during some stage of their business. The general contractor gives emphasis to the cooperative willingness to perform the work according to the schedule. A general contractor may have a stable of subcontractors that have worked effectively.

The competitiveness of the contractor's bid depends on the cost effectiveness of the subcontractor. Assuming each of the subcontractors is qualified to do the work, the general contractor awards the contract to a lower bidder. In effect, the contractor engages the subcontractor in the same fashion that an owner uses to engage the general contractor. Moreover, it is common for subcontractors to engage other specialized subcontractors.

Subcontractors or general contractors use many of the same estimating methods. The unit-quantity method is found in both business arenas. One difference, however, is that the general contractor will include overhead costs of managing the subcontractors in the bid.

Revisit the unit-quantity Table 7.2, where $i = 7$ for Fig. 7.2. Notice that 7.5 yd³ of 4000-psi concrete is necessary for this design, and the row total is $5144, which is the costliest element of the estimate of Table 7.2. The contractor may want to evaluate a concrete pumping truck instead of the crane, bucket, and crew of five for pouring the concrete. The cost for concrete pumping is given by Eq. (7.5). It includes the truck and truck operator.

The labor for concrete loading into the forms is not included with the pumping costs. A crew of four is the manning for this method of concrete delivery. From Chapter 2 and Eq. (2.5) a linear equation for productivity per job is given as *time per job* = $1.182 + 0.017$ (yd^3). For 7.5 yd³, the time per job is 1.3 hr. The unit cost for the crew, called $R_{27} = \$80/\text{hr}$. Subtotal for $i = 7$ of $n_7 H_{72} R_{72} = \$104$. This information is entered in Table 7.12.

TABLE 7.12 Unit-quantity tabulation including subcontractor costs.

Work and Material Takeoff	Quantity, n_i	$j=1$ Materials Unit Cost, R_i	Subtotal, $n_i R_i$	$j=2$ Labor Unit Man Hours, H_{i2}	Total Man Hours, $n_i H_{i2}$	Unit Cost, R_{i2}	Subtotal, $n_i H_{i2} R_{i2}$	$j=3$ Equipment Unit Hours, H_{i3}	Total Hours, $n_i H_{i3}$	Unit Cost, R_{i3}	Subtotal, $n_i H_{i3} R_{i3}$	$j=4$ Subcontractor Concrete Pumping Sub-Total	Row Total
$i=1-6$													
7. Concrete 4000 psi	7.5 yd³	78.50	$2605		1.3	80	$1524				$3	$352	$4132
			589				104						1054
Column total			$3194				$1628					$352	$5186
Cost per yd³			$426				$217				$1	$47	$691
Cost per lin ft			$126				$64					$14	$205
Cost per ft² of form area, two sides			$7				$3					$1	$11

The subcontractor, a concrete-pumping truck, will bill the cost, and assuming a minimum requirement of 15 yd^3, ten-mile trip distance, indirect cost rate of 11%, and a profit markup of 8%, the cost (or the subcontractor's price) as entered on the unit-quantity table is $352 $\big($ = (128.46 + 0.966 \times 15 + 15.09 \times 10)1.11 \times 1.08$\big)$. Row total for $i = 7$ is $1054, which is a reduction over the $5144 line total in Table 7.2. The principle of the unit-quantity method allows the entry of concrete-pumping subcontractor values as $j = 4$. Additional subcontractors can be added as $j = 5, 6, \ldots$. The employment of a subcontractor to pour the concrete with special equipment is seen as advantageous to the cost of the work. The unit-quantity method allows this straightforward comparison. Subcontractor costs can be subsumed into the unit-quantity method.

7.8   DATA-WAREHOUSE EXAMPLES

Prominent in construction-work estimating are data sets, or books, for wide-ranging construction analysis and cost estimating. These publications go back a long time, perhaps to 1906, when the first attempts were made to provide estimating data. These early narrative reports described the design, job, site conditions, and labor, material, and equipment costs and performance. Those authors knew at first hand what had happened and they felt professionally obligated to report their personal experiences. The narrative descriptions were lengthy, and with the construction field becoming large and diversified, brevity became necessary. In the place of these articles, handbooks emerged, filling a significant void and providing convenience to many professional people. From those early first beginnings, pioneering compilers, such as Halbert Gillette, Frank I. Walker, and Robert Snow Means, codified these experiences into tabulations that made sense with parsimony of space.

Publication of these data sets is oftentimes annual, but may be more frequent. Some manuals, such as a "factor adjustment manual," are revised quarterly. At least 25 comprehensive data sets are commercially available, and overall annual sales amount to over 100,000 units. In addition, many trade-estimating manuals exist for the subcontractor or handyman. Topics include carpentry, roofing, HVAC, interior costs, painting, electrical, etc.

These encyclopedias provide an abundance of information. Material, labor, equipment, and subcontractor costs are outstanding and extensive. Crew composition for the many tasks is tabulated. Man hours are listed for the many tasks. Some technical sketches are given. The "unit of measure" for the work and material is defined. Modifications for regional differences are provided. Typically, the 16 CSI MasterFormat divisions are followed in the organization of the information. Hundreds of pages and thousands of entries are presented by these data sets.

7.8.1   Means Building Construction Cost Data[4]

This publisher of estimating information offers several types of data sets to cover a variety of construction designs. Figure 7.3 is a sample of this information.

Using the 16 divisions of the master format system as developed by the Construction Specifications Institute (CSI), the work is initially identified. For example, 03 is the Concrete Division of the CSI numbering format. The number 031100 refers to "Structural Cast In Place Formwork."

---

[4] Phillip R. Waier, ed., *Means Building Construction Cost Data*, Kingston, MA: R. S. Means Company, Inc., 1999. The web address is *http://www.rsmeans.com*

031 | Concrete Formwork

031 100 | Struct C.I.P. Formwork

			CREW	DAILY OUTPUT	MAN-HOURS	UNIT	1995 BARE COSTS				TOTAL INCL O&P
							MAT.	LABOR	EQUIP.	TOTAL	
150	6200	Bulkhead forms for slab, w/keyway expanded metal									
	6210	In lieu of 2 piece form	C-1	1,500	.021	L.F.	1.01	.50	.02	1.53	1.92
	6215	In lieu of 3 piece form		1,380	.023	"	1.01	.54	.02	1.57	1.99
	6500	Curb forms, wood, 6" to 12" high, on elevated slabs, 1 use		180	.178	SFCA	2.15	4.14	.16	6.45	9.15
	6550	2 use		205	.156		1.14	3.64	.14	4.92	7.20
	6600	3 use		220	.145		.80	3.39	.13	4.32	6.40
	6650	4 use		225	.142	▼	.65	3.31	.13	4.09	6.15
	7000	Edge forms to 6" high, on elevated slab, 4 use		500	.064	L.F.	.36	1.49	.06	1.91	2.84
	7100	7" to 12" high, 4 use		350	.091	SFCA	.62	2.13	.08	2.83	4.17
	7500	Depressed area forms to 12" high, 4 use		300	.107	L.F.	.64	2.49	.10	3.23	4.78
	7550	12" to 24" high, 4 use		175	.183		.87	4.26	.16	5.29	7.95
	8000	Perimeter deck and rail for elevated slabs, straight		90	.356		8.50	8.30	.32	17.12	23
	8050	Curved		65	.492		11.60	11.45	.44	23.49	31.50
	8500	Void forms, round fiber, 3" diameter		450	.071		.41	1.66	.06	2.13	3.16
	8550	4" diameter, void		425	.075		.44	1.75	.07	2.26	3.35
	8600	6" diameter, void		400	.080		.77	1.86	.07	2.70	3.90
	8650	8" diameter, void		375	.085		1.26	1.99	.08	3.33	4.64
	8700	10" diameter, void		350	.091		2.11	2.13	.08	4.32	5.80
	8750	12" diameter, void	▼	300	.107	▼	2.78	2.49	.10	5.37	7.15
	8800	Metal end closures, loose, minimum				C	28			28	31
	8850	Maximum				"	146			146	161

Figure 7.3 An example from *Means Building Construction Cost Data*.

The crew is designated as C-1, a reference to the composition for the crew. The reference assigns three carpenters, one laborer, and three power tools. The daily output is the number of units of work produced by the crew in an eight-hour day. The man hours reflects the number of labor hours to produce one unit of work. A useful feature of this line item is the relationship of labor man hours per day, daily production, and man hours per unit. If two of the three factors are known, the third can be found. This relationship is given as:

$$(\text{MH/day})/(\text{Daily output}) = \text{MH/unit} \qquad (7.6)$$

For a crew of four, the man hours per day are 32, and with daily output of 1500, the man hours per unit becomes 0.021.

The unit of work is normally the same as the unit for material, as found from the quantity takeoff. The material cost is expressed in dollars per unit. The labor cost is also expressed in dollars per unit. If equipment is used, then the equipment cost is given in *quantity surveying* units. The sums of these values are called "bare," and the units of this total cost are in units of the line code. The final column gives the total cost, including overhead and profit. The overhead and profit multipliers for this conversion are adopted as "representative" by the R. S. Means Company.

7.8.2   Process Plant Construction Estimating Standards[5]

This system of estimating employs four books, which it calls *Process Plant Construction Estimating Standards*. Figure 7.4 shows one page of this encyclopedia.

The data warehouse provides labor wage rates for many cities, and these rates are codified by individual-worker craft, composite crew, location, and union and Davis-Bacon re-

---

[5] *Process Plant Construction Estimating Standards*, Mesa, AZ: Richardson Engineering Services, Inc., 1997, 4 vols.

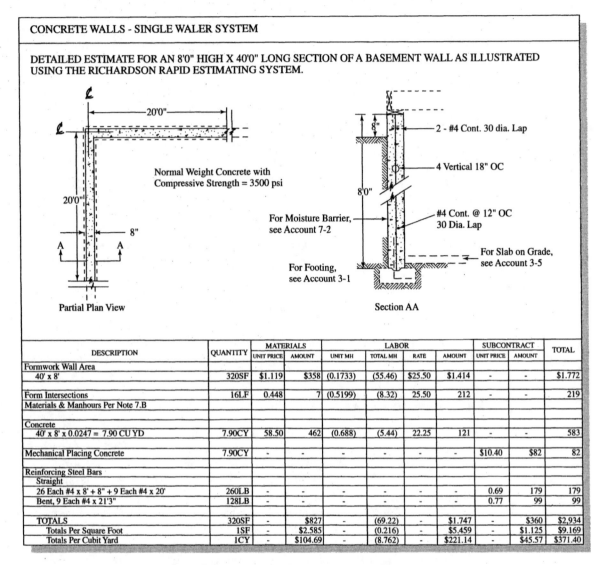

CONCRETE WALLS - SINGLE WALER SYSTEM

DETAILED ESTIMATE FOR AN 8'0" HIGH X 40'0" LONG SECTION OF A BASEMENT WALL AS ILLUSTRATED USING THE RICHARDSON RAPID ESTIMATING SYSTEM.

DESCRIPTION	QUANTITY	MATERIALS		LABOR				SUBCONTRACT		TOTAL
		UNIT PRICE	AMOUNT	UNIT MH	TOTAL MH	RATE	AMOUNT	UNIT PRICE	AMOUNT	
Formwork Wall Area										
40' x 8'	320SF	$1.119	$358	(0.1733)	(55.46)	$25.50	$1.414	-	-	$1.772
Form Intersections	16LF	0.448	7	(0.5199)	(8.32)	25.50	212	-	-	219
Materials & Manhours Per Note 7.B										
Concrete										
40' x 8' x 0.0247 = 7.90 CU YD	7.90CY	58.50	462	(0.688)	(5.44)	22.25	121	-	-	583
Mechanical Placing Concrete	7.90CY	-	-	-	-	-	-	$10.40	$82	82
Reinforcing Steel Bars										
Straight										
26 Each #4 x 8' + 8" + 9 Each #4 x 20'	260LB	-	-	-	-	-	-	0.69	179	179
Bent, 9 Each #4 x 21'3"	128LB	-	-	-	-	-	-	0.77	99	99
TOTALS	320SF	-	$827	-	(69.22)	-	$1.747	-	$360	$2,934
Totals Per Square Foot	1SF	-	$2.585	-	(0.216)	-	$5.459	-	$1.125	$9.169
Totals Per Cubit Yard	1CY	-	$104.69	-	(8.762)	-	$221.14	-	$45.57	$371.40

Figure 7.4 An example from *Process Plant Construction Estimating Standards.*

quirements. It provides indexes for producer prices of materials that might be used in construction. The labor and equipment information, often referred to as "standard," is controlled by a number of design specification statements. The data are compiled from a variety of sources, such as bid openings, news releases, construction newspapers, manufacturers, and builders. Sources and error measurements for the original data are not disclosed.

Information and data are categorized by work or trade discipline into main accounts. The main account numbers usually correspond to the Construction Specification Index (CSI).

The data set uses the word "standard," which is defined as a "forming a basis for comparison." The standard unit man hour involves these considerations: the work is being performed by a contractor who is familiar with all conditions at the job site; the project has the proper supervision; the workers are familiar with and skilled in performance of the work tasks; and there is an adequate supply of labor. There are stated restrictions for the application of the data. For instance, the data must be applied within a 50-mile radius of an urban area.

The data give the direct costs, which are defined as all costs of permanently installed materials and equipment and the labor required to fabricate, erect, and install the permanent materials and equipment.

7.8.3 National Construction Estimator[6]

This data set lists construction costs for general contractors performing the work with their own workers. Overhead and profit are not included specifically. The manual is divided into residential, industrial, and commercial divisions. Figure 7.5 shows one page of the encyclopedia.

The residential division is arranged in alphabetical order by construction trade and type of material. The industrial and commercial division of the manual are arranged by the CSI format.

Material costs are neither retail nor wholesale prices. They are price estimates for contractors, who buy in moderate volume. Delivery costs and sales tax are not included, but waste and shrinkage effects are considered. The labor cost is for installing the material. The labor cost per unit is the labor cost per hour multiplied by the man hours per unit. Labor costs include the basic wage, the employer's contribution to welfare, pension, vacation, and taxes and insurance based on wages.

7.8.4 Computer-Assisted Estimating

Manual estimates, made with pad and pencil and construction estimating information, are time consuming. Computer estimates were first made in the late fifties and early sixties. Then, large-scale central computers were used. Now, desktop and network PCs are the hardware for CAE. Estimating data are available on the internet, avoiding the requirement that manuals and CDs or floppies are placed nearby on the desk.

The use of software is the normal practice. Availability of software ranges from local software outlets to mail order to direct sale of proprietary software. CDs, floppy disks, Internet,[7] and service bureaus are available. Computer on-line services, web assistance, and consulting firms provide estimating. The adaptation of Excel or Lotus spreadsheets is possible for estimates that are constrained to narrow work scope. Engineering construction companies have developed their own internal estimating programs, and these programs receive continued care and nurture. The networking of computers together, where professional people have their own work stations, is a contemporary technological practice.

The extent of computer-assisted cost estimating varies. Simple programs ask for entries for various numbers—material takeoff, labor rates, and the like. Then these programs may calculate the cubic yards, square feet, and do the extensions and other calculations. Printing is a standard feature in these programs. A magnetic pen on the monitor

---

[6] Martin D. Kiley, *National Construction Estimator*, Carlsbad, CA: Craftsman Book Company, 1997.

[7] One source is Timberline Software, and their products can be reviewed on the web at *www.Timberline.com*. Additional locations are *www.WinEst.com* and *www.MC2-ice.com*, and others can be located through a listing given in the references.

	Concrete Formwork

Concrete Form Stakes. Costs per stake.

	13/16" x 13/16"	1/2" x 1"
12" long	1.65	1.65
18" long	1.75	1.75
24" long	1.90	2.20
30" long	2.45	2.65
36" long	2.80	2.80

	Craft@Hrs	Unit	Material	Labor	Total

Concrete Formwork. Multiple use of forms assumes that forms can be removed, cleaned and reused without being completely disassembled. Material costs in this section can be adjusted to reflect your actual cost of lumber. Here's how: Divide your actual lumber cost per MBF by the assumed cost (listed in parentheses). Then multiply the cost in the material column by this adjustment factor.

Board forming. For wall footings, grade beams, column footings, site curbs and steps. Includes 5% waste. No stripping included. Per SF of contact area. When forms are required on both sides of the concrete, include the contact surface for each side.

1" thick forms and bracing, using 1.64 BF of forming lumber per SF (@ $535 per MBF). Includes nails, ties, and form oil (@ $.20 per SF)

	Craft@Hrs	Unit	Material	Labor	Total
1 use	B2@.100	SF	1.12	2.44	3.56
3 use	B2@.100	SF	.61	2.44	3.05
5 use	B2@.100	SF	.50	2.44	2.94

2" thick forms and bracing, using 2.85 BF of form lumber (@$650 per MBF) per SF. Includes nails, ties, and form oil (at $.20 per SF)

	Craft@Hrs	Unit	Material	Labor	Total
1 use	B2@.115	SF	2.15	2.81	4.96
3 use	B2@.115	SF	1.08	2.81	3.89
5 use	B2@.115	SF	.84	2.81	3.65

Add for keyway beveled on two edges, 1 use. No stripping included

	Craft@Hrs	Unit	Material	Labor	Total
2" x 4" (@$580 per MBF)	B2@.027	LF	.41	.66	1.07
2" x 6" (@$610 per MBF)	B2@.027	LF	.30	.66	.96

Driveway and walkway edge forms. Material costs include stakes, nails, and form oil (@$.16 per LF)

Figure 7.5 An example from *National Construction Estimator*.

screen or a mouse with a tablet may be used to find measured quantities from the working drawings.

The higher-level programs receive the computer file from a working-drawing program exchange, then do a material takeoff, develop databases with material, labor, equipment, and overhead costs, subcontractor listings, issue purchase orders, bills of material, provide construction management support, and much more. Engineering construction companies have advanced programs along these lines.

Figure 7.6 shows a screen page of a software program. A mouse is used to make selections, and these are dragged from the database into an interactive spreadsheet.

Prices for commercial software vary from $25 to $250,000. Considerations affecting their value include quality, accuracy, features, serviceability, and time-currency.

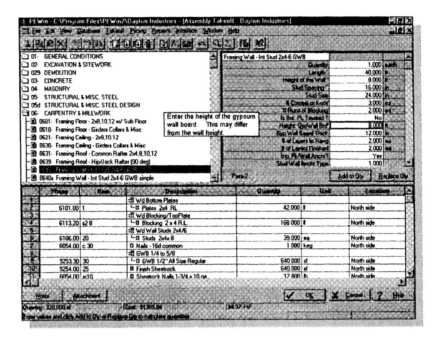

Figure 7.6 Example of a screen page of a computer software program for construction-cost estimating.[8]

7.8.5 Reflections[9] on Data Warehouses

Data warehouses, as popular as they are, need to be evaluated on four counts: accuracy, reliability, verifiability, and convenience.

Accuracy of the data set is not the only issue, as there is also the accuracy with which the data are applied. First, accuracy of the data (as a whole or individual elements) cannot be confirmed, as the original sources of measurement are unavailable, nor are any statistical measures given in the data set, e.g., standard deviation. If the data are accurate, is the estimator using the information as recommended by the compiler? Is there *consistency* (either close or far from the imaginable actual value), and are the applications ever compared to actual performance by the user? The answers to these questions are elusive.

The data set is *reliable*, as it gives the same numeric estimate each time in repetitive situations. Of course, technical and engineering estimating are open to individual interpretation, and the opportunity of comparing the estimates to jobs won provides a limited confirmation of accuracy and reliability. Jobs that are not won cannot be conveniently compared between the estimate and actual cost. Additional discussion of these anomalies is deferred to Chapter 9, "Bid Assurance."

What do the data warehouses say about their own qualifications? Their stated accuracy for these convenient manuals and software programs ranges from "you wouldn't bid the job with these data under any circumstances" to "a source book for preliminary project costs."

---

[8] Timberline Software Corporation, Beaverton, OR.

[9] The author's ideology is tempered by his own estimating data book and DOS software that have been used in commercial practice. A web portal for his cost information is found through www.Bonics.com.

It is the opinion of this author that a firm must develop its own information data set. With uncountable combinations of construction design and engineering, labor skill and effort, material, equipment, and procedures, there is a perception that data sets for estimating are more effective if they are developed and nurtured by the owner or contractor. Estimating data need to be verified by measurement, job tickets, one-cycle and repetitive-cycle time study, and historical experience, and they need to be refreshed continuously. Those data warehouses references are useful only as a secondary opinion if the information matches the future work content.

SUMMARY

In this chapter we considered work-estimating procedures. Important as other motives may be in setting the significant bid price, estimating starts with the work of construction. At this point in time all other purposes for the estimate are secondary. Understanding the elements of the work estimate—labor, material, tools and equipment and subcontract—is critical for the student. The next vital step uses the construction-work estimate as input cost information to the project estimate, which is studied in the next chapter.

QUESTIONS

1. Define the following terms:

   RFE
   Direct labor
   Direct material
   Indirect materials
   Unit-quantity model
   Equipment ownership
   Capital consumption cost
   Average investment life
   Data warehouse manuals
   Allocation of tool cost

2. List ways that a construction-work estimate is used. Give other ways that are not identified by the chapter.
3. What kinds of information are required for a construction-work estimate? To broaden your answer, refer to Chapter 1.
4. What role does the RFQ have in the management of cost estimating?
5. Describe the purpose of the unit-quantity model in cost estimating. How would you alter the model?
6. Do a survey of the items of equipment cost for the following: fixed and rigid scaffolds, tractor trailer, and a pneumatic nailer. Specify whether the cost item is necessary or unnecessary.
7. Find a web base site that lists construction RFQs.
8. Examine a web site on construction estimating. Discuss your impression with respect to accuracy, reliability, verifiability, and convenience.

PROBLEMS

7.1 **(a)** Simple hand labor of clay and rock removal is scheduled for the country of Mexico. The work requires light loosening with a pick prior to shoveling using a long-handled round-pointed shovel. Labor costs for this operation are assumed to be one-quarter of the costs for similar work in the United States, which is $15/hr. Determine the cost in international dimensional units and Mexican currency for volumetric earth removal. (*Hints*: The answer is expressed in units of peso/m^3. Refer to Chapter 1 for currency exchange rates. The work will be performed one year from now. Ignore the cost of simple tools.)

  (b) Repeat for Canadian work, where the wage rates are comparable to those in the United States.

7.2 Determine the cost of materials, labor, and simple equipment to construct forms for a concrete footing, shown as Fig. P7.2. Excavation for the footings is not required. Form ties are not required to satisfy any loading requirement. The crew consists of a carpenter and helper. Portable electric/pneumatic equipment includes saws, nailers, staplers, etc. Ignore the cost of the footing ends, nonpowered tools, reinforcing bars, and concrete. Your answer will be the total cost of the form and the cost per ft^2 of area in contact with the concrete. What is the cost, using the parameters as developed by the unit-quantity method, for similar forms that are 30 ft long? For 12 ft?

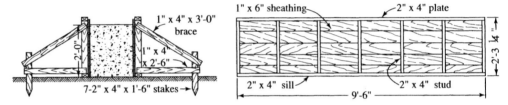

Figure P7.2

(*Hints*: Costs for lumber are $150 per 100 bft. Nail requirement is 1 lb/100 bft and the cost is $0.50 per lb. Portable electric/pneumatic equipment prorated first-cost and operation contributes $1 per hour, and its lump-sum capacity is rated at 400 bft/hr for all lumber.)

7.3 Estimate the cost of materials, labor, and simple equipment to construct a panel-form prefabricated of construction-quality plywood away from the job site. The design is given as Fig. P7.3. Find the total cost of the panel and the cost per ft^2. (*Hints*: Oil-coated plywood is priced at $1/$ft^2$. Costs for lumber are $150 per 100 bft. Corner angles cost $0.75 each. Nails are $0.50 per lb. Portable pneumatic/electric equipment prorated first-cost and use contributes $1 per hour of operation, and its capacity is rated at 400 bft/hr for all lumber. The crew consists of a carpenter and helper. Labor time for the carpenter and the helper is 25% of field time. The panels are reused a total of four times.)

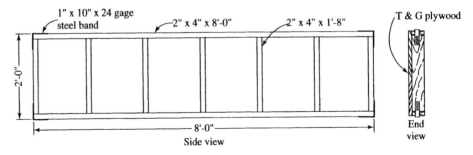

Figure P7.3

7.4 Now reconsider Fig. P7.2 for an operation of setting steel reinforcing bars inside the footing. The horizontal reinforcing steel for the wall consists of pairs of No. 5 bars spaced 4 in. above and below the bottom and top surface. The horizontal bars extend beyond both wall ends 12 in. Vertical No. 5 S-hook bars tie to the horizontal bars and are spaced 24 in. on centers. They extend 6 in. above the surface of the wall concrete after the concrete is poured, and their developed length is 30 in. A crew of a steel setter and common laborer is scheduled. Find the total cost of the reinforcing. (*Hint*: Assume tie wire for the bars is included in the bar cost.)

7.5 Determine the total cost to place and cure ready-mix 3000-psi concrete for the foundation shown by Fig. P7.2. Additionally, find the cost per cubic yard. (*Hints*: Assume truck delivery and chute methods for the placement of the concrete. Two common laborers are scheduled. Hand tools are rented at $65 for the period of work. Let the waste be 2%.)

7.6 Estimate the total cost of a concrete mason and a helper to mix, haul, and cure the concrete required for Fig. P7.2. Additionally, find the cost per cubic yard. Assume low-technology hand-mixing methods. (*Hints*: A 28-day compressive 3000-psi strength concrete is composed of 6 sacks of cement, 39.1 gal of water, and 1260 lb of fine dry aggregate for each cubic yard of concrete. Each sack is valued at $11. Anticipate 10% waste. An allowance for water and aggregate is $125 for the job. Hand tools, buggies, etc. are rented at $205 for the period of work. Weather is ideal. Crew elapsed time is set by longest member of crew.)

7.7 Estimate the direct cost of labor and materials for the construction of a double-sided straight retaining wall 48 ft long, 10 ft high, and having a 10 in. interior dimension for the concrete. Refer to Fig. P7.7. A panel-form, prefabricated of construction-quality $\frac{3}{4}$-in. plywood away from the job site, is designed and available to be used for this requirement. The panel is expected to give four applications. Your answer will be the total cost for the retaining-wall structure, and the cost per ft² of area in contact with the concrete surface. Cost of concrete and any excavation is not required. The wall is resting on a footing that is in place.

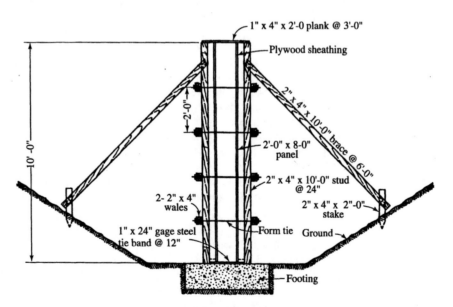

Figure P7.7

(*Hints*: Each panel is 2 × 8 ft and, prorated for the four applications, costs $100 each delivered to the job site. Treated Douglas fir construction-quality studs cost $6.89 and $7.89 each for 10- and 12-ft lengths. Life of the studs is four applications. Form ties and end clamps are installed every 4 ft², and reuse of the tie assemblies is not planned. Galvanized medium-carbon steel tie bands, sheared to length and width, cost $1.28 each. Top planks are $0.85 each. Oak stakes cost $0.65 each and are destroyed in the one application. Provide an allowance of $50 for nails and other fasteners. A scheduled crew consists of a carpenter, helper, and a foreman for 50% time. Ignore portable-tool equipment cost.)

7.8 Determine the total cost for reinforcing bars and 4000-psi concrete for Fig. P7.7. (*Hints*: The 10-in. width of wall uses No. 6 horizontal bars that are 25 ft long. The three horizontal pairs are tied to vertical No. 4 bars, which are 10 ft 9 in. high. Pairs of vertical bars are spaced every 48 inches. A crew consists of a steel setter and common labor. The concrete order is increased by 3% to cover expected slump and waste. This company, which annually uses over 750 yd^3, negotiates a favorable concrete cost. The concrete is poured directly into the cavity from the truck with the aid of a chute A crew of four common laborers and a foreman is scheduled for full-time attention of the concrete pour.)

7.9 (a) Estimate the total cost of labor, equipment, and 3000-psi concrete for a slab floor 100 ft long, 60 ft wide, and 6 in. thick. The floor is 16 ft above ground level. The concrete is delivered to a 1-yd^3 bucket, hoisted with a crane, and deposited into a 1-yd^3 hopper, then hauled in buggies and spread to the thickness. Concrete is delivered at the rate of 15 yd^3/hr. (*Hints*: An additional 1% concrete is ordered. The crew is 1 crane operator, 16 laborers, 1 carpenter, 3 concrete masons, and 1 foreman. Equipment rental and prorated costs are $215/hr.)

(b) Repeat with a concrete-pumping truck replacing the crane-and-buggy method. (*Hints*: The crew is 8 laborers, 1 carpenter, 2 masons, and 1 foreman; average pumping capacity is 125 yd^3. The concrete-pumping truck is faster than the crew, and the labor crew will require an additional two hours after the conclusion of the concrete pumping. The job site is 8 miles away from the home station of the equipment. Indirect costs are 11%. There is no rental of equipment, other than the use of the concrete-pumping truck.)

7.10 A 10,000-ft^2 parking lot is to be estimated from a unit-quantity table. The fine-grading crew and equipment need 8 hours at $280 per hour for the total area. The building of forms using a forming crew and their equipment requires 12 hours at $252 per hour. To complete pouring and finishing, a crew requires 6 hours at $296 per hour. Reinforcing material of 10 × 10 in. mesh and No. 4 bars (12 in. each way) are $0.09/ft^2 and $1.10/ft^2 for material and installation. Requirements and cost of concrete are 200 yd^3 at $61/yd^3. Form material, $\frac{1}{2}$-in. premolded expansion joints, and concrete-curing material costs are $500, $112, and $172. Thirty-nine precast concrete bumpers are $18 each, and 840 ft of 4-in.-wide painted lines are estimated to cost $0.25/ft. What is the job cost? Find the cost per ft^2. (*Hint*: Construct a simplified unit-quantity table considering material and subcontractor costs.)

7.11 Initial cost of equipment is $48,000. The estimated useful life is 5 years, and there is no salvage value. Find the unrecovered capital value at the end of the second year, using an equal yearly capital-consumption cost method. Next let the equipment have a salvage value of 10% of the original cost at the end of year 5, and find the unrecovered capital value at the end of year 4.

7.12 Initial cost of a 2-yd^3 160-hp diesel-engine-powered crawler-type power shovel at the factory and its shipping cost to the buyer is $325,000. It estimated useful life is 6 years, and there is no salvage value. Find the annual yearly capital cost consumption. How much capital value needs to be recovered after two years of operation? Find the unrecovered capital value at the end of the second year. Next let the equipment have a salvage value of 10% of the original cost at the end of year 6, and find the unrecovered capital value at the end of year 2. (*Hints*: Use an equal yearly-capital-consumption cost method.)

7.13 What is the average annual investment cost for a tractor power shovel having an initial full cost of $325,000, and a useful life of 6 yr? There is no salvage value. Repeat for a salvage value of 10%. Use an investment cost of 14% of the average investment value of the tractor.

7.14 What is the average annual investment cost for a 325-hp diesel crawler tractor where the delivered cost is $265,000. Equipment usage is planned for 6 years, after which it will salvaged for 10% of original value. Cost for the average value of the investment is 16%.

7.15 Find the operating cost of fuel consumed per hour in operating a power shovel with a diesel engine. The diesel engine is rated at 160 hp. During a 20-second cycle, the engine is operating at full power while filling the dipper, which requires 5 seconds. During the remainder of the cycle, the engine is operating at 50% of rated power. Also, the shovel is operated 45 min/hr on the average. (*Hints*: A diesel engine consumes approximately 0.04 gal of fuel for each actual horsepower-hour developed. Diesel fuel costs $1.77 per gal at the job site.)

7.16 Find the cost per hour of owning and operating a 2-yd^3 160-hp diesel-engine-powered crawler-type shovel. The price at the factory together with its shipping cost to the buyer is $325,000. Equipment usage is planned for 6 years, after which it will be placed into secondary service. There is no salvage value. Use an equal yearly capital consumption cost method. Cost for the average value of the investment is 15%. Maintenance is assumed to equal 100% of depreciation. Ignore cost of labor and crew operation. (*Hints*: Hours per year, 2000; crankcase capacity, 8 gal; hours between oil changes, 100; operating factor, 0.6; diesel consumes 0.04 gal of fuel per actual horsepower-hour developed; lubricating oil consumed per hour, 0.15 gal; cost of diesel fuel, $1.85/gal; cost of lubricating oil, $7.25/gal.)

7.17 Find the cost for concrete-pumping equipment if the job requires 1500 yd^3 and is 5 miles away from the home station of the equipment. (*Hint*: Use the model developed in the chapter.)

7.18 Find the cost equation for pumping equipment using the chapter information, except with the following changes. The equipment costs $1.5 million, major overhaul is $200,000 and the average pumping rate is 150 yd^3/hr. (*Hint*: Overlook the indirect cost, that being applied to the job cost, once the RFQ becomes known.)

MORE DIFFICULT PROBLEMS

7.19 Estimate the direct cost of labor, materials, and simple equipment for the construction of a double-sided straight retaining wall 125 ft long, 15 ft high, and having a 12-in. interior dimension for the concrete. The wall will be resting on a footing that is in place. Oaks stakes to secure bracing are placed in firm earth. Reinforcing steel bars are required. Horizontally, the design calls for 25 ft 3 in. No. 6 pairs at 2-, 7-, and 13-ft stations. Vertically, No. 5 paired-bars are tied every 10 ft. The vertical bars are 16 ft long. Do not include the cost of the concrete and its placement. Find the total cost for the retaining wall structure, cost per yd^3 of concrete placed, and the cost per ft^2 of area in contact with the concrete surface.

(*Hints*: Use 1-by-6-in. tongue-and-grooved matched pine with rough and smooth sides, 2-by-4-in. lumber for studs and wales. Braces are 10-ft studs and are required every 4 lin ft, both sides. The oak stakes cost $0.65 each. Assume that lumber is used two times, which allows the overlooking of waste for this problem. Costs for lumber are $150 per 100 bft. To overcome the pressure of the concrete, the spacing for the vertical studs is 20 in. and for the horizontal wales is 25 in. Spacing of ties is 30 in., and the area per tie is 5 ft^2. Nails are $0.50 per lb. Portable electric equipment prorated first-cost and use contributes $1 per hour of operation, and its capacity is rated at 400 bft/hr for all lumber. A scheduled crew consists of a carpenter, a helper, and a foreman for 50% time for the forms. A reinforcing crew consists of a steel setter and a common laborer.)

7.20 A 2000 × 25 ft road is to be constructed in a remote region. Cut and fill are necessary, and 5000 ft^3 of roadbed and dirt will be hauled an average 200 ft distance. The road will have a 4-in. aggregate base and 2 in. of asphalt thickness. Excavation will require a bulldozer that can displace 145 ft^3/h, and two are necessary for a total of $230 per hour. Hourly labor rates for this excavation are foreman, $30 per hour; three laborers, $45 per hour; truck driver, $21 per hour; and grade checker, $26 per hour. A water truck and a packer-roller will be used at $68 per hour and $138 per hour for 5 hours. Grading and asphalt laying will require: fine grading and crew, $848 per hour for 12 hour; base equipment and crew for 10 hours at $780 per hour; paving equipment and crew 6 hours at $848 per hour; prime equipment and crew for 5 hours at $446 per hour; and seal equipment and crew for 8 hours at $332 per hour.

Material requirements for 50,000 ft^2 of road are given as:

Material Takeoff	Quantity, tons	$/ton
Asphalt	619	50
Seal oil	4	34
Sand	28	16
Aggregate base	1233	18
Prime	170	34

Determine total cost, cost per 100 linear ft, and cost per ft^2. (*Hint*: Prepare an abbreviated unit-quantity table.)

7.21 Find the cost per hour of owning and operating a 325-hp diesel crawler tractor. The price at the factory together with its shipping cost to the buyer is $265,000. Equipment usage is planned for 6 years, after which it will be salvaged for 10% of original value. Use an equal yearly capital consumption cost method. Cost for the average value of the investment is 16%. Maintenance and repairs are assumed to equal 75% of the annual capital recovery cost.

(*Hints*: Hours per year, 2000; crankcase capacity, 9 gal; hours between oil changes, 120; operating factor, 50% and a 45-min hour; diesel consumes 0.04 gal of fuel per actual horsepower-hour developed; lubricating oil consumed per hour, 0.15 gal; cost of diesel fuel, $1.85/gal; cost of lubricating oil, $7.25/gal).

7.22 Develop the cost equation for concrete-pumping equipment. Use the chapter information, except with the following changes. The equipment costs $1.5 million, major overhaul is $200,000, average pumping rate is 150 yd^3/hr, salvage is 17.5%, and indirect costs are 4%. Include the indirect costs in the cost equation.

PRACTICAL APPLICATION

There are many opportunities for a practical experience in construction work estimating using the objectives of this chapter. Initially, form a team to call and interview contractors and suppliers. Your instructor will give additional and refining instructions.

Your team is to show comparisons of the cost of concrete placement for several technologies. Expand the methods discussed in the chapter. Ask subcontractors for courtesy quotes and visit the job site or plant. Interview the management about these costs. Graph the methods with *cubic yard* specified as the *x* axis and *cost of concrete placement* as the *y* axis. Discuss the economical choice of concrete placement on the basis of your facts. Conclude your practical application by word-processing a team report of your findings.

(*Hints*: Establish your constraints for the survey, such as, does the cost include the material or just the placement? A concrete-pumper truck will have a range of costs, which depend on the amount of concrete pumped. This becomes one curve on the graph. Other technologies will have their own separate cost curves, which will decline with increasing volume.)

CASE STUDY: ESTIMATING A SMALL DESIGN

This chapter concentrates on work estimating. Under the direction of your instructor, select a minor design from your textbooks that is of interest to you, or the instructor will provide a design. Selection of a sketch or problem in this text is a possibility. Use an estimating manual found in the library to estimate the design. Estimate the design for your area. Identify the labor, material, and equipment costs. It is only necessary to find the cost, as profit and overhead will be considered in the following chapter.

(*Hints*: Be thoughtful in selecting the design. Keep it physically small with few materials, as detail estimates tend to become lengthy and time consuming. The estimate may use either "pad and pencil" or a computer. Alas, despite the tedium of doing pad-and-pencil estimates, you will learn more this way than by doing a computer estimate. Moreover, it is unimportant if your estimating manual is not the latest year, as costs can be indexed to the current year. Review the material about indexes in Chapter 4. Your instructor will clarify the means of estimating.)

(*Suggestions for the instructor*: An interesting gambit is the assigning of the same design to the entire class, and then making comparisons among the answers. Each student will have a different answer. For instance, you can statistically calculate the coefficient of determination, σ/μ, which is a measure of the variability to the mean.)

CHAPTER 8

Project Estimating

Techniques for estimating projects are strikingly similar despite large differences in the engineering design. Nor do the principles differ significantly for projects ranging from thousands to many millions of dollars. Further, the principles given in this chapter are germane for either an owner or a contractor.

The student will want to avoid the mind set of data-warehouse estimating, as was stressed in Chapter 7. Our presentation uses a graphical approach to illustrate cash flows, rather than the methods of encyclopedia lookup, and we explain the consequences of those maneuvers. To simplify project-estimating considerations into a short tabular description is not helpful.

We deal initially with the elements of the project cost. Then a work package is introduced. Along the way we estimate direct labor, direct and subcontract material, facilities and equipment, and engineering, which are critical elements. Subsequently, overhead, contingency, interest, and profit are calculated through cost and bid analysis. Once these procedures are complete, we arrive at the bid.

The project engineer establishes the definition and work breakdown structure. Organizational policy may require that other company units assist the engineer in this work. Complicated projects require a team. There is no single best approach, because many styles can be successful.

Project estimating uses specialized software solvers in finding the bid value, as general software may be too limited for a broad range of project types. Some firms have developed proprietary software able to estimate and track project costs.

8.1 PROJECT BID

A project bid is initiated by a variety of sources. An owner may instruct his own engineering and estimating staff to begin preparing a bid on a project, which preliminary studies may have shown to be economically favorable. Similarly, an owner (or his architect under the owner's instructions) may distribute a request for quotation (RFQ) to contractors, which will set their efforts into motion. Various governmental agencies seek construction services. Other opportunities may suggest the development of a business prospectus, entailing a project bid as one step in the overall planning.

Very simply, a *bid* comprises the cost and profit that a project requires. The value is monetary and expressed in the currency of the country. Sometimes a bid is known as a quotation, proposal, investment, price, or even cost, because terms are used loosely in practice. Jargon for a bid is a "grass roots" or "greenfield" estimate.

Cost elements of a project bid vary widely. Their importance may differ between a road, plant, electrical transmission line, a building, etc. Owners or contractors may disagree as to their selection. Thus, we expect diversity in the methods of organizing and selecting the cost elements of a construction bid. Symbolically we define project cost as given by the layers in Fig. 8.1. Not all the layers may be required, depending on the design and the purposes for the bid.

A project-cost estimate includes the Fig. 8.1 layers from construction work to contingency. Once profit is considered, we have the bid. Note that the language of bidding includes the terms *price* or *quote*.

The bottom layer identifies the estimate of construction work. The principles of this estimate were given in Chapter 7. We saw that a *construction work* estimate is composed of direct labor, direct materials, tools, equipment, and subcontractor labor and materials. Direct labor is the work that transforms materials from one state to a value-added condition. Material required for the construction of the design is direct material and is installed by the bidder's work force. Equipment costs for construction work can be calculated specifically for the job or can become a part of the overhead. Subcontract materials, unlike direct materials, entail installment by subcontract labor. The subcontract will include both the material and the subcontract labor. The student may want to review the pertinent material in Chapter 7.

Initially, in a construction-work estimate, *direct labor* is expressed in various time units, such as man hours, man months, or man years. An individual or crew wage multiplies the units of time, leading to direct cost. Previous chapters have dealt with this computation. If those costs are identified as *allowed*, or *standard*, then an adjustment by *efficiency* becomes necessary. The efficiency adjusts allowed time to *actual time*. Productivity factor varies with season, location, worker experience, and so on. Review Chapter 2 for additional principles.

Direct materials are subdivided into raw, standard commercial items and subcontract items. They appear in the design or are necessary for construction. The direct material

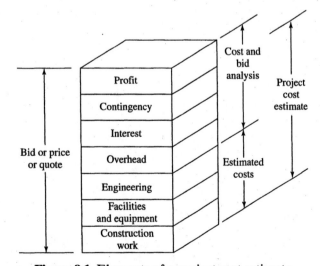

Figure 8.1 Elements of a project cost estimate.

amounts must be increased for losses stemming from waste, scrap, and shrinkage. Chapter 3 deals with material resources and costs.

Facility and equipment cost is a term that can be broadly defined. Examples for an uranium-ore-processing plant would be the pieces of capital equipment, such as conveyors, tanks, and rod mills. A high-voltage transmission line requires a *field office* or *facilities*. Equipment for the construction of the transmission line is used for other lines. In this case, equipment is an overhead charge. Indirect materials and indirect labor are other costs conveniently handled by overhead.

Engineering costs are those incurred for design, specifications, or reports. Included are the salaries and overhead for engineering, administration, CAD, estimating, and drawing reproductions.

Contingency costs are anticipated in situations where prior experience and data are lacking. Projects requiring extraordinary research, development, and design are the best candidates for contingency. Unfortunately, contingency is sometimes a cover-up for poor estimating practices.

Because projects are a large financial undertaking, interest costs are usually charged against the contractor or owner while construction is in progress. Those *interest* charges, which can be substantial, are a part the cost of doing business and are included in the estimate.

The layers in Fig. 8.1, which vary from the construction-work estimate to contingency, are the elements of a project-cost estimate. Profit is calculated upon these items. The sum of cost and profit constitutes a *bid*.

The remainder of this chapter uses the foregoing terms and definitions. The foundations of project estimating rest upon them and upon a special organization for the project. This project organization is called the "work package," and the student needs to know its principles, which we discuss next.

8.2 WORK PACKAGE

A work package is a vital part of a project development. A number of routes may be taken to estimate projects, and Fig. 8.2 is suggestive. One likely way is a synchronized maneuver of several of the steps, though a project engineer develops this style only after experience. Our approach to project estimating coordinates with the modern concepts of the work package.

The proposal plan gives the technical statement, preliminary designs and estimate, and the work breakdown structure. An encouraging preliminary plan leads to additional effort. This results in the RFQ, which initiates the work. Specifications and working drawings are the principal technical information for construction-work estimating, as described in Chapter 7. A project, however, may have significant nonhardware costs. The project design, by itself, is insufficient for estimating; other information must be consolidated with it.

The collected project information is called a *work package*. "Forms"[1] included within a work package are the definition, designs, work-breakdown structure, schedule, and estimates. If the project is large, then the amount of digital information in a work package is very extensive. Even small projects require a thick notebook or large computer filing.

---

[1] This textbook does not show various samples of "forms." Current practice dictates that organizations adopt *paperless* operations and take advantage of the digital computer wherever possible. Our reference to "forms" does not imply *paper* forms, but rather the frames displayed on the monitor of the computer.

PICTURE LESSON
Interstate Highway System

The Dwight D. Eisenhower System of Interstate and Defense Highways is the most successful and impressive highway system ever constructed. President Eisenhower, impressed by the German autobahn, signed the Act in 1956 creating the highway system. With over 42,000 miles constructed, it set superhighway standards.

One of the last segments brought into the system is the Glenwood Canyon of I 70. Located 160 miles west of Denver, Colorado, Glenwood Canyon is a rocky gorge carved by the Colorado River, in some places 2000 feet high. Before construction, the river shared its banks with a railroad line, the old two-lane U.S. Highway 6, and a small hydroelectric generating plant.

The beauty of the canyon compelled special treatment of the construction and design. Environmental impact statements, public hearings, and surveys beginning in the 1960s eventually involved more than 64 federal, state, and local agencies. Requirements emerging from the numerous hearings were that the canyon be revegetated to restore its former pristine natural state and that the highway blend well with the mountain backdrop and have pleasing aesthetics. Rules for the contractors included fines for needlessly destroying vegetation.

Approximately half a billion dollars were spent for engineering design, right of way and environmental work, construction, and rest areas and revegetation. Overall, about 91% of funding came from the federal government, who in turn taxed the purchase of gasoline, with the money flowing into a trust fund set up for this very expenditure.

The picture shows an overhead gantry crane moving a road section into location. The advantage of the gantry crane is placement of road sections from the top, unlike other lifting methods which hoist from the river bed, or cantilever techniques which add successive lengths from the roadway. The divided highway feeds into twin 4000-ft tunnels. The tunnels use 24-hour television monitoring, variable message signs, AM/FM rebroadcast system, weather sensors, and roadway-ice, heat, and fuel-oil-leak detectors. Viaducts, bridges, and tunnels comprise over half of the 12-mile length.

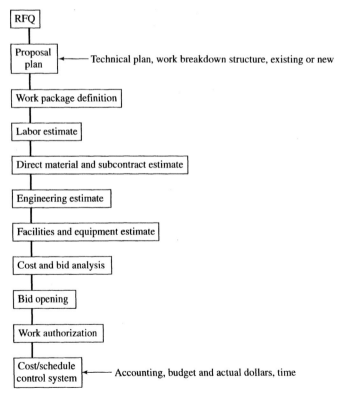

Figure 8.2 A flow chart for estimating an integrated project within the work-package concept.

A *project definition* is a planning form that identifies what is to be done, when, and by whom. Typically, a numbering system connects the definition to other forms. The definition indicates a baseline initiation date. The project definition, then, gives a workable scheme to achieve the project goal.

Design engineering is responsible for designs that satisfy the project goals for a cost, which is to be found in the WBS process. Design efforts assist the definition that translates performance objectives into a project design. While this sounds like a chicken-egg process, it is the pattern. There is variety to these schemes, and we give a glimpse of the possibilities. So a project engineer will take an overall project definition and assign packages of the work breakdown structure to other cognizant engineers of the proposal team.

The definition subdivides the project into smaller-scaled tasks, and eventually it defines construction for physical items. For example, a reinforced concrete wall may be one task, or the task may be divided into erect outside forms, tie reinforcing steel, erect inside forms and bulkheads, pour concrete, remove forms, and clean. This redefinition depends on the size, complexity, and effort. Simply, "It all depends." Giving full weight to these three words, we recognize that flexibility, knowledge, and skill are key factors in project development. For the remainder of this chapter we assume that the definition phase is complete and that the cost estimate remains to be prepared.

The *work breakdown structure* (WBS) is a graphical display of the project. It results from subdivision of the project into logical components, or *work packages*, that are arranged into a treelike chart to allow visibility and analysis of a single component. Work packages can also be grouped into larger tasks or the end item itself. A construction project may be viewed as a hierarchy of work packages, the lowest or most detailed level being called an elemental work package or element. The WBS is used for estimating, planning, and performance measurement and control. In some cases, accounting charges are linked to the design by means of WBS.

Application of the WBS is practiced by owners, contractors, and agencies of the government. Indeed, it was the Department of Defense that developed the WBS concept into the styles that are prevalent today in construction. However, subdivision of a project into smaller, less complex, and more manageable tasks is not new, having been basic to construction for a century or more.

The WBS differs from a bill of material (BOM). The BOM identifies part numbers of hardware. The WBS also shows hardware, but not to the detail of a BOM. The WBS will show nonhardware costs. The work breakdown structure is composed of hardware, software, technical and management services, and other work tasks. In some instances, it is useful to have specifications and drawings conform to the numbered WBS with drawing notes that are placed directly on the title block of the drawing.

Figure 8.3 is an example of a six-level WBS. The title of the project is Creston 500 kV Transmission Line. Levels are used to specify the WBS. Level 1 is the entire project. Level 2 is the major elements, such as land, engineering, transmission line, etc. Each lower level adds another digit, and 1322 is a fourth-level "transformers and control panel." Numbering of the WBS corresponds to the definition and the estimates. The definition is the source of the original WBS number system.

After the project summary WBS is formulated and numbered, individual contract WBSs are used for procurement actions for subcontractors and vendors. For example, the owner may issues RFQs for 1321, "Building," as it can be handled independently of other WBS items. The subcontractor may extend the WBS to lower levels as the basis of another RFQ.

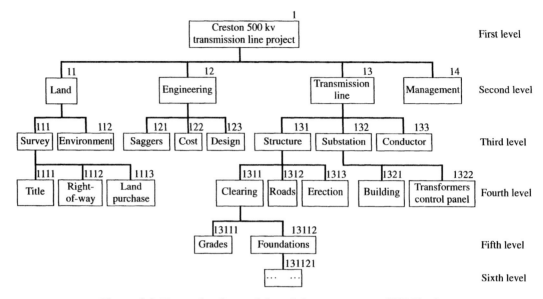

Figure 8.3 Example of a work breakdown structure (WBS) of a project to the sixth level.

When the WBSs for the subcontractors are attached to the summary WBS, the project WBS is created. The project-summary WBS and its derivatives are used throughout contract definitions, design, construction, and operation for technical and engineering activities. Reports of progress, performance, and financial data are often based on the project WBS. These thoughts are amplified in Chapter 9.

Schedules are an important part of the work package. Though the definition is the planning document and notes what must be done, scheduling determines the calendar dates for the start and conclusion of the WBS activity. Details of scheduling, such as CPM (critical path method), PERT (program review and evaluation technique), or other network methods, are considerable, and for further information[2] we refer the reader to the many excellent texts available.

Instead, look at Fig. 8.4, which symbolically shows the integration of the WBS, schedule, and costs. In Fig. 8.4(b), open and solid triangles indicate the start and conclusion of the major task. Each task can be similarly scheduled. The length of the bar indicates the time requirement and not the cost. The fourth level of the WBS is estimated at $5, $10, $5, $10, and $5 for each task. The third-level summary is $35. Two other third-level tasks are estimated at $20 and $25 for a second-level summary of $80. The first-level summary is $300.

If we add period designation to the horizontal axis, we match the costs for levels of the WBS as shown by Fig. 8.5. In Fig. 8.5(a), the $300 bid is broken down for the WBS. The work packages are homogeneous from a project point of view. For example, work package #1 can be owner-built, while work package #2 can be let for contract, and they will have technical and management oversight. In turn, Fig. 8.5(b) shows the *baseline* defined as the time-scheduled cash flow for the estimate. This WBS cash flow can be for the owner, major contractor, or subcontractor. The contract cost of $300 is shown. The *baseline estimate* is graphically similar to a smoothed ogive curve and is useful for estimate assurance and cost control—topics covered in Chapter 9.

---

[2] Software for project planning: www.PRIMAVERA.com, www.USCost.com, www.COSTTRACK.com, and www.ESTIMATINGSYSTEMS.com.

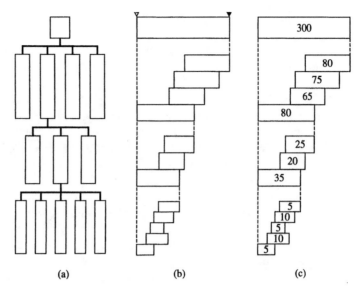

Figure 8.4 Integration of (a) WBS, (b) scheduling of work, and (c) estimation of costs for tasks.

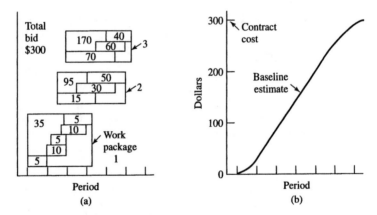

Figure 8.5 Definition leading from (a) estimated costs of tasks from work packages to (b) scheduled cash flow, called the baseline estimate.

While the baseline estimate is important to project estimating, the work package gives other information that is useful for cost analysis. Indeed, it would be inappropriate to think that the WBS preparation ends with only that feature. As projects are large dollar undertakings, the money required for the work may exceed that which an owner has on the assets side of the balance sheet. Similarly, the contractual arrangements between the general contractor and subcontractors can exceed money on hand, so that short- or long-term loans are needed to continue financing the work.

Two concepts are introduced in financing projects, and they emerge from the WBS. They are *commitment* and *expenditure*. These concepts are more appropriate for materials, as labor costs are required to be paid weekly to monthly. Labor costs, for the most part, are funded with money on hand or short-term loans. Project costs, on the other hand, are longer-term obligations.

The *ordering* of material is the commitment stage, while the *payment* for the materials is the expenditure stage. Commitment dates are not the same as delivery or payment dates. The commitment date allows a time lag for delivery, construction, and payment.

Materials are placed on order long before they are received. Depending on the contract between the buyer and seller, payment may be made immediately as COD (collect on delivery) or on a deferred schedule. Of course, other arrangements are possible. But we are keeping our explanation of commitment and expenditure simple, overlooking contractual complexities at this point.

We digress a moment to consider triangular cash flows, in order to simplify large money behavior as found in some projects. Because our project is large and there are many materials, suppliers, and contractors, a continuous cash-flow model is considered reasonable. To simplify the discussion of cash flows for this chapter we use graphical models, such as triangles. Imagine cash flows of such magnitude and frequency that a triangle or other geometrical feature can represent their behavior. Refer to Fig. 8.6 for an illustration.

In this example we illustrate two types of materials: long-term equipment and short-term materials, where long and short term are indicators of the lead time. If our project is a chemical processing plant, equipment is reactors, boilers, etc., which are ordered early in anticipation of their design, complexity, and elapsed time for fabrication. Short-term materials may be structural material, which is readily available.

The vertical axis is cost per period, or $\$10^6$/period. Consider the horizontal axis. The period $E = 0$ is the moment of the estimate, or the *benchmark* period, which was defined with the discussion on indexes. See Chapter 5 for that. The horizontal axis is designated "period." That may mean any ordinary period, but here years are implied. For (a), cash flows grow to 8 units at period 2 and then decline to $0 at period 6. For (b), initial materials are committed at period 2 and grow to 10 units, then decline back to $0. These triangles, we contend, can model cash flows.

Using simple algebra, the 10^6 cost for the triangles is $\frac{1}{2}$ base $\times$ height, or $20 and $30 for (a) and (b). As the components of a project come together, those costs are added to determine the total project cost. In our little illustration, the sum of the triangles (a) and (b) results in (c), which gives $50.

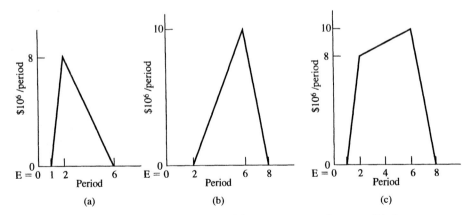

Figure 8.6 Triangular cash flows: (a) long-term equipment, (b) short-term materials, and (c) sum of equipment and material costs.

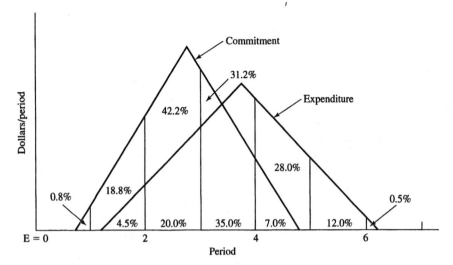

Figure 8.7 Matching commitment and expenditure to period.

We return now to our discussion of commitment and expenditure.

Consider Fig. 8.7, which shows the approximating isosceles triangles for commitment and expenditure. The y axis is designated dollars per period. The numbered x axis is labeled "period" with the origin E set equal to zero. The numbered period axis implies fiscal half-years, but other calendar periods are suitable. For our simplified example, the areas under the two triangles are equal. Each period area of cost is determined. For example, 42.2% of the material costs is committed in period 2–3. For expenditure, 28% of material costs is paid during period 4–5.

Note Fig. 8.7 again, where the commitment triangle is five periods long and is divided into fiscal half-year percentages by using geometry. Similarly, the expenditure triangle is six periods long, and its amount can be synchronized to the fiscal half-year percentages. The half-year percentages shown on the figure are repeated in Table 8.1. Notice in the table the obligation of the contractor and the owner with respect to cash flows.

Total project funds can be distributed by using those percentages. If the project materials are estimated as $11,750,000, then the committed and expenditure dollars for each

TABLE 8.1 Converting material commitment and expenditure to cash flows

Fiscal Half-Year	Percent Commitment	Percent Expenditure	Contractor Dollars Commitment	Owner Dollars Expenditure
0	0.8	0	$94,000	0
1	18.8	4.5	2,209,000	$528,750
2	42.2	20.0	4,958,500	2,350,000
3	31.2	35.0	3,666,000	4,112,500
4	7.0	28.0	822,500	3,290,000
5	0	12.0	0	1,410,000
6	0	0.5	0	58,750
Total	100.0	100.0	$11,750,000	$11,750,000

half-year are given by the product of the percentage and the total project funds. With those ideas and continuity of periods, it is straightforward to find monthly commitment and expenditure percentage and dollar amount.

Curves approximating the actual commitment and expenditure other than the isosceles triangle are used. Rectangular, trapezoidal, right-triangular, and bell shapes are additional approximations of the cash flows of the project.

If the committed or expenditure dollars are progressively accumulated for the periods and a smooth line drawn through the points, then an *ogive curve* or an *S-curve* results. This is shown in Fig. 8.8. A third line, available funds, is included.

Available funds represent the money received periodically from the owner, bank, investor, etc., and used by the contractor to pay subcontractors or itself. The available-funds line is herky-jerky as shown in this figure, because discrete step payments are received.

Those three lines are important in financing a project. At the period midpoint, commitments lead available funding dollars, and if the graph had divisions, then this financial lead time could be measured. Expenditures lead available funds by so many periods, although less than committed funds. *Current liabilities* represent the difference between the expenditure and available-funds line. *Current liabilities* for material can be defined as total dollars that must be assumed if the contract were terminated. *Unliquidated commitments* comprise the unpaid balance between commitments and expenditures. Those three lines are used in cost/schedule control.

Now we turn to cost scheduling of direct labor.

The discussion so far in cost scheduling within the WBS has been concerned with materials, but many of those ideas are effective with direct labor. There are several ways to schedule direct labor cost. The work package is a contributing document to this activity.

The labor estimate is used for time-scheduling direct labor cost. Recall that the labor estimate "form" provides entry opportunity for labor craft, period and year, total hours, average hourly rate, and total dollars. The labor hours are also totaled for the period. For any modest or major project, those data from the labor estimate are already period-scheduled. This scheduling is in either hours or dollars. Hours are more useful for work planning, though dollars are necessary for financial planning and control. Hours are not subject to *escalation* because of inflation effects, although dollars are.

The labor estimate integrates the elements of the estimate, learning theory if it exists, schedule, and proper lead times to accomplish the tasks in their order. After summing the labor estimates for various periods, as in Fig. 8.9, an ogive curve can be drawn.

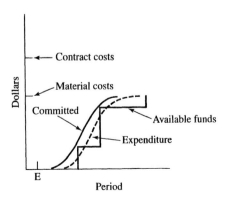

Figure 8.8 Collection of many cost estimates to form smooth committed and expenditure curves.

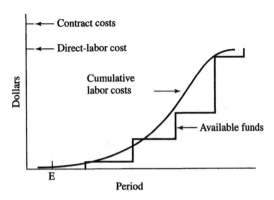

Figure 8.9 Plotting of cumulative total labor costs and available funds.

Estimates are a major part of the work package. Estimating is done concurrently with the work-package development. The estimates are the most important documents of the project bid.

8.3 ESTIMATING

A project estimate is composed of direct labor, direct material and subcontract, facilities and equipment, and engineering. Procedural detail for estimating vary because of application or dollar amount. The larger-valued projects may have addenda to these forms. Distinctions exist in construction—building, road, plant, and so on. An owner or contractor has needs that call for variety. But for instructional purposes, these four forms are sufficient.

The labor estimate deals with the bidder's labor or work force. This assumes that labor will be direct hire for the duration of the project rather than a subcontract arrangement.

Direct-hire labor is hired and paid on a per-hour basis to the worker. Consequently, the labor estimate may involve effective gross hour wages, payroll overhead, and productivity. In those situations where union labor is used, the term "off the bench" implies that the hiring is though a union-controlled hiring office and that the labor is waiting for a "call." Where union labor is not used, the firm will conduct its own hiring.

Whether the labor is union or nonunion, the company pays the workers directly. *Subcontract labor* works for a subcontractor, and the costs are estimated and quoted on a fixed unit price or lump sum, often including material costs. Subcontractor direct labor is not the same as direct labor for the purposes of these estimates.

A brief labor estimate is shown in Table 8.2. The labor estimate is cross referenced to the WBS and the definition. Additional information may include start period, duration, conclusion and labor types, hours, and wages. Some estimates will include a check-off that indicates how hours were estimated (i.e., opinion, comparison, standards, cost-estimating relationship, etc.). The project engineer determines if the work is recurring or nonrecurring. *Recurring work* for project estimating is cyclic; that is, it is done and estimated with repetition in mind, such as the erection of transmission towers.

The estimated time is adjusted for efficiency. The same job may require more time in Alaska than in Texas because of weather conditions. Though a labor standard for a job

TABLE 8.2   Labor estimate for work package

			Period							
WBS	Description	Total Hours	3	4	5	6	7	8	Gross wage	Cost
1311	Clearing	4,000	2,000	2,000					$30	$120,000
1312	Roads	8,000		2,000	4,000	1,000	500	500	35	280,000
1313	Tower erection	55,743		22,297	16,723	7,861	7,862	1,000	40	2,229,720
1321	Substation building	24,286		5,000	5,000	5,000	5,000	4,286	35	850,000
	Total	92,029								$3,479,720

Project: High-voltage transmission line

is consistent, effects of location, crew skill, native or green, and so on, are factors that the engineer weighs.

Productivity for the bidder's and subcontractor's work force can vary, and some experience suggests that subcontractors have more favorable productivity.

Various methods of labor estimating have been discussed, such as the unit-quantity method of Chapter 7. The preferred approach is to estimate separately the allowed man hours (or man days) and their productivity, rather than posting a lump sum that is the product of the two quantities. Those features are a part of the usual labor estimate, but are not demonstrated in Table 8.2.

The direct material and subcontract estimate requires title-block information similar to that for the labor estimate. It is cross identified to the WBS and labor estimate.

There is a difference between direct material and subcontract. *Direct materials* end up in the project design and are usually installed by the bidder's work force. Those direct materials do not include the contractor's work force in their cost.

The WBS and the definition may indicate that a subcontractor is hired. *Subcontract materials* may be specially designed or standard commercial materials. Especially important is the notion that subcontract materials use subcontractor labor, unlike direct materials, which use the contractor's work force. A subcontract may include both the material and the subcontractor's labor.

A quantity of the material is posted. Additions for scrap, waste, shrinkage, and spares are included for direct material. The source of the direct material estimate may be opinion, comparison, take-off for shape and rate, or statistical relationship.

The subcontract materials may be item-estimated or handled as a lump sum, or a *quotation* may be available. Selection of the subcontractor may be based on low bidder, technical competence, or best delivery schedule. Table 8.3 is an abbreviated example of the direct material and subcontract estimate.

Facilities and equipment are also direct materials but of a different character—that is, delivered and erected equipment, such as large storage tanks or field-fabricated vessels involving both material and labor as a single lump sum. Facilities and equipment are specified in detail, are custom fabricated by vendors for the project, and may be produced and estimated like a product. Those job shop manufacturers are sensitive to demand and adjust prices to accommodate demand. Price fluctuations can be expected in those costs, especially for external suppliers.

TABLE 8.3 Direct material and subcontract estimate for work package

			Project: High-voltage transmission line							
			Period							
WBS	Description	Direct material or sub-contract	2	3	4	5	6	7	8	Cost
1312	Road	Direct material	5,000	5,000	5,000					$15,000
1313	Tower	Direct material	500,000	940,625	300,000	100,000	60,000	30,000	10,000	1,940,625
1321	Substation building	Direct material			90,000	45,000	45,000			180,000
1322	Substation transformer, controls	Subcontract		100,000	150,000	50,000	50,000	50,000		400,000
133	Conductor	Subcontract		117,000	117,000	117,000	117,000	117,000		585,000
	Total									$3,120,625

It is also necessary to estimate any raw or bulk materials that may be required with the facilities and equipment. For example, an electric transmission tower will require concrete foundations. Will the foundations be estimated with the transmission towers? If not, then a cost connection must be made between the WBS, definition, and estimate. The concrete foundations are estimated as direct material by using the contractor's work force.

Because facilities and equipment are custom designed, a quotation from suppliers becomes necessary. Often a predesign estimate is made. Thus, the source for those data may be external to the estimating team or may be based on internal information.

Facilities- and-equipment estimates can include land and building and processing equipment. Equipment supporting a subcontractor is included in the subcontractor's quotation. Rental charges for construction support equipment can be included in the labor estimate but usually are included in project overhead. Table 8.4 is an abbreviated example of a facilities-and-equipment estimate.

Engineering costs are often significant for a project design. Their separate consideration points to this importance. Table 8.5 is an example of an engineering estimate for project designs. If engineering costs are not covered by an estimate, they are treated as an overhead cost.

Productivity can vary and should be forecast. Overtime, job size, and specific working conditions can affect productivity. The engineering costs used to develop proposals that do not lead to new jobs are collected into this conglomerate hourly rate.

Cases exist where an engineering effort is significant in proportion to a contract cost. A firm will be hired to do the engineering and, say, manage a large-scale construction job. Estimating those costs may become a competitive bid among engineering firms. If major equipment is installed, say, as a result of this bid, and is in the range of $500,000 to $50 million, then engineering costs for complex pilot and chemical plants range from 20% to $7\frac{1}{2}\%$. As project values increase, fees paid for engineering typically decline. In repetitive types of construction, engineering costs vary from $3\frac{1}{2}\%$ to 13% of total installed cost.

TABLE 8.4 Facilities-and-equipment estimate for work package

						Project: High-voltage transmission line					
						Period					
WBS	*Description*	*Type*	*1*	*2*	*3*	*4*	*5*	*6*	*7*	*8*	*Cost*
111	Land, survey	Facility	75,000	11,575							$86,575
112	Environmental	Facility		6,425							6,425
1321	Substation building	Facility						50,000			50,000
1322	Substation transformer	Equipment				200,000	300,000	200,000	50,000	50,000	800,000
	Total										$943,000

TABLE 8.5 Engineering estimate for work package

					Project: High-voltage transmission line			
					Period			
WBS	*Description*	*Hours*	*Hour Rate*	*Overhead Rate*	*1*	*2*	*3*	*Total Cost*
12	Engineering	750	$44.67	25%	35,000	5,000	1,875	$41,875
121	Saggers	150						
122	Cost	80						
123	Design	520						

Engineering costs may be negotiated on a lump-sum turnkey basis where the cost of engineering is included in the erection package for an entire plant. Sometimes the engineering contracts are negotiated on a *cost-plus basis*.

Variations of engineering professional contracts include

- Cost plus a negotiated fee or profit for the engineering contractor.
- Cost plus a fixed-fee contract with a guaranteed maximum.
- A contract for engineering design manpower to be supervised by the client's engineering staff.

No matter what the contract type, engineering estimates contain these elements of cost:

$$C_e = \Sigma S + \Sigma E + \Sigma OH + \Sigma F \tag{8.1}$$

C_e = engineering costs, dollars
S = salaries, dollars
E = variable expenses such as travel, living away from home, communications, and computers, dollars
OH = overhead such as rent, depreciation, heat, light, clerical supplies, workmens compensation, etc., dollars
F = fees paid to other specialists and engineers, dollars

These WBS estimating procedures have similar advantages. The uniformity created by a work package encourages consistency for later estimate assurance. The source of information, auditing, a standardized communication format, cross reference, and the central WBS engineering document make these estimating methods of value throughout the project.

The four kinds of estimates provide the factual basis for the project. If the estimates are poorly done, then no amount of superficial analysis will improve the estimates. But we separate the task of estimating from its later cost and bid analysis. For it is important that the bid be competitive and compatible to the firm's ability. We now consider this next step in project estimating.

8.4 COST AND BID ANALYSIS

More information may become available to the engineer during the estimating and analysis period. Tips may be found in the local newspapers, budget disclosure, or the owner or major contractor may indicate boundaries for the bid. Rebidding may occur in large projects, and first bids may be known. Typically, past winning bids are open knowledge for public works. Business magazines and trade newspapers publish information regularly. Even rumors may be available.

Competitive bidding is usual for projects. In the simplest case, the offeror will announce a deadline date for the bid, and sealed envelopes containing the bid and other information are opened and the winner announced. In technical projects, the bidder provides a design along with the bid. In many cases, evaluation may take a long time before the winner is selected.

When formal advertising and competitive bidding are impractical, a bargaining process begins between the parties, each having its view and objective. This is termed *negotiation*. Cost and bidding analysis are different for each of those situations. The estimating procedures should be identical, however. Eventually the bid is "laid on the table," so to speak, and its acceptance or denial depends on many factors.

As a *pro forma* document, the bid is the center of much interest. If the project estimate is a public document subject to audit by an owner, major contractor, or the government, and this depends on the contract, then analysis is handled by a contracting officer, negotiator, or engineer on an "arm's-length" basis (i.e., each side having a competitive and self-serving interest). The audit is from several vantage points: engineering, accounting, purchasing, and estimating. Thus, the cost and bidding analysis must satisfy many objectives. First and last, it must be professionally done.

After the estimating stages are completed the bid and cost analysis finds the following:

- Overhead
- Contingency
- Interest
- Profit

8.4.1 Overhead

Overhead costs were discussed in Sec. 4.9. There we learned that overhead for projects is of two types: office and job. Overhead costs exclude costs of direct labor, direct and subcontract materials, and facilities and equipment. Items appearing in overhead must not be

PICTURE LESSON
Worker Safety

The picture shows an insulation worker clamping the external stainless steel covering for the insulation prior to the erection of the fractionater at a chemical process plant. Notice that the worker is wearing a body harness and he is tethered by strong strap to the hauling ropes. Safety helmet, sun-protecting goggles, gauntlet gloves, warm clothing, safety shoes, secured hand tool, woven hauling rope, and a separate crane for supply of materials are incorporated in the contractor's workplace for this occupation.

In spite of precautions, tragedy does happen. Fortunately, accidents and death are declining in the United States. In a recent year, the incidence rate of injuries and illnesses for every 100 workers declined to about 7, the lowest on record. Concomitantly, this results in lower worker compensation costs.

There are many reasons for this positive reduction. As one example, engineering studies are performed on hauling ropes to determine the desirable performance properties and standards. Operation in daylight or in industrial atmosphere has differing influences. Too, knots at intersections form a source of weakness. Daily, weekly, and monthly inspections of numerous items are checked off. A great collection of knowledge has been codified into standards, regulatory requirements, and practice.

In addition to the strong leadership by contractors, the insurers, unions, and the Associated General Contractors of America have contributed to safety and health. Since 1970 the Occupational Safety and Health Administration has been committed to saving lives and preventing injuries in American workplaces. OSHA conducts unannounced inspections, develops regulations, levies fines and legal penalties, and assists contractors, unions, and employees with various programs.

included in those estimates. Perhaps, certain aspects of engineering may appear in the overhead accounts, if it is reasonably convenient and the work applies to more than one project.

Office overhead includes general business expenses, such as home office rent, office insurance, heat, light, supplies, furniture, telephone, legal expenses, donations, travel, advertising, bidding expenses, and salaries of the executives and office employees. Those charges are incurred for the benefit of the owner's or contractor's overall business. In office overhead, the final cost objective is multiple (i.e., several projects), and those indirect costs cannot be pinpointed as specific estimating amounts.

Each project requires a special analysis to determine its own overhead items. Typical charges of *job overhead* include permits and fees, bid or performance bond, job office expenses, insurance, electricity at job location, storage and protection, etc. Those charges, their basis and rates were discussed in Chapter 4, and the student will want to read that material. With the rates it becomes possible to find the applied overhead that is added to the project cost using the following:

$$C_{op} = C'_p(R_o + R_j) \tag{8.2}$$

where C_{op} = overhead charged to future project, dollars

R_o = overhead rate for office costs, decimal

R_j = overhead rate for job costs, decimal

C'_p = direct labor, direct and subcontract materials, facilities and equipment, and engineering costs of future project, dollars

8.4.2 Contingency (Optional)

Contingency is another cost element evaluated in the cost and bid analysis. Sometimes this element is called *engineering reserve*. Uncertainties in developmental projects are included with this self-insurance. Especially in the early stages of a project where technology detail and its cost are absent, contingency estimating is preferred to careless or unjustified padding of cost elements. Contingency-cost estimating is practiced for major projects, albeit by various techniques.

As the project design matures and as more information becomes available, the contingency estimate becomes less in absolute value and percentage of the total bid. In large-scale project-cost planning, contingency amounts are more important during early estimating. Eventually, an estimate is determined by using a work breakdown structure. Thus, contingency dollars abide by the rule: more information and less contingency.

In high-risk high-dollar projects, contingency is usually present, and it may appear as a line item in the bid. As the project spends money for materials, direct labor, subcontract agreements, and so on, the dollars of budgeted contingency are available for the payment of unexpected problems that arise.

Two methods of assessing contingency are considered:

- Experience and opinion
- Analytical treatment

In discussing an informal method of assessing contingency, we assume that the dollar amounts for the major WBS levels are already estimated. By using opinion, the engineer estimates an additional dollar amount for major elements. For instance, a project for developing commercial laser electrical power is risky, and prudent evaluation would determine an incremental cost for possible overrun.

But contingency analysis is also possible with the estimating of the major elements of the WBS. It is an extension of range estimating discussed in Sec. 6.10.2. Refer to Fig. 8.10 for the steps.

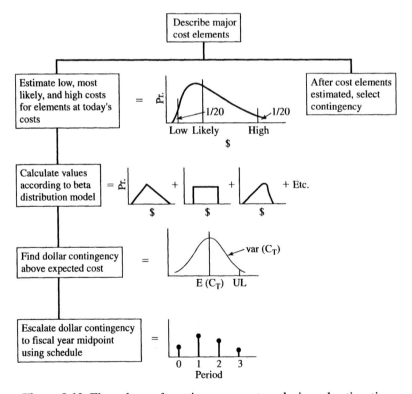

Figure 8.10 Flow chart of contingency-cost analysis and estimating.

Contingency-cost estimating begins with estimating a lowest value, L, most likely value, M, and the highest value, H, of the cost elements of the WBS. The most likely value is the modal or most common value that would be repeated in an unlikely repetition of this cost element. Similarly, the lowest and highest cost values are not to be exceeded either downward or upward with a probability of $1/20$.

Seldom, if ever, are those instructions exactly executed, but three values are necessary to allow the use of the beta probability distribution. The form of this general distribution has properties that make for simple calculation, which is important for early estimates of the project. It is unnecessary to specify the probability distribution of the individual cost elements, which is fortunate, because their behavior is nonnormal and unknown. After many elements are added, according to central limit theory and practice, the resulting probability distribution is normal. The normal distribution for the mean cost can be manipulated by using conventional probability rules.

The three values for each major WBS element are estimated at today's costs, even though it is future expenditure that is desired. If historical or raw information is available, then it is often at the current time, such as $E = 0$. Further, in conference and round-table estimating, current values are easier to estimate intuitively than future values.

Now, assume that the expected cost of Table 6.6 $E(C_T) = \$10,600$, and its variance $\text{var}(C_T) = 22,300$ dollars^2, are to be analyzed for contingency. Recall Eq. (6.26):

$$Z = \frac{UL - E(C_T)}{[\text{var}(C_T)]^{1/2}} \tag{8.3}$$

where Z = standardized normal variate

UL = upper limit cost, arbitrarily selected, dollars

$E(C_T)$ = expected cost of range estimating model, dollars

$\text{var}(C_T)$ = expected variance of range estimating model, dollars^2

$E(C_T)$ and $\text{var}(C_T)$ were calculated by using Eqs. (6.24) and (6.25). The square root of the variance is called the standard deviation.

Values of the upper limit are selected by the project engineer and cover the potential range of overrun or underrun cost of the project. The range of those values depends on the risk of the project. With several values picked, Table 8.6 is completed by using Eq. (8.3) and Appendix 1, standard normal distribution tables.

Appendix 1 provides probabilities as measured from the reference $Z = 0.00$. Overrun and underrun are defined with respect to this reference. At this symmetrical location, 50% of the probability is above and below $Z = 0.00$. For example, if $Z = 1.00$, then the table reads 0.3413, which is the area or probability from $Z = 0.00$ to 1.00. If we want the value to the right of $\Pr(Z = 1.00)$, then the upper-tail area is given by $0.5 - 0.3413 = 0.1587$. If we are interested in the probability that cost exceeds \$10,300, then we would have

$$Z = \frac{10,300 - 10,600}{(22,300)^{1/2}} = -2.00$$

Because we are interested in $\Pr(\text{Cost} > 10,300)$, the appendix would be used as $0.5 + 0.4772 = 0.98$, which is observed as the first entry in Table 8.6.

Observe Fig. 8.11, which gives the probability of exceeding the expected project cost. The expected value, \$10,600, gives a 50% probability. The probability that cost will assume

TABLE 8.6 Calculation of probability given that a cost upper limit exceeds a mean value

UL	Z	Pr(cost ≥ UL)	UL	Z	Pr(cost ≥ UL)
$10,300	−2.00	0.98	$10,750	0.99	0.16
10,400	−1.35	0.91	10,800	1.32	0.09
10,500	−0.68	0.75	10,850	1.66	0.05
10,600	0	0.50	10,900	2.00	0.02
10,650	0.32	0.37	10,950	2.33	0.01
10,700	0.66	0.25	11,000	2.67	0.004

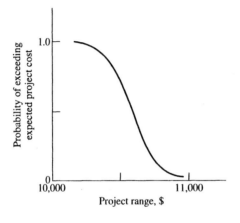

Figure 8.11 Risk graph of contingency for commercial laser electrical power.

values less than $10,600 is greater than 50%, naturally. From this analysis, the engineer gains an opinion for the risk with this curve.

Several contingent maximum costs may be considered for the bid, keeping in mind the uncertainty, design, technology, and construction conditions. Eventually, one upper-limit cost is selected as the *contingent maximum* cost. For example, suppose that engineering selects $10,800 as this value, which, as we see from Fig. 8.11 or Table 8.6, will be exceeded with a 9% probability.

The next step is to schedule the increment of contingency of $200 (= 10,800 − 10,600) over the project periods. The contingent maximum is at current cost, meaning that the time is $E = 0$. If inflation or deflation is suspected, then this value is adjusted by means of indexes relating to the project. Indexes are now used such as those provided by Table 5.6, which assumes declining costs for laser electrical power owing to active technology. The index at time $E = 0$ is 1.000, and the indexes for periods 1, 2, 3, and 4 are found from analysis of indexes, as discussed in Chapter 5.

Additionally, expenditure scheduling, given in percent terms, is known. This information is shown as Table 8.7. The scheduled contingent amount for period 2 is $44.8 (= 200 × 0.25 × 0.896).

The total of $177.95 is posted to the project summary as a line item for contingency. Similarly, the $10,600 is entered as the project estimate amount. Note that contingency is shown for the total project rather than for single-cost elements. Even though it is element variability that gives project contingency, it is as likely that other cost elements may be overlooked, and this contingent amount becomes available for unspecified requirements.

TABLE 8.7 Scheduling the contingency amount of $200 for percent expenditure and indexes

Period	1	2	3	4	Total
Expenditure (%)	15	25	40	20	100%
Period Cash Flow	30	50	80	40	$200
Index	0.941	0.896	0.876	0.871	
Contingent Cash Flow	28.23	44.80	70.80	34.84	$177.95

Opinion and technical experience are indispensable in estimating contingency amounts. On the other hand, algebraic refinements, such as those presented, do not substitute for effective methods in project estimating. Unfortunately, in practice, contingency is a device that replaces qualified cost-estimating methods.

Depending on the contract type, if unjustified contingency is added to the cost, then the bid becomes noncompetitive. In some cases, should a surplus of contingent dollars be available at the conclusion of the project, they are returned to the owner, according to provisions of the contract.

8.4.3 Interest (Optional)

Interest is an element of the project-cost estimate. As projects may exceed cash-on-hand, bank or investor loans are necessary to meet the ongoing cash-flow requirements. Various methods are employed to determine the interest for the project. Both the owner and contractor need to understand these procedures. The basic steps are as follows:

- Project cash-flow model adoption
- Net cash-flow determination
- Interest calculation

Figure 8.12 shows three geometrical models for project cash-flows: triangular, rectangular, or the trapezoid. The triangular approximation has been previously discussed. In view of its constant-period cash flow, a rectangular likeness, for example, may mimic a high-voltage transmission line or a pipe-laying project. Other approximations are found, but the trapezoidal model is the more common.

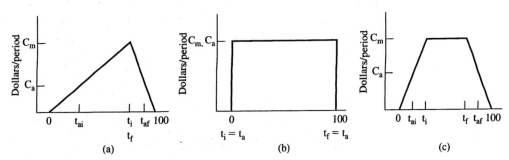

Figure 8.12 Cash-flow models: (a) triangle, (b) rectangle, and (c) trapezoid.

Again, because projects are large dollar undertakings, we assume that the continuous cash-flow models meet the needs of estimating. In these models of Fig. 8.12, the x axis is the contract time t scaled as a percent. The y axis is dollars per period. Let

t_i = initial time when dollars per period is maximum of contract time, percent

t_f = final time when dollars per period is maximum of contract time, percent

C_a = total value of contract divided by frequency of progress payments, average dollars per period

C_m = maximum baseline dollars per period

Geometrically, for the trapezoid, Fig. 8.12(c),

$$\frac{C_m}{C_a} = \frac{200}{100 + t_f - t_a} \tag{8.4}$$

TABLE 8.8   Values of C_m/C_a for the trapezoid where $0 \leq t_i \leq t_f \leq 100$

			t_f	
t_i	50	60	70	80
20	1.54	1.43	1.33	1.25
40	1.82	1.67	1.54	1.43
60	—	2.00	1.82	1.67

An early step in the analysis for the project cash flow and interest is to assume the geometrical approximation, t_i, t_f, and then calculate C_m/C_a. Other project and contract facts need to be known to find the amount of the interest.

The period of contract time during which the baseline dollars per period is equal to the average value can be shown by similar triangle geometry to be

$$t_{ai} = \frac{t_i}{\dfrac{C_m}{C_a}} \times \frac{N}{100} \tag{8.5}$$

and the final instance of the time for an average cash-flow rate is found when

$$t_{af} = \left(100 - \frac{100 - t_f}{\dfrac{C_m}{C_a}}\right)\frac{N}{100} \tag{8.6}$$

where N = progress dollar payments for project, number

t_{ai} = initial period when average cash flow is reached

t_{af} = final period when average cash flow is reached

As the project progresses, the cumulative cash flow appears as an *ogive* or *S*-curve. These cumulative curves are the integral of the curves given in Fig. 8.12. But it is also pos-

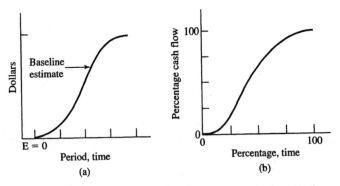

Figure 8.13 Cumulative baseline curve plotted as (a) period versus dollars and (b) percentage time versus percentage cash flow.

sible to have the cumulative baseline value as a percentage of the total contract amount, y axis, to the percentage of the total contract time, x axis. This is shown in Fig. 8.13. Both the triangle and trapezoid of Fig. 8.12 give the S-shaped curve. The rectangular model, Fig. 8.12 (b), will give a straight line for the cumulative cash flow.

For a trapezoid cash-flow model, we have after integrating

$$CBV = \frac{C_m t^2}{2 C_a t_i}, \quad 0 \le t \le t_i \tag{8.7}$$

$$CBV = \frac{C_m}{C_a}\left(t - \frac{t_i}{2}\right), t_i \le t \le t_f \tag{8.8}$$

$$CBV = \frac{C_m}{2 C_a}\left[2t_f - t_i + (t - t_f)\left(1 + \frac{100 - t}{100 - t_f}\right)\right], t_f \le t \le 100 \tag{8.9}$$

where CBV = cumulative baseline value, percentage

Application of those formulas requires t to be a percentage number. With Eqs. (8.7), (8.8), and (8.9) and arbitrary $t_i = 25\%$ and $t_f = 62.5\%$, we determine sample values, as shown in Table 8.9:

We now return to the high-voltage transmission-line project. Previously, we illustrated the estimated elements of direct labor, direct material, subcontract material, facilities, equipment and engineering. Project and office overhead is applied to the subtotal of

TABLE 8.9 Illustrative table for cumulative baseline value percentages

t, %	CBV, %	t, %	CBV, %
0	0	62.5	72.8
12.5	4.5	75.0	87.9
25.0	18.2	87.5	97.0
37.5	36.4	100.0	100.0
50.0	54.6		

direct costs. A trial value for interest is entered as \$110,000. A contingency of \$40,896 is indicated in Table 8.12. Profit is added to these costs. The project amount is found to be \$9,118,175. Look ahead to Table 8.12 for a summary of these costs.

The terms of the contract stipulate a construction schedule of 8 months and, equivalently, 8 payments. Projects involve periodic payments between the owner and contractor, or between the contractor and subcontractors. Those payments are known as *partial* or *progress payments* and reimburse for work.

The trapezoid cash-flow model with $t_i = 25\%$ and $t_f = 62.5\%$ is useful for a transmission-line project. Using Eq. (8.4), then

$$\frac{C_m}{C_a} = \frac{200}{100 + 62.5 - 25} = 1.4545$$

as the ratio of maximum to average cash flow.

$C_a = 9{,}118{,}175/8 = \$1{,}139{,}772$ per period average cash flow

$C_m = 1.4545 \times 1{,}139{,}772 = \$1{,}657{,}800$ maximum cash flow per period from t_i to t_f

$t_i = 0.25 \times 8 =$ second month when maximum cash flow starts

$t_f = 0.625 \times 8 =$ fifth month when maximum cash flow concludes

$t_{ai} = \dfrac{25}{1.4545} \times \dfrac{8}{100} = 1.375$th month, when initial average cash flow rate is reached

$t_{af} = \left(100 - \dfrac{100 - 62.5}{1.4545}\right)\dfrac{8}{100} = 5.9$th month, when final average cash flow rate is concluded

The cash-flow trapezoid for these assumptions is given as Fig. 8.14(a).

Steps for an algorithm to find the net cash flow are demonstrated by Table 8.10. For our purposes, notice that the columns are numerically identified. Other and more elegant algorithms and software are available, and the reader will want to review the References for possible programs.

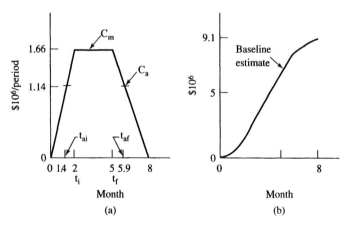

Figure 8.14 Cash-flow approximation for high-voltage transmission line: (a) trapezoid, and (b) baseline estimate.

TABLE 8.10   Cumulative baseline value, profit, net cash-flow, retainage, and expenditure for the transmission-line project.

(1)	(2)	(3)	(4)	(5)	(6)	(7)	(8)	(9)
Month	Contract Period, %, t	CBV, %	CBV, $	Profit, $	Expenditure, $	Available Funds	Retainage, $	Net Cash Flow, $
0	0							
1	12.5	4.5	414,592	37,690	376,902	–	414,592	(376,902)
2	25.0	18.2	1,658,368	150,761	1,507,607	352,403	1,305,965	(1,155,204)
3	37.5	36.4	3,316,736	301,521	3,015,215	1,409,613	1,907,123	(1,605,602)
4	50.0	54.6	4,975,104	452,282	4,522,822	2,819,226	2,155,879	(1,703,596)
5	62.5	72.8	6,633,472	603,043	6,030,429	4,228,839	2,404,634	(1,801,591)
6	75.0	87.9	8,014,876	728,625	7,286,251	5,638,451	2,376,424	(1,647,799)
7	87.5	97.0	8,844,630	804,057	8,040,573	6,812,644	2,031,985	(1,227,928)
8	100.0	100.0	9,118,175	828,925	8,289,250	7,517,935	1,600,240	(771,315)
9	112.5	100.0	9,118,175	828,925	8,289,250	7,750,449	1,367,726	(538,801)
10	125.0						0	828,925

In column (1), each month is a progress payment, which ends with the total number of periods and contract payments, plus those periods to receive all monies, including retainage, by the contractor and paid by the owner.

The cumulative baseline value percentages, (3), are entered by the steps that have been described. Column (4) is the product of column (3) and the project contract value. For month 1, the CBV is $414,592($\approx$ 0.045469 $\times$ 9,118,175).

Column (5) is the product of (4) and 9.09% and is identified as "profit." The 9.09% is the markup rate. It is prudent business policy to recognize that profit is a concurrent opportunity in construction. A different profit would result if the rate were a markup on cost.

Expenditure, column (6), is the difference between columns (4) and (5).

There may be a delay from work concluded to money paid. Available funds depend on contract terms. For our example, there is a 1-month delay for the owner to certify that the work is done and is paid on the basis of 85%. For example, at $t = 2$, column (7) is found by multiplying the cumulative baseline one period earlier by 85%, or $352,403 (= 0.85 $\times$ 376,902).

Often the owner will hold back an amount and not immediately pay for work concluded. This amount is called *retainage*. The contract may require that retainage be placed in an escrow account with a bank, thus assuring the contractor of its availability. An escrow account removes any unfair advantage and the motivation is contractual. Eventually, the owner will pay all monies, but with a sufficient delay to assure that the project design is concluded according to the terms and specifications of the contract. Retainage, column (8), is found by subtracting available funds from the cumulative baseline value (4) for each period. For period 2, column (8) is $1,305,965 (= 1,658,368 − 352,403).

The last column describes the contractor's net cash flow and is found as the difference between columns (7) and (6) or (5) and (8). For period 2, (376,902) denotes a negative value, and a loan equal to this amount is required. It is only in the last period, $t = 10$, that the apparent profit of $828,925 is available to the contractor. Note that column (9) is a cumulative deficit or surplus. Values in this column are a guide to the amount to which the contractor must finance the project with a bank loan. Interest must be paid for this interim financing.

The cumulative baseline value column (4), when plotted against the month, provides the *S* curve. Note Fig. 8.14 (b). The *y* axis can be identified in dollars per month in either actual or percentage terms.

Because projects are large dollar undertakings, the owner or contractor may have insufficient cash on hand to meet all financial requirements. Borrowing from banks and other lending institutions then becomes necessary. An owner may have a surplus, as indicated on the firm's balance sheet under net worth, that needs to be liquidated and converted to money for payment. Once the project is concluded, the contractor is paid by the owner, but typically the owner has incurred a long-term debt.

Before interest is calculated for the estimate pro forma, it is necessary to establish a prepayment plan. The difference between *payment for work* concluded and *available funds* is the contractor's amount required for bank loans and is the principal on which short-term construction interest is charged. Both an owner and a contractor establish this cash-flow stream.

For project analysis there are two components to owner's interest. The first is during the period of the construction phase, and the second is following the conclusion of construction. The owner deals with both, but the contractor is concerned with project construction cash-flow requirements only, which involves only the interest for operations of the job, unless the contractor has sufficient working capital to cover the costs of the work, materials, subcontractors, and so on.

The owner will use a cash-flow analysis such as that in Table 8.10. In particular, the owner will pay attention to column (7). Initially, there will be some cash on hand available for progress payments. If we assume that the owner has 25% available money to finance the project, then from period 5 on, a loan for the remainder is required. This is construction financing and terminates at project end. It is usually more expensive (i.e., concerning interest) than ordinary long-term debt for which collateral and other securities (such as the project) are pledged. The short-term loan is converted into long-term debt. The student is encouraged to study the material about the liabilities account of the balance sheet in Chapter 4.

In a similar way, the contractor needs money to meet obligations. The contractor may need short-term financing, which is measured by the difference between columns (6) and (7) up to period (10), or column (9). The money out of pocket is the same as indicated by column (9). Repayment by the contractor to the lender would be concluded following final payment by the owner. The amount of interest paid by a contractor to a lender also depends on initial working capital.

Consider Table 8.11, which demonstrates the calculation of interest for short-term financing by a contractor. In this evaluation we assume that external financing is obtained by the contractor and that column (9) of Table 8.10 reflects the net cash-flow requirement to be borrowed monthly.

Column (1) is the month for the project, and is also a point of a period payment. Column (2) is the loan requirement at the start of the month. At the beginning of the project, there are no costs of construction. During the first month there are costs requiring a loan. Construction operations during month 1 are $376,902; this amount is removed from Table 8.10, column (9). The $376,902 is entered in column (3).

Outstanding loan balance is given in column (4), or the sum of columns (2) and (3). The total loan is the quantity on which interest is charged. For this example, we use a 1% monthly interest rate. The interest calculation gives $3,769 (= 0.01 × 376,902). The end-of-period balance is the sum of columns (4) and (5), or $380,671 (= 376,902 + 3,769).

TABLE 8.11 Schedule of interest payments for short-term loan to finance construction as required by net cash flow for transmission-line project

(1) Month	(2) Loan, start of month, $	(3) Monthly net cash flow, $	(4) Total loan, $	(5) Interest amount, $	(6) End-of-period balance, $
0					
1	—	376,902	376,902	3,769	380,671
2	380,671	778,302	1,158,973	11,590	1,170,563
3	1,170,563	450,398	1,620,961	16,210	1,637,171
4	1,637,171	97,994	1,735,165	17,352	1,752,517
5	1,752,517	97,994	1,850,511	18,505	1,869,016
6	1,869,016	(153,791)	1,715,225	17,152	1,732,377
7	1,732,377	(419,871)	1,312,506	13,125	1,325,631
8	1,325,631	(456,613)	869,018	8,690	877,708
9	877,708	(232,513)	645,195	6,452	651,647
10	651,647	(1,367,726)		6,516	
Total				$119,361	

The end-of-period balance is the loan quantity at the start of the next month. Notice that $380,671 is entered as column (2) for period 2. The monthly cash flow of $778,302 is found from Table 8.10 or (1,155,204 − 376,902). Deficit quantities are considered positive for the purpose of a loan requirement.

Note in Table 8.11 that the outstanding loan is paid off in month 10. The total of the monthly interest, $119,361, would be posted to interest on the pro forma, Table 8.12. The original estimate of the interest that was posted was $110,000. Inasmuch as profit and other calculations are based on this trial value, the algorithm recalculates the loan requirement until there is convergence in the interest amount, as shown between Table 8.11 and Table 8.12.

The owner would use a similar procedure, but interest is based on column (7) of Table 8.10. The amount of a short-term note to cover shortage of payments to the contractors depends on the cash availability of the owner. A short-term note liability is posted to the balance sheet, which is converted to a long-term mortgage, if required.

With the work package estimated and calculated, from direct labor to interest, we now turn to the next step, pricing and bid. Even so, the estimating and calculation of these pro forma elements are tentative, because the steps include the setting of profit to reach the bid. Backtracking may be necessary to reevaluate the design and elements, if the bid is not competitive. Or there may be a greater opportunity for profit. Our understanding of the process of project estimating is not yet complete.

8.4.4 Pricing and Bid

Pricing for project designs differs from consumer product pricing in several ways. Consumer products are sold on the open market to an unknown buyer. In the bid market, however, a contractor constructs an end item for a known buyer. For products, price adjustments are routinely based on market or technology or cost movements. This choice is made every so often for the same or similar products.

Projects are procured by contract. Methods of pricing depend on the type of contract selected by the owner or buyer, and pricing is in accordance with the contract. A bid requires a unique analysis where opportunity for repetition is unlikely.

The winning or losing of a bid award may give no hint about how well the project was estimated and priced. A competitor may "low-ball" a bid to keep a project work force busy. Contrariwise, the winning of a bid may be against uninterested competition, who submit high bids. In the product market a 3% increase in price may reduce the number of units sold by 1%, for example. A similar increase in the bid may lose the contract.

Contracts are broadly divided into two groups. In the first, risk is borne by the contractor, and the contract is called *firm price*. The contractor may or may not determine a contingency, which is added to the estimate. Second, an owner or buyer may assume all economic risk, and the contract is called *cost plus*. In this type, costs are reimbursed by the owner. A profit percentage or a fee is usually agreed to between the parties, which increases the total amount. In view of these contract complexities, we defer their discussion to Chapter 11.

We recommend that estimating methods be identical for either type of contract. Preparation of the price and bid analysis, however, depends on the contract type.

Some view project pricing as an interpretive art: "Price can be anything you want to make it." Others would change this art into a routine calculation where price is found by formula. In engineering and estimating, though opinion is unavoidable, several pricing strategies are important, but none guarantees success. Pricing success, when achieved, is a consequence of technical and estimating factors of moderate importance. Seldom is it a result of brilliant strategy.

But to evaluate potential bid prices, analysis and the use of formulas are required. We, however, stress that pricing proceeds once cost is finally estimated. The estimated cost is the most significant and recognizable component of price.

The most common approach to pricing is to use *cost plus a markup*. A formula would be

$$P = C_t + R_m(C_t) + C_m \tag{8.10}$$

where P = project bid, dollars

C_t = total cost of project, dollars

R_m = markup rate on cost, decimal

C_m = miscellaneous costs that may be inappropriate for markup (e.g., contingency), dollars

Sometimes C_t is called *full cost*, implying that all elements "are in" the estimate, and there are no unintentional omissions.

Another variation would use various markup factors on the several cost elements. Simple as Eq. (8.10) appears, there are many variations. For instance, C_t can be separated into material, direct labor, burden, and engineering with each term having its own markup. Some companies use the same markup year after year. Others use markups reflecting the preceding year's actual percentages. For the most part, these percentages vary with business success, and this feature offers ease of understanding as its best advantage. As business falls, the add-on is reduced and vice versa.

Sometimes contractors need to understand *conversion pricing*. In this case, the owner will buy all materials and then employ a contractor to install the materials.

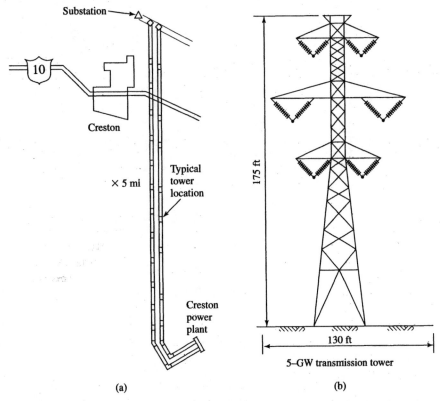

Figure 8.15 Illustrative design of Creston 5-GW transmission line:
(a) double-circuit transmission path, (b) transmission-line tower.

Throughout this chapter we have used the example of a 500-kV transmission line that is to be estimated. Retrospectively, information about this project is more detailed than can be printed in the space available here, and we can only furnish the map of the 5-mile route and an elevation-sketch of the towers. Refer to Fig. 8.15.

The owner will formally invite contractors to bid, and RFQs are issued to qualified contractors. The specifications will state that the transmission line is double circuit from a power station bus wall to the Creston substation. "Double circuit" implies two parallel transmission conductors and hardware that are constructed side by side. The aluminum conductor is 3 × 1.6 in. O.D. per circuit. Performance is 0.28 voltage drop per mile at 525 kV.

Table 8.12 is the bid summary for a 500-kV voltage transmission-line project. Information from the estimates, Tables 8.2–8.5, is posted to the bid summary. The estimating analysis for overhead, contingency, interest, and profit is conducted after this posting. Eventually, a bid value of $9,159,072 for 5 miles of a high-voltage transmission line is found. A markup of 10% on all components, except contingency, is calculated and a price of $9,159,072 (= 0.10 × 8,289,251 + 8,330,147) is *tentatively* set as the bid. The bid is called tentative, because there are other considerations before the bid is "laid on the table," so to speak. Those topics are developed in Chapter 9.

TABLE 8.12   Project bid summary

Project: High-voltage transmission line
Customer: Bonneville Power Administration

Work Package	
Direct labor	$3,479,720
Direct material	2,135,625
Subcontract	985,000
Facilities	143,000
Equipment	800,000
Engineering	41,875
Overhead	
Office, 4% of direct costs, $6,600,345	264,014
Job, 5% of direct costs, $6,600,345	330,017
Subtotal	$8,179,251
Contingency, $\frac{1}{2}$% of subtotal	40,896
Estimated interest	110,000
Total	$8,330,147
Total without contingency	$8,289,251
Profit, 10% (on cost, less contingency)	828,924
Bid (without contingency)	$9,118,175
Bid	$9,159,072

SUMMARY

We have seen the elements of direct labor, direct material and subcontract, facilities, equipment and engineering form the organization for the estimate. Once those estimates are concluded, a bid and pricing analysis for overhead, contingency, interest and profit is conducted. There are no simple tricks to estimate projects, for the size, diversity, and complexity of construction designs make this professional work demanding in terms of thought and skill. Our approach has been to introduce the important work breakdown structure to this organization.

QUESTIONS

1. Define the following terms:

Overhead
Bid
Contingency
Productivity
Work package
Definition
Conversion pricing
Full cost
Project interest

Retainage
Baseline estimate
Commitment
Geometrical cash-flow models
Available funds

2. What is the purpose of a project estimate? Why are these estimates important?

3. What kinds of information are necessary to estimate a project?

4. Illustrate direct and indirect costs.

5. Distinguish between direct material and subcontract in a project bid.

6. Give the purpose of a work package. What is the definition part?

7. How do commitment and expenditure differ? What causes this difference?

8. Describe and construct a new labor estimating form.

9. Why is the bid document important for project designs?

10. Examine a web site for project planning and survey their approach to planning. Write a report giving your impression.

PROBLEMS

8.1 A project requires 17,000 allowed hours. Location of the work is above the Arctic Circle, and the efficiency is estimated as 45%. If the gross hourly cost is $50, then what is the estimated cost? If the work is in Texas, hourly cost is $30, and an efficiency of 85% is assumed, then what is the cost of direct labor?

8.2 Five similar bridges are constructed under the direction of one field office. Facility cost for the field office is $250,000. Engineering for the bridges is $350,000. Overhead, which is calculated on the basis of direct labor and material cost, is 5%. Additional data for each bridge are as follows:

Task	Hour Rate Wage	Hour Rate Equipment	Estimated Time (Months)	Estimated Material
Masonry and concrete	$45	$50	15	$1,160,000
Forms, scaffolding	35	40	6	320,000
Asphalt	30	90	3	415,000
Grading	25	50	2	110,000
Surveying	40	20	3	

Develop a simple project-cost estimate by using the items in Fig. 8.1. Find the project cost and the cost for one bridge. (*Hint*: Each month is a 173.3-hour period.)

8.3 A satellite-repeater-link construction project has the following definition:

WBS	Work Task Title	WBS	Work Task Title
1	Satellite repeater	15	Equipment
11	Design	16	Training
12	Construction	151	Preamplifier
13	Documentation	152	Mixer
14	Integration	153	IF, amplifier, filter
		1511	Other hardware

Construct a graphical WBS. How many levels are necessary?

8.4 A project is defined as follows:

WBS	Cost, 10^6	Starting Milestone, Yr.	Ending Milestone, Yr.
1		0	11
11	$165	6	11
12	225	0	8
13	140	7	10
14	160	6	9
121	55	3	8
122	95	0	6
123	75	3	7
1221	15	3	6
1222	25	2	5
1223	15	2	3
1224	25	1	4
1225	15	0	3

Sketch a work breakdown structure to four levels. Plot the cumulative baseline costs. What is the total cost? Chart a graphical schedule of work. (*Hint*: Have the WBS number along the vertical axis for the schedule of work.)

8.5 A project definition, WBS, and the cost estimates for the design, construction, and manufacture of a kerosene-heater product line, are given as follows:

WBS	Work Task Title	Start Period	Ending Period	Cost Estimate, 10^3
1	Kerosene project	0	10	
11	Design	0	2	$1,000
12	Model assembly	2	8	750
13	Plant construction	1	3	8,000
14	Reliability	4	10	1,600
121	Body assembly	2	6	400
122	Tank assembly	2	6	200
123	Burner assembly	2	5	150
1211	Lower body	2	5	300
1212	Upper body	3	5	100
1221	Upper tank	2	4	150
1222	Lower tank	3	6	50
1231	Glass cylinder	2	3	100
1232	Net	3	4	50

Prepare a graphical work breakdown structure. Sketch a schedule of work similar to a bar chart. Plot the cumulative baseline costs. Determine project cost.

8.6 Prepare a cash-flow plan for $6.5 million, matching dollars commitment to expenditure.

Period	0	1	2	3	4	5
Percent commitment	10	18	36	28	8	0
Percent expenditure	0	8	25	39	21	7

8.7 A project is estimated to cost $2,000,000 and the isosceles triangles in Fig. 8.7 are an approximation for the cash flow plan. If the periods now represent quarters, then prepare a cash-flow plan matching dollars commitment to dollars expenditure. When is the approximate midpoint of the commitment and expenditure triangles? What is the difference in money at this time? Discuss the reasons for wanting to know the differences between commitment and expenditure cash flows. (*Hints*: Eyeball the isosceles triangles for the midpoints. These midpoints give the cash flows, and their difference is a ball-park requirement for cash-flow loans or other monies to meet obligations. If bank loans are necessary, then this is the loan requirement.)

8.8 A developmental project has an expected cost of 25×10^6 and a variance of 9×10^{12} dollars^2. Plot a risk graph of contingency. Find the upper-limit cost for 25% overrun; for 25% underrun. (Hint: Allow the variable upper limit to range from $18 to 31×10^6.)

8.9 Financially, a project is considered risky. Management asks for a contingency estimate as insurance for the risk of the estimate. The range estimating method provides a mean cost and standard deviation of $100,000 and $50,000.

   (a) Sketch a risk graph of contingency. What value is added to the mean cost if management desires to operate at +25% risk? (*Hints*: Provide various values for the upper limit, both above and below the mean cost, and find the probability of exceeding this value.)

   (b) Project life is 4 years, and yearly money flow is scheduled as 25, 50, 15, and 10%. Escalation indexes for those years as related to benchmark time of $E = 0$ are 1.000, 1.005, 1.015, 1.050, and 1.100. Find the total contingency cash-flow amount necessary to add to the project estimate where the contingent maximum cost is selected at +25% risk.

8.10 A construction estimate is $200,000 for a contract period of 5 months. The cash flow is trapezoidal, and $t_i = 40\%$ and $t_f = 80\%$. Find the average and maximum cash flows and the time at which they occur. Sketch the cash-flow trapezoid, which is similar to Fig. 8.7(c). Earned profit is 10%, and there is a 1-month delay for payment by the owner. Payment is monthly. Retainage is 20%. Construct a table similar to Table 8.10. Find the net cash flow for the contract.

8.11 A construction bid is $8 million and it has a contract period of 5 months. The assumed continuous cash flow is trapezoidal, and $t_i = 20\%$ and $t_f = 60\%$. Find the average and maximum cash flows, and when they occur. Sketch the cash-flow trapezoid, showing the significant cash-flow epochs and the dates they occur. The profit earned by the contractor is 20%. Payment is monthly. Retainage is 10%, and there is a 1-month delay in payment by the owner to the contractor. Calculate the project cash-flow table.

8.12 Calculate the interest amount for each period, where the interest rate is 2% per month for the net cash flow. Also, find the total interest for the loan.

Month	1	2	3	4	5	6
Net cash flow, 10^6	(1.92)	(2.10)	(1.83)	(1.05)	(0.64)	1.60

(*Hints*: The net cash flow is cumulative. A bank loan of for $1,920,000 is necessary for the first month, and we simplify that the month acts as a single period despite the number of days in the month. The cost for the loan is the interest rate times the monthly remaining balance. For the second month, $2,100,000 - 1,920,000 = \$180,000$ is the additional amount necessary. Starting for month 3, loan repayment begins, for which it is noted that the cumulative amount is less than the preceding month, as $270,000 (= 2,100,000 - 1,830,000)$ is available above any loan requirements, and during the subsequent periods, the loan is repaid, but interest is charged for any remaining balance. Use the following table headings for computation.)

Month	Loan, start of month, $	Month net cash flow, $	Loan end of month, $	Interest, $	Remaining balance, end of month, $

8.13 Calculate the interest amount for each period, where the interest rate is 1.5 % per period for the project net cash flow. Also, find the total interest for the loan.

Period	1	2	3	4	5
Net cash flow, 10^6	(6.1)	(6.5)	(0.5)	2.0	3.5

8.14 A construction estimate gives the following costs:

Work Package Item	
Direct labor	$10,000,000
Direct material	20,000,000
Subcontract	1,400,000
Job and office overhead	600,000

(a) Find the bid, if the markup of full cost is 5%.

(b) Find the bid, if we disregard the cost of direct material for a markup of 20% of *conversion-cost* components. (*Hints:* This pricing policy results when the owner chooses to write the contract to have the owner buy the material and then employ a contractor to "convert" the material. Terms of the conversion contract require that the bid excludes direct material, but includes everything else.)

8.15 Sketch the cumulative baseline value percentage against percentage contract period for the high-voltage transmission-line project using the work package estimates given by Tables 8.2–8.5. Compare this *S*-curve against the results given by Table 8.10, columns 2 and 3. How do the curves compare? What do you suggest to improve the trapezoid assumptions as directed by t_i and t_f? (*Hints:* Consider using a spread sheet for this problem. Increase the estimates for overhead, contingency, interest, and profit by a constant ratio to give a total cash flow of $9,159,072.)

MORE DIFFICULT PROBLEMS

8.16 Consider a project where the materials are divided into four categories: (a) facilities and equipment, (b) long-lead-time materials, (c) standard commercial materials, and (d) raw materials and stock items. In Fig. P8.16 you will notice that the estimates are shown as triangles.

The commitment and expenditure cash value per period can be geometrically modeled against period number. $E = 0$ is the point of the estimate, and 16 is the period at which the project is turned over to the customer. (*Hint:* Assume that cash flows are instantaneous because the project is large and there are many suppliers, so a continuous cash-flow triangle is considered reasonable.)

(a) Plot the total commitment dollar ogive beginning at period 1 and concluding at period 12. Describe the method. (*Hints:* Set up a table with an entry of period number, starting at $E = 0$. Progressively find the intersection of the period with the *y*-axis line. Consecutively sum this value for the periods. This is the ogive baseline curve.)

(b) On the same curve, plot the total expenditure ogive beginning period 2 and ending period 13. Show the method.

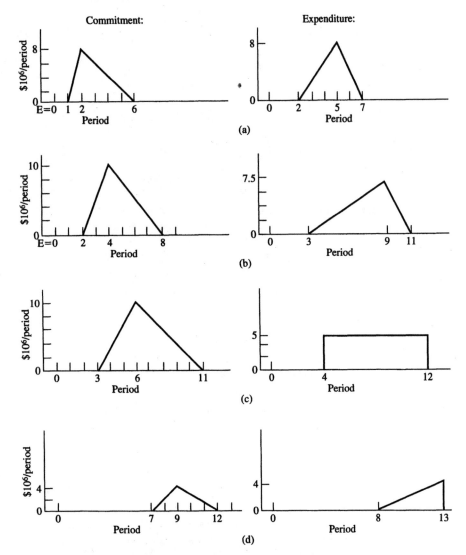

Figure P8.16 Estimates of materials for project. Commitment and expenditure cash flows.

(c) Graphically or analytically determine the effective retainage for the major contractor for periods 2, 7, and 13. In those curves, title the commitment and expenditure axes as *budgeted cost material committed* and *budgeted cost material expended*.

(d) Using the same curves, redefine the range from 1% to 100% for both the *x* and *y* axes.

8.17 An owner requests that a contractor create a *bias contingency* in her estimate–that is, artificially calculating a high estimate to avoid having the owner ask for additional funding from a bank, and thus giving the impression of a project under cost control. Assume that an expected cost of $100,000 and a standard deviation of $50,000 were determined using the method of range estimating at the time $E = 0$. The owner requests a bias contingency of $+ 10\%$ for the 25% over-run risk level. Sketch a risk graph. What is the amount of money needed above the original estimate to meet the owner's request? Discuss the pros and cons of this manipulation. Four-year project cash-flow percentages are 25%, 50%, 15%, and 10% and escalation indexes are 1.005, 1.015, 1.050, and 1.100. Find the period cash flows adjusted for inflation. (*Hint*: Use $110,000 as the mean cost. Thus, a positive bias of 10% increases the mean to an artificial $110,000.)

8.18 A contract of $15 million is estimated. A trapezoid model of cash flow is assumed with $t_i = 30\%$ and $t_f = 80\%$. Project life is 15 periods. Other assumptions are as follows: profit = 10%, owner retainage = 5%, and money as paid from the owner to the contractor is 1 period late.

(a) Find the net surplus or deficit for each period from the contractor's viewpoint. (*Hint*: Use a spreadsheet for the calculations.)

(b) Plot the cumulative baseline dollars (y axis) versus the periods (x axis).

(c) Plot the cumulative baseline percentage against the cumulative contract percent. When does the cumulative baseline percentage equal the elapsed contract percentage?

8.19 A supplier is in the business of producing decorative plating of architectural garnishes and ornamental filigree for commercial projects, such as shopping centers and large buildings, where electroplated embellishments are used extensively. This specialized industrial operation is dedicated totally to the architectural design for the period of time until the order is complete. The sales price depends on the quality and the quantity of the product. Thus, the supplier sells its plating line operation, and the model for the business is given as

$$\text{Plating line profit} = \text{revenue} - \text{cost}$$

$$z = nS_a - (V_c + F) + b_i - m_i$$

n = saleable product per period, tons per day

S_a = sales price, dollars per ton

V_c = variable conversion cost, dollars per period

m_i = raw materials entering the process, which appear as a part of the architectural product, dollars

b_i = by-product waste materials, dollars per period

F = fixed cost, dollars per period

The electroplating line uses special alloys for the anode, which amount to 100 pounds per day. The line is able to operate at three conversion efficiencies of 65, 75, and 80 %. The cathodic output from any level is 25 tons of plated product. Any plating raw material not appearing on the product is fully salvaged following recovery procedures by the firm. The variable conversion costs differ with respect to the level of efficiency. The anode material cost is $75 per pound.

Conversion Efficiency	65%	75%	80%
Plating material usage, lb	100	100	100
Revenue, $/ton	$470	$480	$485
Daily variable cost	$1,825	$2,280	$2,320
Daily fixed cost	$2,250	$2,350	$3,500
Credit for recoverable material	$700	$500	$400

What are the revenue and net total costs for each conversion efficiency? Which one is the most profitable conversion? (*Hint*: Use one day as the basis of analysis. Incidentally, the labor costs are a part of the fixed costs, which includes the usual items but also includes direct and indirect labor.)

PRACTICAL APPLICATION

Develop a project definition, and provide a work breakdown structure showing and describing the principles used in its development. Consider the possibility of a real project known to you or to your instructor. You will need to take a field trip to understand the project.

Decompose the project, where one block in each level is further reduced to a lower level. Indicate the numbering and title for each block. Prepare project assignments for the WBS elements. It is not necessary to estimate the tasks. Form a team for this application. Your instructor will define the limits and operation of the team.

CASE STUDY: HIGH-VOLTAGE TRANSMISSION-LINE PROJECT

An RFQ is packaged for a high-voltage transmission line, similar to the Creston project. The owner's drawings and specifications include the following: 525 kV single circuit, 1.5 GW capacity, 3 × 1.5 in. O.D. conductor per phase, 30 towers, length of transmission line, 8 miles; other data remain the same as given in the chapter example. The work schedule calls for completion in 8 periods. The contractor's database is shown by Fig. C8.1 and Table C8.1

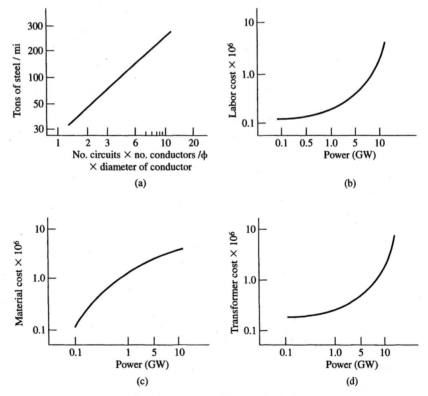

Figure C8.1 Estimating information for high-voltage transmission-line projects. (a) WBS 1313; (b) WBS 1321; (c) WBS 1321; and (d) WBS 1322.

TABLE C8.1 WBS information for estimating high-voltage transmission-line projects

WBS	Type of Estimate	Principle of Estimating	Information
111	Facilities	Unit	$C_a = \$17,315/mi$
112	Facilities	Unit	$C_a = \$1285/mi$
12	Engineering	Linear regression	$a = 600$ hr; $b = 50$ hr/mi;
1311	Labor	Comparison, opinion	6400 hr
1312	Labor	Comparison, opinion	10,800 hr
1313	Labor	Average learning	$K = 3000$; $\phi = 90\%$ for no. of towers
1321	Labor	Curve	Fig. C8.1(b)
1312	Direct material	Opinion	$24,000
1313	Direct material	Curve	Fig. C8.1(a), $I_c = 1.15$; cost/lb = $0.75
1321	Direct material	Curve	Fig. C8.1(c)
1321	Facilities	Historical quote, comparison	$60,000
1322	Subcontract	Power law and sizing	$Q_r = 1.2$ GW; $C_r = \$200,000$; $m = 0.9$; $I_r = 1$; $I_c = 1.18$
1322	Equipment	Curve	Fig. C8.1(d)
133	Subcontract	Cost-estimating relationship	$22,800x_1^{2.39}x_2^{1.15}$; $x_1 = $ dia. of conductor; $x_2 = $ no. of conductors per phase

(*Hints:* For information that is not identified in the case study, use chapter facts. Use the work breakdown structure to identify the WBS tasks. Period designation for estimating is ignored. Use the comparison method to estimate the amount of interest in the bid. Ignore contingency in the estimate. Markup for profit is 7.5%.)

Determine the direct labor, direct material and subcontract, facilities and equipment, and engineering estimates. Conclude by finding the bid, which is similar to Table 8.12.

CHAPTER 9

Bid Assurance

The engineer's interest in his or her estimate continues beyond its initial preparation. The next important step is the fine tuning of the bid, because there is a grooming of the bid before it is laid on the table.

It is an important dogma of this author that the cost estimate is more important for engineering and the design than the actual cost. Competitive bids win new business—actual costs do not. The teaching objective of this chapter presents methods for optimizing the bid, and then regulating costs to match the estimate. This defines the scope of bid assurance.

The estimate's accuracy, reliability, and quality are important. If the bid leads to winning a job, then there is an opportunity to verify its general goodness. This chapter describes the forensic analysis of estimates when compared to a so-called actual value. There are lessons to be learned, for instance, in improving the capture rate, understanding the importance of human behavior, and other techniques.

9.1 BIDDING STRATEGY

Understanding bidding practice is meaningful. For example, we may bid high with a corresponding high markup, which may mean that no jobs are won; or we may bid low and get many jobs at no profit. The higher the bid, the lower are its chances for obtaining work. The lower bid gives better chances for obtaining the work, but perhaps at a loss. This is called the *winner's curse* dilemma. It is the in-between range of the high and low bid where skillful analysis is vital.

The first consideration is "to bid or not to bid." Intelligent preanalysis by the contractor is necessary, and a number of factors need to be addressed, such as

- Strength of company
- Scheduling with existing and future workload
- Bonding capacity
- Market situation
- Risk

If the decision of the firm is "to bid," then the contractor analyzes many relationships to assure his bidding strategy. There is an art and science to competitive bidding.

The spread between the bid price and the cost depends on several factors, such as the contractor's need for work, minimum acceptable markup, and the maximum that the contractor believes is possible. In Chapter 8, markup is defined as a percentage of the job cost, and in this discussion it remains as profit only. Choosing the markup is an inexact, feely kind of thing, wherein experience and opinion are important. Experience would be influenced by the estimate, location of the job, history of success with a range of markups, engineer/architect, competition and so forth. These variables may indicate whether a low or high markup is effective. This intuitive feel is the most common manner of establishing markup. The contractor will attempt to have the largest markup and still win the job. The markup is only so large as to minimize the difference between his firm and the next lowest bidder.

Typically, the approach is to consider each job independent of past or future work. Moreover, there are technical methods that evaluate the history of the bidding pattern to increase the markup percentages and profit over the long run of business. The methods of evaluation of this history in a manner to optimize the value of the bid laid on the table, so to speak, are known as *bidding strategies*. These strategies are adaptable for both fixed-price and unit-cost types of bids, but this discussion is pointed to lump-sum contracts. The per-unit bid prices are the contractually obligated prices for each of the bid items. Bidding in a low-price auction of the type considered now involves two or more contractors whose value is not the same for each bidder, since the value is established by the individual firm's bid.

Previously, we had defined profit as the difference between the bid price and the estimated cost. At this juncture, we mention that our interest is in *expected* profit, defined in the following familiar way:

$$\text{profit} = B - C_e \tag{9.1}$$

$$\text{expected profit} = P \times (B - C_e)$$

where B = bid price, dollars

C_e = estimated cost, dollars

P = probability of event $(B - C_e), 0 \leq \text{Prob} \leq 1$

The bid includes the markup for the profit along with *full* cost. The contractor does not win every job she bids, and in the case of bids that are not won, she receives a zero profit. This is assuming that the estimated cost is exact, without any error. A long-term profit is a function of her winning bids and the profit within each job, and it is also a function of her capture rate. Typically, the contractor will raise or lower the markup rate, depending on the competition, project, and so forth.

If the contractor submits a bid with a high markup, his chance becomes smaller in any competitive bidding arena. As the markup is reduced, the chances for winning the bid improve. It is possible to assign probabilities to various bids that are feasible, and then expected profit becomes realized. Expected profit involves an application of probabilities. These probabilities are numerical judgments of future events. The techniques for deriving probabilities include:

- Expert *opinion*
- Analysis of historical data to give a *relative-frequency* interpretation
- Convenient *approximations* such as the normal distribution function

These techniques were discussed in Chapters 5 and 6, and the student will want to review that material.

TABLE 9.1 Expected profit for one job with a $200,000 cost estimate

Bid, B, $	P(B < next lowest bid)	Expected Profit, $ $P(B - C_e)$
175,000	1.00	$1.00(175,000 - 200,000) = -\$25,000$ (loss)
200,000	0.96	0
205,000	0.80	$0.80(205,000 - 200,000) = \$4,000$
210,000	0.45	$0.45(210,000 - 200,000) = \$4,500$
215,000	0.15	$0.15(215,000 - 200,000) = \$2,250$
220,000	0	0

Assume the simple illustration where the contractor is able to enumerate the probability of being the lowest bidder. An estimate is $200,000 for a project, and there are prospective bids, B, as shown in Table 9.1.

Notice that if the contractor chooses to bid a price less than cost, there is the certainty (in the mind of the contractor) that she will win the bid. At the higher value, $220,000, there is no chance of winning the bid and earning the maximum profit, because competition will have precluded that opportunity.

In examining the expected profits, the optimum bid is $210,000. Expected profit is not the profit that a contractor will make on any one job. Assume that the bid of $210,000 wins, then the job profit is $10,000. Contractors have the objective of maximizing total long-term profit, and expected profit is the meaningful idea for bidding a number of jobs during an indefinite and future time. Repeated bidding for the same job does not occur, of course, but the idea is valid over a variety of projects. These concepts are what might be called "playing the odds."

Before bidding strategies can be discussed further, it is necessary that "our" contractor evaluate his costs against one or more competitors. For simplicity, consider competition against one competitor known as A. This identification infers that our contractor is in a two-person contest, and future competition is in the same economic catch basin.

In public works, bids frequently are opened from their envelopes and read aloud to those attending the bid opening. The contractor is able to gather information at these bid announcements, some of which is useful for future bidding. Also, trade magazines publish the results of bidding. Even in private construction some of this information is legitimately disclosed. For this analysis to continue, we assume that past bidding records are available, which allows the development of Table 9.2.

In this post-mortem analysis, 74 incidents are arranged systematically in a histogram of competitor's bid B_A divided by the cost of our estimate, C_e. See Fig. 9.1(a). These 74 comparisons are assumed recent and will be like the bids of contractor A in the future. It is recognized that we know only that information which is public knowledge, for we do not know the estimated cost of contractor A's bids. For example, if B_A/C_e is 0.98, it implies that our contractor, if he bids this ratio, will bid 2% less than contractor A. In the interval, $0.98 \leq B_A/C_e < 1.00$, the historical data reveal that there was one occurrence and, cumulatively, it is 73/74, as our cost has a 99% probability of being less than the ratio. The cumulative probability is shown as $P(\text{our bid } B < B_A)$ in Fig. 9.1(b).

Continuing, if our contractor bids within a $1.00 \leq B_A/C_e < 1.02$ interval ratio, there is a 93% chance that he will be lower than contractor A. Further, if our contractor bids a *markup* of 2%, then this same probability of being less than contractor A persists. The ratio of B_A/C_e identifies the amount of the markup, which is allowed to become the variable

TABLE 9.2   Bidding history of competitor A against "our" contractor

No. of Comparisons	B_A/C_e	$74 - \Sigma$ Comparisons	$P(\text{our bid } B < B_A)$	B/C_e	Expected Profit, \$ $P(\text{our bid } B < B_A)$ $\times (B - C_e)$
0	$B_A/C_e < 0.98$	74	$74/74 = 1.00$	0.98	$1.00(0.98C_e - C_e) = -0.02C_e$
1	$0.98 \leq B_A/C_e < 1.00$	73	$73/74 = 0.99$	1.00	$0.99(1.00C_e - C_e) = 0$
4	$1.00 \leq B_A/C_e < 1.02$	69	$69/74 = 0.93$	1.02	$0.93(1.02C_e - C_e) = 0.019C_e$
6	$1.02 \leq B_A/C_e < 1.04$	63	$63/74 = 0.85$	1.04	$0.85(1.04C_e - C_e) = 0.034C_e$
14	$1.04 \leq B_A/C_e < 1.06$	49	$49/74 = 0.66$	1.06	$0.66(1.06C_e - C_e) = 0.040C_e$
22	$1.06 \leq B_A/C_e < 1.08$	27	$27/74 = 0.36$	1.08	$0.36(1.08C_e - C_e) = 0.029C_e$
16	$1.08 \leq B_A/C_e < 1.10$	11	$11/74 = 0.15$	1.10	$0.15(1.10C_e - C_e) = 0.015C_e$
6	$1.10 \leq B_A/C_e < 1.12$	5	$5/74 = 0.07$	1.12	$0.07(1.12C_e - C_e) = 0.008C_e$
3	$1.12 \leq B_A/C_e < 1.14$	2	$2/74 = 0.03$	1.14	$0.03(1.14C_e - C_e) = 0.004C_e$
2	$1.14 \leq B_A/C_e < 1.16$	0	$0/74 = 0$	1.16	$0(1.16C_e - C_e) = 0$
$\underline{0}$	$1.16 \leq B_A/C_e$				
74					

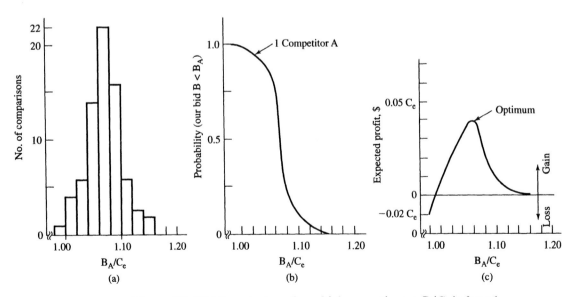

Figure 9.1 Bidding strategy for which an optimum B/C_e is found: (a) histogram of comparisons of cost to competitor A bids, (b) conversion of histogram to probability $(\text{our bid } B < B_A)$, and (c) optimum location for bid-to-cost ratio.

that optimizes our chances of winning the bid. This strategy raises and lowers the markup concerning competitor A. Review of the past data and its arrangement finds the optimum markup for a job, for which we estimate the full cost. We have assumed that our costs and competitor A's costs for the future project are identical, which for the moment we accept.

We find expected profit, or $P(\text{our bid } B < B_A) \times (B - C_e)$, which is traced against the ratio B/C_e. For one competitor, the plot of expected profit against B/C_e is shown as Fig. 9.1(c).

The x axis represents the ratio of the bid to the cost estimate, while the y axis is the expected profit. At the y value 0, there is the crossover point of loss-to-gain for profit, and the y-axis values are in units of the cost estimate, C_e. The peak value is found at about $B/C_e = 1.065$. This suggests that we multiply the cost estimate by a markup of 6.5%, whenever we are in a two-contractor bidding match with contractor A. The expected profit will amount to about $0.04C_e$.

Notice that optimum markup percentage is independent of the estimated cost and it may be determined in advance of bidding. The optimum markup remains the same for any size of estimate, either small or large.

In the next case our contractor bids against two other contractors, called A and B. The post-mortem analysis is identical as in the case of one competitor, except for contractor B. There would be two cumulative probability distributions. These results are shown as Table 9.3.

At the B/C_e ratio of 1.08, there are $P(\text{our bid } B < B_A) = 0.36$ and $P(\text{our bid } B < B_B) = 0.65$. With this markup of 8%, our contractor has the probability of 36% and 65% of being less than A or B. Since these are independent cases, the product of $0.36 \times 0.65 = 0.23$ is the joint probability that our contractor is less than both simultaneously. What we are seeing is that more competitors will invariably reduce the optimum markup. Refer to Fig. 9.2.

The previous two cases dealt with one or two competitors where identified information was available. The next case is that of the average competitor, in which some or all of the bidders are unknown. We label this case as the *average competitor*, or B_{AVE}. This mythical contractor is a composite of the bidders, and these facts are lumped into one distribution. Then if our bidder knows that he will be bidding against three other contractors, the probability is the joint product. For example, if the markup is $B/C_e = 1.04$, then $P(\text{our bid } B < B_{AVE}) = 0.85$. If three competitors are anticipated, then $0.85^3 = 0.61$ is the likelihood that our contractor will be less than the three contractors simultaneously. Table 9.4 shows

TABLE 9.3 Bidding strategy with two competitors, A and B and expected profit

B/C_e	$P(\text{our bid } B < B_A)$	$P(\text{our bid } B < B_B)$	Expected Profit, \$, $P_{A \times B}(B - C_e)$
0.98	1.00	1.00	$1.00 \times 1.00(0.98C_e - C_e) = -0.02C_e$
1.00	0.99	1.00	$0.99 \times 1.00(1.00C_e - C_e) = 0$
1.02	0.93	0.95	$0.93 \times 0.95(1.02C_e - C_e) = 0.018C_e$
1.04	0.85	0.84	$0.85 \times 0.84(1.04C_e - C_e) = 0.029C_e$
1.06	0.66	0.70	$0.66 \times 0.70(1.06I) = 0.028C_e$
1.08	0.36	0.65	$0.36 \times 0.65(1.08C_e - C_e) = 0.019C_e$
1.10	0.15	0.25	$0.15 \times 0.25(1.10C_e - C_e) = 0.004C_e$
1.12	0.07	0.09	$0.07 \times 0.09(1.12C_e - C_e) = 0.0008C_e$
1.14	0.03	0.05	$0.03 \times 0.05(1.14C_e - C_e) = 0.0002C_e$
1.16	0	0.04	$0 \times 0.04(1.16C_e - C_e) = 0$
1.18		0	0

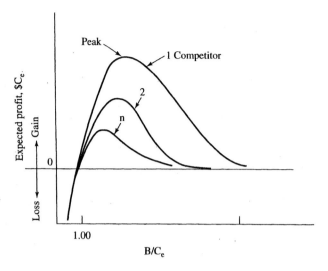

Figure 9.2 Strategies for bidding against 1, 2, and n competitors.

this case. The optimum location is somewhere near 1.04, and the expected profit is about 2.5%. Greater accuracy in the computation, and increasing the decimal places to a finer resolution of the data, are necessary as the pool of competitors increases.

The expected profit and location of the optimum bidding point decline with increasing number of competitors, as is seen in Figure 9.2. The peak is lowered as more competitors enter the bidding contest, which is favorable from an owner's viewpoint. Notice that the optimum moves to the left. Both the lowering peak and movement to the left are to less profit and tighter bid-to-cost ratios for the contractor, and the results approach pure competition.

It should not be assumed that smart bidding replaces competent and qualified estimating. For it is conjectural whether a bidding technique and strategy are not countered by a competitor having superior design and construction engineering practices. Further, there are a number of refinements that were not discussed here. For example, we over-

TABLE 9.4 Bidding strategy with average competitor, and expected profit.

B/C_e	$P(\text{our bid } B < B_{AVE})$	Expected Profit, \$, $P_{3\,AVE}(B - C_e)$
0.98	1.00	$1.00 \times 1.00(0.98C_e - C_e) = -0.02C_e$
1.00	0.99	$0.99^3(1.00C_e - C_e) = 0$
1.02	0.93	$0.93^3(1.02C_e - C_e) = 0.016C_e$
1.04	0.85	$0.85^3(1.04C_e - C_e) = 0.025C_e$
1.06	0.66	$0.66^3(1.06C_e - C_e) = 0.017C_e$
1.08	0.36	$0.36^3(1.08C_e - C_e) = 0.004C_e$
1.10	0.15	$0.15^3(1.10C_e - C_e) = 0.0003C_e$
1.12	0.07	$0.07^3(1.12C_e - C_e) = \text{negligible}$
1.14	0.03	$0.03^3(1.14C_e - C_e) = \text{negligible}$
1.16	0	$0 \times 0.04(1.16C_e - C_e) = 0$

looked such considerations as size of the contract such as small vs. big, giving differing weight to later experiences with the competitors, cash-flow timing, required return, capital investment, and different types of construction.

There are other bidding modifications. The divergence of the actual from the estimated cost can be considered a statistical distribution, and its variation can be entered into the process. Proprietary and fuzzy logic models are software enhancements that various firms employ.

Strategies do not have to involve probabilities, as some contractors adopt other maneuvers that are successful. Consider this stratagem. On a unit-price project, a *balanced bid* is one where each bid item includes its own direct cost, overhead, bond, other costs, and markup. Each bid item has its own fair share of these costs that are necessary in the performance of the task. In a way, all the elements of construction work are consolidated, but for a unit-price project. This differs from the lump sum type of a contract.

For a variety of reasons, a contractor will occasionally increase the unit costs on selected bid items while decreasing the unit costs on others. This is done in such a way that the jiggling of unit prices leaves the total bid price unchanged, as if the unit prices were as calculated on an unbiased estimating basis. This deliberate adjustment is called *unbalanced bidding*.

The reasons for this unbalanced bidding are summarized as

- Offset early mobilization costs
- Errors by owners in material take-off quantities
- Perceived productivity advantages or disadvantages

Unbalanced bidding may allow early payments to cover the higher cost of startup, where, if flat rates were used instead, the cash flow recovery for startup would be delayed. This type of cash flow could reduce the need for bank loans for expenditures. Earlier reimbursement is possible with unbalanced bids.

A contractor may detect errors in the determination of the take-off quantities that the owner submits for each contractor's bid. During the progress of the work, actual quantities will be certified, and the contractor is paid as the work progresses. If there are errors in the estimation of the quantities, and the contractor has noted these contradictions, greater profit is possible.

On some tasks the contractor may have an advantage over competitors and be able to increase the markup on those bid items. Correspondingly, the contractor may be aware of inefficiencies and may choose to reduce markup on those tasks. The unbalancing may be to adjust markup rates on work items.

The unit-quantity procedure, as described in Chapters 6 and 7, is the estimating procedure that is selected for the unbalanced bid strategy. The unit-quantity method consolidates labor, material, overhead, equipment, subcontractor, and markup into a single unit price. It is this cost that is posted to the column "balanced unit estimate." Table 9.5 shows the techniques.

Finding the required quantity is the responsibility of the owner. The contractor is promising to perform the work to specifications for the unit estimate, which is a calculation that is proprietary to the contractor. Adjustments are made to the unbalanced unit estimate, which reflect the strategy of the contractor. The contractor submits the bid showing the modified unit estimates and the total.

TABLE 9.5   Balanced and unbalanced bidding strategy

Work and Material Take-off	Unit	Owner's Quantity	Balanced Unit Estimate	Balanced Total Bid, $	Unbalanced Unit Estimate	Unbalanced Total Bid, $
Lumber	bft	1,300	$2.10	$2,730	$2.25	$2,925
Reinforcing steel	lb	395	4.10	1,620	4.25	1,679
Concrete	yd^3	7.5	112.50	844	78.70	590
Total				$5,194		$5,194

TABLE 9.6   Payments between balanced and unbalanced bid

Work and Material Take-off	Actual Quantity	Balanced Payment, $	Unbalanced Payment, $
Lumber	1,375	$2,888	$3,094
Reinforcing Steel	408	1,673	1,734
Concrete	7.4	833	582
Total		$5,394	$5,410

Once the job is completed with the quantities known, final payment is made to the contractor. Table 9.6 shows the results of unbalanced bidding. A net gain between the two approaches can be determined.

When submitting an unbalanced bid, the contractor assumes the risks and disadvantages of this strategy. Further, it is possible that unbalancing can result in losses to the contractor. Finally, the contracting authority may disqualify the bid as unsuitable if it recognizes these maneuvers. Owners, who are aware of these practices may permit a startup fixed cost by the contractor, giving the owner an opportunity to realize these mobilization costs. Permissive clauses of this nature reduce the need for unbalanced bidding.

Unbalancing bidding is not a new strategy in the twenty first century, for contractors have used this stratagem for over 100 years. But it remains an intriguing business gambit.

More information may become available to the engineer during the estimating and analysis period. Tips may be found in the local newspapers, budget disclosure, or the owner or major contractor may indicate boundaries for the bid. Rebidding may occur in large projects, and first bids may be known. Typically, past winning bids are open knowledge for public works. Business magazines and trade newspapers publish information regularly. Even rumors are sought. Eventually, the bid is packaged and presented according to the dictates of the owner or architect or project manager representing the owner, oftentimes following a formal procedure.

9.2   PRESENTING THE BID

Bidders need to be aware of owner instructions, as the contractor will frequently follow guidelines in the preparation of the proposal. These instructions usually include such things as

- Qualification of the bidder
- Site examination, survey, subsurface tests, etc.

PICTURE LESSON
Steel

The picture shows the erection of the structural steel frame for a power plant. Portable-motorized and tower cranes are used for lifting. One of the basic and most important commodities for contemporary heavy construction, steel is the sinew of engineering design for buildings.

Steel has its beginning with pig iron, the molten product of the blast furnace, which is obtained by smelting iron ore, coke, and limestone with air or oxygen. Steel is a crystalline alloy of iron, carbon, and several other elements. Carbon is an important constituent because of the ability to increase the hardness and strength of steel. More tons of steel are used than all other metals combined. Steels are further classified into carbon steels and alloy steels. The major elements for alloying the iron after carbon are manganese, molybdenum, chromium, vanadium, tungsten, etc. The alloying elements added during steel making increase corrosion resistance, hardness, strength, and other physical properties. An enormous number of distinct properties can be contrived by selection of alloying elements in the recipe. Unlike wrought iron, steel contains no slag, and it may be rolled, cast or forged.

Molten steel is either cast into large ingots or continuously strand molded. Hot working is the process of converting these steel-mill intermediate products into the structural steel products for construction. The metal deformation process of rolling the structural shapes exploits an interesting fact of metallurgy: the ability of metals to flow plastically above the recrystalline temperature (but below the softening temperature) in the solid state without the accompanying deterioration of properties. Moreover, in rolling the hot metal into a desired shape there is little or no waste of material. The term "hot finished" refers to steel bars, plates, expanded metal grate, beams, columns, and other structural shapes that are purchased in the as-rolled condition from steel suppliers or fabricators.

- Bid security
- Bid forms
- Date and place of bid submission
- Bid modification and withdrawal
- Legal requirements
- Opening and award procedures

Competitive bidding is usual for projects. In the simplest case, the owner will announce a deadline date for the bid, and sealed envelopes containing the bid and other information are opened and the winner is announced. In technical projects, the bidder provides a design along with the bid. In many cases, evaluation may take a long time before the winner is selected. The details of the submission procedures are important. The owner will have a variety of stipulations.

The owner may reject any bid not prepared and submitted in accordance with the provisions or reject any or all bids. Bids received after the time and date specified may not be considered. Generally bidders are required to use an owner's bid forms. All blank spaces must be filled and prices must be in both numbers and written words. The sealed bid should contain the bid form, bid security (check or bid bond), and power-of-attorney (if bid bond is used as bid security).

The owner may have a requirement for minority business opportunity, or insist on contractor and subcontractor qualifications, such as years of experience or jobs that are valued in excess of some floor cost. General contractors shall include in their bid envelope statements, references, and information showing experience in projects of similar scope and complexity to this project.

If the bidder is doing business as a corporation, the bid shall be signed by an officer, i.e., president or vice president. If the bidder is an individual or a partnership, the bid shall so indicate and be properly signed and witnessed.

The owner specifies the nature of the selection, such as lowest responsible bidder. The determination of the lowest responsible bidder is made on the basis of the base bid, unless alternates are requested with the bid.

The bid security must be accompanied by cashier's check, certified check, or a bid bond executed by the bidder. If a surety company is used, then the owner may require approval of the surety company. Such checks or bid bonds are returned except one to three of the lowest bidders after the opening of bids, and the remaining checks or bid bonds are returned after the accepted bidders have executed contracts.

The owner may have requirements prohibiting the services of a bid registry or any other such bid service, where for example software that is issued by companies specializing in this service is considered to be anticompetitive and illegal restraint of trade.

The owner may insist that the contractor provide evidence of being able to be bonded up to some designated value. The successful bidder, upon failure or refusal to execute and deliver the agreement and performance payment within a specified period of the notice of award, shall forfeit to the owner the security deposited with the bid as liquidated damages.

Once the deadline is passed for submission of the bid, the owner (or his architect for buildings) proceeds with the opening of the bids. Bids for public works are read aloud before the public. Most bids for private projects are not released as public information and remain proprietary. For public bid disclosure the contractor or her representative is typically in attendance, taking notes, and determining if the selected contractor is nonresponsive and fails to meet the RFQ requirements, and if the government is appropriately discharging its legal obligations. A losing contractor may attempt legal redress if the government acts against its RFQ instructions. Additionally, the information that is tallied at these bid openings is useful for analysis of estimates and optimization of bids—topics previously covered.

Though it would seem that the bid opening and selection of the winning contractor conclude the owner's attention to the bid value, they do not. The owner may choose to return to the best several candidates and negotiate on "best-and-final offers." The owner fine-picks the contractors' details and suggests alternatives that may lower costs. Changes in the owner's design allow this negotiation to continue.

9.3 ANALYSIS OF ESTIMATES

Cost estimating is seldom done without an effort to check its success. A common method of verification is to find the *so-called actual costs* and compare those with the original estimates. This analysis was done in an actual study of 157 cost estimates, as shown in Fig. 9.3. In this study the deviations range from a low of 50% to a high of 450%, but the sum of the estimated amounts is less than 4% above the total actual cost. Notice that the curve is non-symmetrical and is called "positively skewed." Nonsymmetrical data can be normalized for additional statistical analysis, but those techniques are beyond the scope of this text.

Obviously, all the jobs that were estimated low did not yield as much as expected. The others brought in more than expected, because they cost less than estimated.

For simplicity, the deviations in cost estimates are expressed by the ratio C_e/C_a. This does not change the shape of Fig. 9.3, which shows the new scale C_e/C_a. This ratio equals one where actual cost equals the estimated cost and is the important *break-even* point. Those points to the left on the horizontal axis are operationally undesirable, because the firm is operating at a loss. Depending on bidding policies and objectives, the preferred location of estimates is to the right of the break-even point. Long-range survival depends on this.

Curves such as Fig. 9.3 are seldom found, because actual costs may not match neatly with the design and estimate. Inasmuch as estimates precede their actual costs, time passes between the estimate and the historical determination of the actual value. In some situations, data may never be gathered to allow even a nominal comparison. At the other extreme, an abundance of data may be on hand. Due to accounting effects, a side-by-side

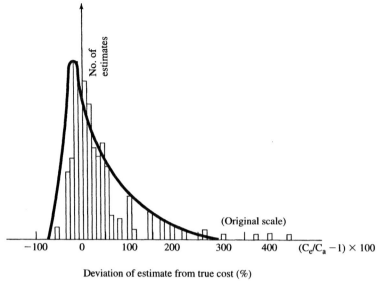

Figure 9.3 Distribution of estimates plotted against actual costs.

comparison may be impractical. For instance, depreciation, deferred costs, loans and insurance, refunds, and the like cloud the collection of actual costs. In practice, a reconciliation between estimate and actual measure, though very desirable, is difficult to achieve.

In further discussion, the estimated cost, C_e, is understood to be the total cost for a job. This includes the usual items of cost, direct and indirect, subcontract materials, equipment, overhead, engineering, and contingencies. The contingency allows for unassignable extra costs that may occur. The actual cost, C_a, includes those same factors of cost and contingencies, because they are inherent in the performance of each job.

A natural question is, "What can be learned from an analysis of this sort to help improve the cost-estimating practice?" Should some amount be added to each estimate to reduce the losses from the low ones? If so, then how much? To do so would raise the high estimates and diminish their chances in competition. The low estimates cannot be identified beforehand or there would not be any low estimates.

Though analysis of this sort is useful, it does not give information about job proposals that were not selected. So, what about the performance of losing bids? The actual costs of unsuccessful bids were never determined. Study of estimating policies must consider all estimates, not just the ones that have produced orders.

An important objective for the estimating team is to produce estimates that are exact. For estimates will deviate from the actual cost. This thought introduces the *error of the estimate*, which is defined as

$$E = \left(\frac{C_e}{C_a} - 1 \right) \times 100 \tag{9.2}$$

where E = error of estimate, percent
        C_e = estimate of cost, dollars
        C_a = actual cost, dollars

Exact estimates are desired. While this is a commendable purpose, it is more realistic to say that the goal is to have the estimate value fall within some acceptable range. This advances the notion of a *tolerance*. Factors such as cost of preparation, time available, impact on the organization, and data requirements bear on the selection of an estimate tolerance.

In making an estimate, we perform a combination of computations—usually addition, multiplication, exponentiation, and so forth. The data for these computations are from a variety of sources, each with its tolerance or error. In any system, say a chemical plant composed of tanks, heat exchangers, columns, and so on, each of the individual components is estimated at the center of expectation, or the mean. There is a range of values, usually what is called *plus or minus*. If the constructed system is composed of a random and independent selection of the equipment, the overestimated actual cost tends to compensate for the underestimated actual cost components of the system. The total cost of the system is the sum of the expected estimates or means of the individual components. The estimated percentage error of the system is less on a percentage basis than it is for the individual components of the system taken one at a time. An example illustrates, as shown in Table 9.7:

TABLE 9.7   Illustration of addition of estimating costs with tolerance

Equipment	Estimated Cost, $	Tolerance, ±	Cost Tolerance, $	Squared Cost Tolerance, $^2 \times 10^6$
Reactor	3,000,000	±20	±600,000	360,000
Direct materials	9,250,000	±10	±925,000	855,625
Engineering fee	2,000,000	±5	±100,000	10,000
Site development	3,000,000	±5	±150,000	22,500
Process building	2,100,000	±10	±210,000	44,100
Total	$19,350,000		$1,985,000	1.292225×10^{12}

The square root of the sum of the squared cost-tolerance is $1,136,760. If this is divided by the sum of the expected costs of the components of the system, we have ±5.9% (=1,1367,760/19,350,000). When independent estimates add or subtract, the uncertainties in those quantities always add. This illustration does not extend to propagation of errors for multiplication, and so forth.

Costs are *estimate*, *actual*, or *standard*. Actual costs are incorrectly assumed to be the more important. However, it is infrequent that actual costs are accurately known. Further, actual costs are not always indicative of future values. Sometimes actual costs are more expensive to uncover than estimates, and they are not available until after the project is complete. Then actual costs are too late, except for analysis and reestimating. Contrariwise, actual costs, or their approximations, are necessary for accounting analysis in the income statement and balance sheet, as we learned in Chapter 4.

Standard costs are hypothetical and in a sense are "should be" costs. Standard costs are fashioned on work-measurement or accounting principles and are useful for construction-work estimating, income and expense recording, tax finding, and other financial reports. Of course, standard costs are never "true" except in the sense of a definition.

It is a dogma of this author that the estimate value, properly determined, is superior to either actual or standard values. While actual costs for large-scale projects (public projects in particular) may never be determined, it is interesting to speculate on the deviation from estimates that would be revealed if actual costs were known. This difference is due to the error of the estimate. The three kinds of error weeds in estimating are

- Mistakes
- Policy
- Risk

Mistakes result from imprecision, blunders such as $2 + 2 = 5$, and omissions. Typically, mistakes pass unnoticed, but if they occur, "nature" may be kind due to compensating effects of offsetting mistakes. Prevention results from uncompromising arithmetic and strict attention to methods that inspire faultless computation. The use of computers is a popular solution for overcoming mistakes. As a tireless machine, it removes the burden of routine calculation. Many estimates are handled not by a computer, but by pad and pencil. Cross checks by other engineers or the stapling of a calculator's paper tape to the estimate, or row and column arithmetic simultaneously agreeing for the final value, are simple ways to reduce mistakes.

Preventing omissions in cost elements is encouraged by pro forma procedures, cross talk within the estimating team, and checklists. Typically, most management views errors as inferring only mistakes.

Policy errors are errors of belief made through ignorance or inadvertence. Simplified illustrations include failing to recognize material price breaks for quantity purchase, or overlooking a planned contractual increase in direct labor cost. Excessive or low values for overhead ratios, for instance, or first cost or cost of operation are typical mischoices. A cost-estimating relationship may have higher statistical correlation with a nonlinear relation instead of a linear model. For those simple situations of estimating with construction encyclopedias, a policy error would be using unconfirmed data. Errors of policy are prevented by well-thought-out policies, data collection, and analysis practices.

Risk errors are the least understood. Assume that in a competitive bidding situation your value is "perfect," and yet a competitor submits a lower value. The lower value may be the result of their more productive business, and thus, your failure to win is a consequence of your unproductive business. On the other hand, the competitor's bid may be a "low-ball," a bid simply to keep the work force busy. Alas, your bid still lost. Contrariwise, your bid may win because your business is more productive, or competitors are not really interested and submit high bids, or they make mistakes and overbid. Maybe your bid is intentionally low to keep your work force busy. Profit margins may be overlooked also. *Risk error* is the difference between "your" estimate and the winning value.

The winning bid for public works construction may be known, where the law dictates open disclosure of information. In private business, some competitors' estimates may be proprietary and are unknown. Thus, the deviation between the estimate and actual cost for losing estimates may never be known. It is important to improve business by increasing the yield of the estimates made.

Equation (9.2) finds the error by comparing actual costs to estimates and overlooks policy and risk errors. Unfortunately, the comparison is not possible until after the actual costs are reported. If there are excessive overruns, they may not be apparent until after the job is complete, which is too late. An improved measure of error, Eq. (9.3), compares a preliminary and a detail estimate, and it also tends to consider policy errors:

$$E = \left(\frac{C_p}{C_d} - 1\right) \times 100 \tag{9.3}$$

where C_p = preliminary estimate of cost, dollars
        C_d = detail estimate of cost, dollars

Naturally, the preliminary and detail estimates are made for an identical design. The preliminary estimate is made quickly and independent of the detail estimate. A detail estimate uses a pro forma procedure and emphasizes comprehensiveness. There are pitfalls to avoid. It is necessary to prevent the preliminary estimate from becoming a self-fulfilling value, where a detail estimate is guided to match a preliminary estimate. Two groups or individuals or alternate methods may make those estimates separately. The value of C_p should be restricted on a need-to-know basis so as not to influence the value of C_d Chapter 6 discussed a number of approaches and the philosophy of preliminary methods.

Any discussion of the analysis of the estimate needs to consider the cost of estimating, which is a labor-intensive activity. The amount of estimating detail and the effort that go with the preparation contribute to overhead cost.

Imagine the following picture: An estimator leans back in his chair, he glances at the drawings, looks at his navel, closes his eyes, and a moment later he says "Bingo! The cost is eighty-three thousand, seven-hundred dollars, and yes, thirty-nine cents." Now imagine another picture: A different estimator, examining the same drawings, looking to his computer screen with its minutia of data for the umpteenth time, and scribbling on an extensively used computation pad, enters the value of $118,763.25. Which is right: "personal guessing or analysis-paralysis?"

The answers to this question are debated in the construction-estimating field. Rightly so. For this activity needs to be conscious of its own costs of operation and its effectiveness. There is a cost of making a cost estimate and relying on the value of that estimate. In some businesses there is evidence that 90% of the estimates do not lead to contracts and sales.

This book stresses that quality and quantity of information are crucial to estimating. If information is lacking, then estimating methods accommodate to the precondition that data are limited. Even so, many hard-dollar decisions are made with little information. As more information becomes available, estimating methods earn the appearance of certified tools, and faith and acceptance, justifiably so, improve. Decisions get better, errors are reduced, and a positive factors become apparent. If this is true, then with what level of detail should the estimating activity operate? Should we employ the detail of the data warehouses, or use preliminary methods with less information? Should we insist on an abundance of data, or are tried-and-test rules of thumb acceptable? We do not give specific answers to these questions, but, broadly, the issues are made clearer by examining Figure 9.4 (a).

The axes are not scaled in this figure, other than to show increasing or decreasing magnitude. The amount of estimating detail increases left to right. The detail of estimating information influences the costs of database development, training, methods improvement opportunity, computer advantages, and so forth.

We have discussed the errors of the estimate, and these errors can cause estimates to gain or lose business on their own dismerit. In Fig. 9.4(b), $C(E)$ is the monotonic decreasing cost of the errors of the estimate as the amount of detail increases. Alternately, $C(M)$

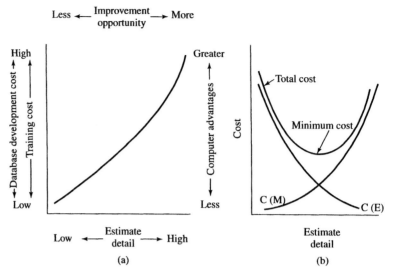

Figure 9.4 Illustrative arguments on estimating detail: (a) tradeoff of information detail, and (b) optimum cost for estimating detail.

is the monotonic increasing cost of making the estimate as the detail increases. Then it is possible to write an expression as

$$C_T = C(E) + C(M) \tag{9.4}$$

where C_T = total cost of making estimate, dollars
         $C(E)$ = functional cost of the error in estimate, dollars
         $C(M)$ = functional cost of preparing the estimate, dollars

$$= C_f + C_v(n - 1)^a, \quad n = 1, 2, 3\dots \tag{9.5}$$

where C_f = fixed cost of preparing the estimate, dollars
         C_v = variable cost of preparing the estimate, dollars per element n
         n = elements or detail of estimate, number
         a = experience exponent

Equation (9.5) and its empirical coefficients are uncovered by regression methods. (See Chapter 5 for development.) The fixed value C_f includes space, utilities, training, documentation, and so on. C_v is related to second-level estimates, as the function assumes that a first-level estimate is always made. If there is only one item or detail, then the cost of making the estimate is C_f. More advanced techniques are necessary to understand the cost of the error term. There is an end-of-chapter exercise for the student to find the cost of preparing the estimate.

Consistency is another quality of estimates and of the estimating group. There is a criticism that engineers are seldom consistent one day to the next and that different people will estimate the same design differently. True, the nature of estimating requires a pinch of judgment and experience. Technical work, by its very nature, has variances between competent and trained engineers. Estimating is no different. One standard statistical model, useful for evaluating consistency, is to construct the noise-to-signal ratio, or

$$SD/C_{ave} = \text{noise-to-signal ratio, number} \qquad (9.6)$$

where C_{ave} = average of several cost estimates, dollars
        SD = standard deviation, dollars

C_{ave} is the average estimate of several engineers of the same design under the same procedures. The SD is the standard deviation of the number of estimates. Usually, if there is a database, and estimating rules are followed in making estimates with the database, then the noise-to-signal ratio is smaller than for situations where judgment plays a greater role.

Ultimately, estimates either win or lose in the sense that business is obtained. Contractors submit cost estimates to potential customers. Winning and losing business is functionally related to cost-estimating performance. This analysis is coined the *capture rate*, and is defined as:

$$\text{capture rate} = \frac{C_w}{C_t} \times 100, \text{ percent} \qquad (9.7)$$

where C_w = cost estimates that win, number
        C_t = total cost estimates, number

This capture rate may be determined periodically. Some subcontractors make hundreds of estimates per month, others make only a few yearly. The capture rate may differ between new and repeat business. Some firms adjust their costs or profit margins, depending on the direction of the capture rate. If the capture rate falls, then profit margins are reduced, and vice versa. Some firms operate successfully with 10% capture rate and others are unsuccessful with a 90% rate. This equation alone does not give the entire picture of successful estimate assurance. The equation can be modified to indicate capture on the basis of dollars that win to dollars that were bid.

Direct labor costs deviate from estimates because of inefficiency. Adjustments to the cost estimates are made by using an overall efficiency factor. Engineers usually think of efficiency as analogous to output/input, which is a number less than 1. We recommend the reverse ratio. An output/input factor would divide the new estimate to indicate a "realized estimate", or "adjusted estimate."

$$\text{efficiency} = \frac{\Sigma C_a}{\Sigma C_e} \qquad (9.8)$$

For instance, if ΣC_a = \$515,758 and ΣC_e = \$471,507, then efficiency = 1.094. If those data were representative of future work, the total estimated cost for a future design would be multiplied by 1.094 to anticipate the cost actually reported. The efficiency factor may be found from a single sample or a time series of experiences. It is doubtful if a constant efficiency factor is ever achievable. Ultimately, the engineer will use opinion and experience to state whether a historical value can be used for the future.

Despite a variety of possible ways to make an estimate, resources for the preparation are always restricted, because time, money, and the technical staff are limited. Eventually, a point is reached where the objectives must be satisfied with what time, money, and intelligence are available. The ideal policy has the estimate coincide with the reconciled actual cost.

Actual cost is, perhaps, never known. No procedure, mathematical technique, or policy employed in the engineering world is without its flaws and shortcomings or is able to guarantee perfect estimates. Although flaws in estimating may be obvious, those procedures and techniques are used for the simple reason that they are the best means at hand. Imperfection seldom deters usage.

The behavior of the business and its people has a good deal to say about the quality of the bid, as we shall see next.

9.4 BEHAVIORAL CONSIDERATIONS

Another goal of estimate assurance is to develop and sustain business systems and to inform management about the extent, type, and status of expenditures. The computer has given impetus to systems by enabling them to be more effective. Computerized business systems should lead to improvement of estimate assurance, but it is recognized that this is only the routine part. Though those enamoring features are traditionally thought to be the essence of estimate assurance and despite their development, there remains a general lack of estimating success.

Little heed is given to the possibility that those systems operate in an organization. Behavioral principles need to be considered, since they encourage the better preparation of estimates. (A detailed discussion of behavioral principles is beyond the scope of this text.) Once the estimate is "secured," so to speak, and the business is in hand, the assuring of the bid continues.

A general "control" model consists of actual and standards, measurement of actual results, comparisons of actual results to standard, and engineering action. Figure 9.5 describes a simple network.

Those control models that do not respond to external social or economic influences are known as *closed*. Systems that, for example, respond to poor labor efficiency, late material delivery, engineering changes, competition, and price changes are known as open.

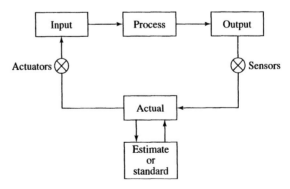

Figure 9.5 Simple control model for bid assurance.

PICTURE LESSON
Proud Moment

This 1937 photo is of a turbine rotor assembly, prior to its installation in the Bonneville Dam's first powerhouse. The rotor weighs 13 tons. At the time of completion, Bonneville's powerhouse was the largest in the world.

Turbomachines are classified partially by the flow path through the passages of the rotor. When the path of the through-flow is wholly or mainly parallel to the axis of rotation, the design is termed an axial flow turbomachine. The cast-steel blades sweep through a circle 23 feet in diameter. The blades were designed for the Kaplan-type hydraulic turbine. The bullet-shaped nose allows laminar flow of water, minimizing turbulence on the diffuser or discharge side of water flow.

Two main categories of turbomachines are those that absorb power to increase the fluid pressure or head (ducted fans, compressors, and pumps) and those that produce power by expanding fluid to a lower pressure or head (hydraulic, steam, and gas turbines).

Open systems do not lend themselves to automatic regulation and are more difficult, and hence challenging. Less management time and finesse are necessary for a mechanistic or closed system.

The "sensor" measures the output transmitted for comparison with a standard. If an unacceptable difference is noted, then action is taken by an "actuator."

Let us relate those model components to engineering and estimating. Estimates or standards deal with performance, time, and cost. To be meaningful, unambiguous, and useful, the estimate must be quantitatively stated in measurable operational units. Measurement of results, records, and actual cost relates to the cost code and work breakdown structure, and so on.

Work and cost standards are noticeably absent in the construction business. As such, the ineffectualness of the construction industry to increase its productivity can be partially attributed to its inability to establish work and cost standards methodically. Undoubtedly, the uniqueness of the construction jobs gives an impression that standards cannot be established. Estimates that are engineered can in some respects be considered as standard. The development of these work standards was discussed in Chapter 2. There is not the labor efficiency in construction that is found in manufacturing. Alas, arguments about high-volume manufacturing versus construction uniqueness are too shallow to let us ignore the lack of worker performance in the construction industry.

Comparison of actual results to standards is done by sundry means. There are many excellent texts that tell how. A simple example is the side-by-side comparison of estimate and actual costs for material, labor, and overhead, broken down in as fine detail as required. Much of this is done routinely in conjunction with results management. Trending and charting of milestones are other techniques for this purpose.

Engineering actions encourage performance to match the objective. If there is agreement between the actual result and the bid, then action is unnecessary. A match of this kind is unlikely. Because estimates are "open" systems, change is commonplace, and management is needed in order to encourage conformance. This is a crucial component in the cost-estimate assurance program.

A management objective may choose to have the bid's accompanying labor targets made "tight or loose." Behaviorally, it can be argued that tight labor productivity will motivate attainment and lower cost, as Fig. 9.6 suggests. A counterargument is that loose labor productivity which is easily attained will probably generate good feelings. Here psychological motivation is needed for self-esteem. Research findings are biased toward tight estimates for productivity attainment. How tight the cost estimate should be is unresolved, because if set too tight, then the labor standard may demotivate and cause lower performance. The student will want to reread Chapter 2 for more information on construction labor measurement.

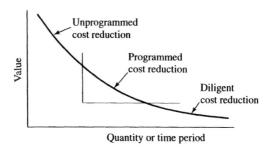

Figure 9.6 Management reactions to meet cost estimate objectives.

The cost estimate should reflect the anticipated *money out of pocket* that the task will require. Thus, if a diligent cost-reduction program is planned, lower cost is anticipated by the cost estimate.

Successful bid assurance operates in the regime of human activity, and behavioral models, such as the simple one displayed here, become important to achievement. The construction-work estimate is the point at which principles of cost departure from the estimates are first applied. These departures are called "variances."

9.5 WORK-ESTIMATE ASSURANCE (OPTIONAL)

Construction work is a value-added conversion of direct material by direct and subcontract labor. Comparison of actual to estimate values is important in the cost assurance of construction work.

The unit-quantity method of construction estimating is significant in bid assurance. For example, a standard unit cost of construction work is a predetermined cost computed even before construction is started. In estimating the construction cost of a design, it is necessary to study the kind and grade of materials that should be used, how labor jobs ought to be performed, how much time each labor job should take, how the indirect services should be best administered, and what the entire specifications are for the complete and total construction. The aim, of course, is to specify the most effective method of constructing the design and then, through adherence to the specifications in the actual jobs, achieve the lowest practical unit cost. This planing occurs whenever the unit-quantity estimating method is used. For information about the unit-quantity plan, see Chapters 6 and 7.

The word "standard" is often used synonymously with "estimate." Accountants are involved with standard costs. The engineer, when referring to standards, thinks in terms of a rigid specification, but there are differing viewpoints. A standard cost, as discussed here, provides a dollar amount that is a "should be" amount. It is not an immutable natural law. In the present discussion we do not differentiate between standard and estimate.

An attitude is necessary in viewing standard costs. Standard costs may be classified as perfection-level standards, and engineering encourages attainment of those standards as goals. Some businesses contend that perfection standards are preferable to attainable standards, because perfection standards provide a stimulus to workers and engineering to achieve the best possible performance.

The unit-quantity method to estimate direct material cost uses quantity, requirements, and the material unit cost. Actual material cost may have a *difference* from the estimate because of a quantity, or material unit cost, or both. Understand that the word difference means *variance*. In considering estimate-assurance principles, we are not discussing statistical variance.

The reported actual costs are very likely to deviate from the estimates or standards. These differences are called *variances* and are expressed as dollar amounts or percentages. They are *favorable variances* when the actual costs are less than the standard costs and *unfavorable* when actual costs exceed standard costs. It should not be interpreted that excess of actual cost or of standard cost is detrimental to the firm. Similarly, not all favorable variances represent actual benefits to the company. The terms favorable and unfavorable, when applied to variances, only indicate the direction of the variance from standard cost.

Some businesses use variance analysis to understand inflation, deflation, schedule delays, or engineering change orders. Those variances are the result of an owner or contractor action. The variance analysis allows for an explanation of the cost increases or decreases.

Information for variance analysis is found from the unit-quantity estimating method and the report about the actual costs. Consider an example for material:

	Number of Units	$/Unit	Total
Estimate	200	17.38	$3476.00
Actual	210	17.05	3580.50

The actual cost is greater than the estimate, but why? There are two possibilities–the number of units and the cost per unit. An unfavorable variance of $104.50 resulted in part from an excess of 10 units and in part from an increased unit cost of $0.33. The analysis is pictured in Fig. 9.7.

Figure 9.7 starts with actual material cost per unit times actual units. We change one of the factors (cost or quantity) in the estimate. The difference gives a variance due to the factor that was changed. The remaining difference is the variance for the other factor. In Fig. 9.7(a) the quantity factor is changed first, and for Fig. 9.7(b) the material cost factor is changed first. Those procedures result in an ambiguity in the quantity or cost variance, de-

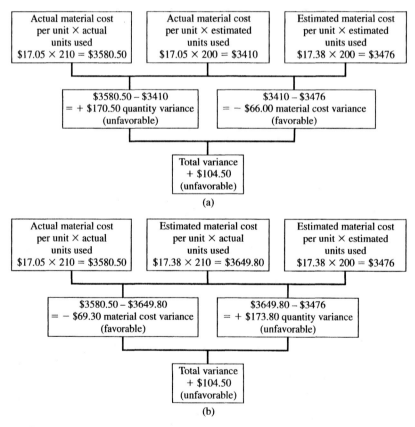

Figure 9.7 Finding material cost variance: (a) quantity first, and (b) unit cost first.

pending on the order of calculation. While the total variances are identical by either method, the intermediate steps and their variances are not.

This ambiguity can be significant in owner-contractor or contractor-subcontractor disputes. Terms of the contract may stipulate the manner in which variance analysis is undertaken. A typical resolution is as follows:

$$V_m = (N_a - N_e)C_e \qquad (9.9)$$
$$V'_m = (C_a - C_e)N_a$$

$$\text{net material variance} = V_m + V'_m$$

where

V_m = variance for material due to quantity change, dollars
V'_m = variance for material due to cost-per-unit change, dollars
N_a = actual quantity, number
N_e = estimated quantity, number
C_e = estimated material dollars per unit
C_a = actual material dollars per unit

Those calculations are shown in Fig. 9.8(a). Both favorable and unfavorable variances are possible. By using Eq. (9.9) we find the solution to Fig. (9.8) as V_m = \$173.80 and V'_m = −\$69.30. The net material variance is an unfavorable \$104.50 (=173.80 − 69.30).

Labor costs can be analyzed for both man hours and the labor dollar rate. Figure 9.8(b) illustrates the notation.

$$V_l = (MH_a - MH_e)R_e \qquad (9.10)$$
$$V'_l = (R_a - R_e)MH_a$$

$$\text{net labor variance} = V_l + V'_l$$

where

V_l = variance for labor due to difference from estimated man hours, dollars
V'_l = variance for labor due to difference from estimated hourly labor rate, dollars
MH_a = actual total hours for job

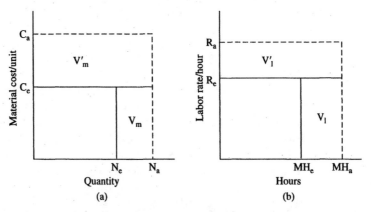

Figure 9.8 (a) Variance for material; and (b) variance for labor.

MH_e = estimated total hours for job

R_a = actual labor hourly rate, dollars per hour

R_e = estimated labor hourly rate, dollars per hour

Consider the example for labor wage rate:

	Wage Rate, $/hr	Hours	Total
Estimate	$18.67	283	$5,283.61
Actual	17.32	325	5,629.00

The unfavorable variance owing to changes in hours is +$784.14 $[=(325 - 283)18.67]$, though the favorable rate variance is −$438.75 $[=(17.32 - 18.67)325]$. The unfavorable net labor variance is +$345.39 (=784.14 − 438.75).

Man hours are the product of a labor estimate and the number of units. A three-dimensional variance analysis is possible for labor jobs if we consider quantity, labor hourly rate, and the estimated unit time. Define

$$V_{dl} = (N_a - N_e)H_e R_e$$
$$V'_{dl} = (H_a - H_e)N_a R_e \qquad (9.11)$$
$$V''_{dl} = (R_a - R_e)N_a H_a$$

$$\text{net direct labor variance} = V_{dl} + V'_{dl} + V''_{dl}$$

where V_{dl} = variance for direct labor due to difference from estimated quantity, dollars

V'_{dl} = variance for direct labor due to difference from estimated wage rate, dollars

V''_{dl} = variance for direct labor due to difference from estimated wage rate, dollars

H_e = estimated hours per unit

H_a = actual hours per unit

	Hours/Unit	Quantity	Wage Rate, $/hr
Estimate	1.4	200	18.67
Actual	1.5	210	17.32

The following variances are calculated: V_{dl} = +$261.38, V'_{dl} = +$392.07, and V''_{dl} = −$425.25. The unfavorable labor variance is +$228.20.

Indirect expenses are important for construction. Methods of their regulation and cost assurance are not as straightforward as those for direct material and direct labor, where variances are the focus of attention. Indirect costs, since they are in support of direct costs, are assumed to vary proportionately. A ratio of indirect to direct costs is one way to monitor those costs. Overhead can be analyzed for variances also.

Behavioral considerations are important for construction cost of labor and materials. Direct labor hours can be reduced by better management, training, harder work, improved methods of construction, material changes, and so forth. Contrariwise, strikes, morale problems, unforeseen design changes, or bad weather can increase the cost of construction jobs.

PICTURE LESSON
Power House

The Bonneville Dam, put onstream in 1937, was the first federal dam among 55 major hydroelectric projects on the Columbia River that constitute the largest hydroelectric system in the world. At Bonneville there are 20 generators, and for a head of 50 feet, the generation capacity is 1,075 MW.

9.6   PROJECT-ESTIMATE ASSURANCE (OPTIONAL)

The money required for a project is large compared to the available resources. Because of the importance of projects to the long-term financial health of the firm, considerable effort is devoted to assuring the success of estimate. Projects are scheduled over a long duration, and opportunities exist to regulate the actual costs. Both the owner and the contractor desire those objectives. However, an estimate-assurance program is more than a mere objective: It defines details, provides a strategy, and indicates the steps. Owners sanction the project because of a profitable time-value-of-money concept (see Chapter 10), and money and time (or schedule) are major factors of concern.

The engineering team, in preparing the work breakdown structure (WBS) package, provides details, strategy, and steps. The WBS package is discussed in Chapter 8, which developed the scope and definition, designs, work-breakdown structure, schedule, and estimates. The WBS is useful for estimating, planning, and performance measurement and to assure that actual costs achieve the estimate. Now we extend those principles for estimate assurance.

The purpose of cost assurance is to guarantee the project profit. Though estimate assurance deals with overall project cost, it is defined as the procedures where actual costs are controlled to the level of the elements. It makes little difference whether the budget is a scheduled commitment or a scheduled expenditure for the elements. Nor do values of dollar magnitude alter the basic approach.

Our intent now is to supply a few rules and some understanding of work-package control.[2] Integration of data for estimate assurance requires a single unit of measure that combines cubic yards, for example, and cost performance with schedule performance. Cost and schedule variances are reported in dollars, while time variances are reported in periods. While the preferred denomination in actual construction is days, we suggest a *period* can be a day, week, month, or other unit, and therefore a period is a general dimension for time.

At this point it is useful to consider the principles for project estimating that were first introduced in Chapter 8. There we saw the importance of the work packages of the work breakdown structure, and how estimates were determined. The elements of the project estimate are direct labor, direct material and subcontract, facilities and equipment, engineering, overhead, interest, contingency, and profit. Contingency allows for unassignable extra costs that occur, and its eventual outcome is not known beforehand. Actual cost includes those contingencies because they are inherent in the performance of the cost element.

The *S* curves are developed first by plotting cumulative baseline value (CBV) against the planned period. The subjects for the plot are the major elements of the project. The baseline values were determined by using trapezoidal or other models or by using the WBS schedule. This ideal baseline expenditure is now called the *budgeted cost for work scheduled* (BCWS) and is available at the moment of the project start. This vertical axis may also be titled BCWS%, which is equivalent to cumulative baseline percentage. The contract-period percentage is an alternative *x* axis for cost/schedule performance reports.

Once the project is under way the contractor's reporting methods will provide the *actual cost of work performed* (ACWP), and by cross checking actual work to the WBS we find the derivative *budgeted cost for work performed* (BCWP). Additionally, the cost/schedule status report will provide a forecast cost at completion and budgeted cost at completion. Those concepts are illustrated in Fig. 9.9. Excellent assurance software (see References) exists that eases the burden of project estimate, but it is educationally important to have the student perform manual sketching and understand these relationships.

At the close of each period the engineer graphically forecasts the remainder of unfinished WBS elements for ACWP and BCWP. The extension depends on a graphical ability, and more importantly, upon extra knowledge and opinion about delivery promises, potential strikes, bad weather, and so on.

There is a problem, however, with work-in-process tasks having a long time period. If the reporting cutoff period slices into work-in-process for some task, then the percentage of completion must be subjective. Estimating the percent of task complete may be done by the person responsible for the task.

Work packages vary. For example, fabrication work packages tend to be short and discrete. Engineering work packages are difficult to plan, since the work is variable, making it difficult to judge for percentage completion. Contrariwise, work completion can be better

---

[2] For more information, see Robert I. Carr, "Cost, Schedule, and Time Variances and Integration," *Journal of Construction Engineering and Management*, Vol. 119, No. 2, June 1993, p. 245–265.

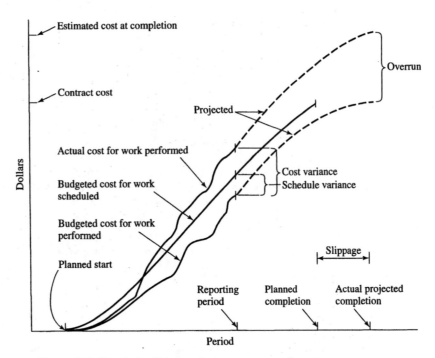

Figure 9.9 Cost/schedule performance reporting.

judged over several reporting periods, because single-period judgment of percentage completion is tricky.

A related problem deals with reporting cost and schedule variance of fixed-price subcontracts, since some subcontractors may not report internal progress on work. For those who do not, perhaps, the engineer will do "subcontractor talk" and learn informally of their progress in order to keep the cost/schedule report informed. On the other hand, the subcontractor may be contractually obligated to supply information as a condition for progress payments.

Performance measurement is a comparison of budgeted versus actual accomplishment. Comparing the budgeted cost for work scheduled to the budgeted cost of work performed produces a dollar-schedule variance. If the BCWP exceeds BCWS (i.e., is higher on the vertical axis), then more work is accomplished than was planned, and the favorable variance reflects the dollar (or percentage value) of the extra work. If BCWP is less than BCWS, then less work is accomplished than was planned, and an unfavorable schedule variance is indicated. The horizontal difference between BCWS and BCWP is the schedule variance expressed as periods or period percentage.

Contract cost performance is related to work done versus work planned. The BCWP when compared with BCWS indicates a schedule variance, but when compared with ACWP yields a cost variance. If the BCWP exceeds ACWP, then a favorable variance is noted. If ACWP is greater than BCWP, then a negative cost variance, or an overrun, is noted. Cost and schedule variances are indicated in Fig. 9.9.

The dashed projection of BCWP to the contract cost level indicates the conjectured schedule completion. Project slippage is the number of calendar days or periods between the actual and planned project completion. The dashed projection of ACWP in Fig. 9.9 indicates the projected dollar overrun when compared with BCWP.

Consider the following relationships:

$$\text{cost variance} = BCWP - ACWP \tag{9.12}$$

where $BCWP$ = budgeted cost of work performed, dollars
                $ACWP$ = actual cost of work performed, dollars

$$\text{schedule variance} = BCWP - BCWS \tag{9.13}$$

where $BCWS$ = budgeted cost of work sheduled, dollars

Schedule variances are the difference between the earned value of work performed and that which is scheduled. Schedule variances are positive when actual output exceeds scheduled output. Therefore, a positive schedule variance is a favorable one.

If the estimate assurance program reduces the magnitude of the actual expenditure, then project attractiveness is enhanced. If the project decision is based on optimistic cost expectations, then a cost-assurance program should reveal this fact early enough for reassessment. Cost information must be available so that significant items of the project cost may be watched. The project budget is tabulated to have this record correspond to the WBS, definition, and estimates. It has happened that projects having overly optimistic estimates, poor performance, and schedule delays have been canceled after reassessment, or reduced in magnitude by the owner.

Consider the example of a construction bid of $8 million having a contract period of 5 months. Cash flow is trapezoidal, where t_i = 20% and t_f = 60%. Profit earned by the contractor is 20%. The estimated cost total is $6,400,000. The owner has a retainage policy of 10% with a 1-month delay in the owner's payment. Table 9.8 provides the expenditure calculation as BCWS. Actual data have been provided for periods 1–4 as the work progresses.

At period 4 there is a current unfavorable cost variance of $2,400,000 (=7,000,000 − 4,600,000) and a time delay of 1 month. The BCWP is extended to the contract cost. A horizontal difference between BCWS and BCWP gives a project slippage of three-quarters of a month. The extended ACWP and BCWP indicates a project overrun of $1,350,000. The cost/schedule performance of the construction bid is shown as Fig. 9.10.

In our view cost assurance of projects continues with the knowledge of variances between actual measured costs and progress on the one hand and target budgets and schedules on the other, to determine if the project is being managed as required. Understanding these principles is important for student learning.

TABLE 9.8   Cost/schedule performance report where costs are $1000

	Report			Reported			Variance			Forecast		
Period	Period (%)	BCWS (%)	BCWS	ACWP	BCWP	Schedule	Cost		BCWP	ACWP	Variance
1	20	14.3	$9150	$500	$1250						
2	40	42.9	2745	2000	2250						
3	60	71.5	4575	5500	3500						
4	80	92.9	5950	7000	4600	1	$2400				
5	100	100	6400			1.1	2250		$5600	$7500	−$1900
6	120	100	6400			.7	1800		6400	7750	−$1350

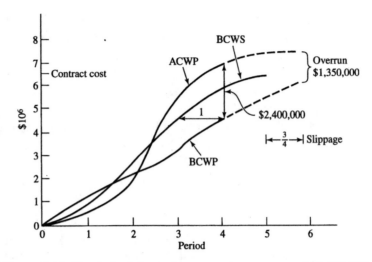

Figure 9.10 Cost/schedule performance for contract of $6,400,000.

SUMMARY

Jobs are obtained in a competitive bidding arena, and thought and planning are necessary before the bid is submitted. An important purpose of an estimate is to improve business opportunity. Not all bids succeed, but those that do are matched against a comparison. The comparison could be an actual or preliminary value. Actual costs suffer from inaccuracy or lateness and do not always provide a valid future lesson. Nonetheless, a variety of techniques, including behavior considerations, variance, productivity, and trend charting, guide the actual cost values to become estimated values.

QUESTIONS

1. Define the following terms:

Estimate error	Variance
True value	Capture rate
Contingency	Closed estimate
Mistake error	ACWP
Policy error	Assurance systems
Standard	Bid laid on the table
Unbalanced bidding	Control model

2. Why are estimates compared to actual cost or standard cost?
3. Describe the philosophical differences between an estimate-assurance and a cost-control program.
4. Engineering usually stresses the abolition of mistakes in cost analysis and estimating. Name practices to achieve this objective.
5. Describe management actions for (a) a declining capture rate, and (b) an improving capture rate. Why is a corrective policy based on a capture-rate strategy alone defective?

6. Develop a list of behavioral principles that would be useful for cost analysis and estimating.

7. Give different formulas for variance calculations. Can ambiguity be completely overcome?

8. Why are project estimates difficult to cost track? Cite some cost overruns from local or national circumstances. List the factors that cause these overruns.

9. Define a capture rate based on dollars won versus dollars estimated. What advantages does this have over Eq. (9.7)?

10. Discover a web site that features project estimate assurance. Report on the approach.

PROBLEMS

9.1 Sketch the *expected profit* for a range of bid values where the bid, B, is allowed to range over several values. Select the optimum bid. Cost for the project, C_e, is $500,000. The *opinion* probability of winning the project is given as $P(B < \text{next lowest bid})$ and is:

Bid, B; $	P(B < next lowest bid)
490,000	1.00
500,000	0.95
510,000	0.65
520,000	0.45
530,000	0.15
540,000	0.05
545,000	0.00

9.2 Two contractors bid the same job opportunities. Information about contractor A's bids is uncovered from public records. A summary of the ratio of the contractor's A bid against our cost is given below:

No. of Comparisons	B_A/C_e
0	$B_A/C_e < 0.8$
1	$0.8 \le B_A/C_e < 0.9$
2	$0.9 \le B_A/C_e < 1.0$
7	$1.0 \le B_A/C_e < 1.1$
12	$1.1 \le B_A/C_e < 1.2$
21	$1.2 \le B_A/C_e < 1.3$
18	$1.3 \le B_A/C_e < 1.4$
7	$1.4 \le B_A/C_e < 1.5$
2	$1.5 \le B_A/C_e < 1.6$
0	$1.6 \le B_A/C_e$

Sketch the expected profit function against B/C_e and find the peak point. Recommend a bid value if the cost of the future project is $725,000.

9.3 Two contractors compete in a two-company arena of construction. Bid knowledge about contractor A is available from public information. Using these records, our contractor compiles a ratio of the bid of contractor A to his estimate, which is identified as B_A/C_e. Sketch the expected profit for a two-company bidding strategy. From the sketch, find the bid-to-cost ratio for an optimum strategy. If our estimated cost is $400,000 for an upcoming bid against contractor A, what is our expected bid? The information is given as

No. of Comparisons	B_A/C_e
0	$B_A/C_e < 0.98$
1	$0.98 \le B_A/C_e < 1.00$
3	$1.00 \le B_A/C_e < 1.02$
5	$1.02 \le B_A/C_e < 1.04$
13	$1.04 \le B_A/C_e < 1.06$
18	$1.06 \le B_A/C_e < 1.08$
14	$1.08 \le B_A/C_e < 1.10$
5	$1.10 \le B_A/C_e < 1.12$
2	$1.12 \le B_A/C_e < 1.14$
1	$1.14 \le B_A/C_e < 1.16$
0	$1.16 \le B_A/C_e$

9.4 The construction competition for business is a threesome of contractors. Enough is known about the bids of competitors A and B, along with facts from your firm, to give the following information:

B/C_e	$P(\text{our bid } B < B_A)$	$P(\text{our bid } B < B_B)$
0.8	1.00	1.00
0.9	0.99	1.00
1.0	0.96	0.98
1.1	0.86	0.80
1.2	0.69	0.70
1.3	0.53	0.60
1.4	0.13	0.25
1.5	0.03	0.09
1.6	0	0.05
1.7	0	0

Sketch the expected-profit function. Find the optimum bid location in terms of your cost, C_e. If your cost is $3,725,000, what is the bid to be laid on the table?

9.5 Historical information is collected for an undifferentiated group of competitors. Sketch the expected-profit function if the recent information from the owner indicates that a total of three bidders is planned for. From the sketch determine the better point of operation. What does the relatively flat dome of the expected-profit function say about bidding in this situation? What is the bid for a cost of $1,885,000? Discuss the consequences of more bidders from the vantage point of the model.

B/C_e	$P(our\ bid\ B < B_{AVE})$
0.98	1.00
1.00	0.97
1.02	0.93
1.04	0.84
1.06	0.64
1.08	0.34
1.10	0.12
1.12	0.07
1.14	0.05
1.16	0.03
1.18	0

9.6 A contractor is submitting a unit-price contract for construction work. The contractor is planning to unbalance the bid in two of the three items, as noted below. Determine the payment gain in this strategy.

Work and Material Take-off	Unit	Owner's Quantity	Balanced Unit Estimate	Unbalanced Unit Estimate	Actual Quantity
Excavation	yd^3	15,000	$7.25	$8.75	16,313
Footings	L.F.	7,500	11.50	8.50	7,319
Gauges	Ea.	115	295.00	295.00	118

Discuss the advantages and disadvantages of this tactic. What do you think the owner will say about this, if she becomes aware of it?

9.7 (a) A concrete contractor bids a job of steps and risers for an apartment. Her estimate is as follows:

Work Element	Amount	Rate
Direct labor	125 hr	$25/hr
Concrete	32 yd^3	$75/yd^3$
Other materials	$150	
Overhead rate on basis of direct labor	22.5%	

Her pricing policy is 20% as a markup on full cost. Determine the bid for the job. After the job, which she wins, a reconciliation of all costs by the bookkeeper gives a cost of $7,250. What is the variance? Find the error of the estimate. What is her final profit for these actual costs?

9.8 (a) The contractor submitted 298 quotations for the winning jobs of Fig. 9.3. What is the capture rate?

(b) Of the winning jobs reported by this survey, 63 were for jobs required by an engineering design change resulting from 109 quotes. The contractor had provided the original work, which were not reported by this survey. Find the capture rate for the new and repeat business.

(c) Discuss ways to improve his repeat business, considering that profit margins that are 20% on repeat business and 2.5% on new business, versus profit margins that are 2.5% on repeat business and 20% on new business.

9.9 A full-cost of a job is estimated as $400,000 and $500,000 for direct material and direct labor. Additionally, the work-package definition specifies the following range of accuracy:

Item	Accuracy
Quantity of material	±4%
Unit cost of material	±2%
Unit labor rates	±3%
Labor productivity	±5%

What is the possible error as measured in dollars? Find the range of the estimate.

9.10 A chemical plant system is composed of tanks, pumps, and heat exchangers. Estimates and their tolerances are shown below. The equipment is to be installed in a unified system. Find the percentage tolerance error of the installed construction.

Equipment	Estimated Cost, $	Tolerance, ±
Tanks	40,000	±20
Pumps	35,000	±10
Heat exchangers	60,000	±25

9.11 A contractor reviews last year's performance of "bids left on the table," and the following information is uncovered.

Bids Left on the Table	Above Winning Bid, %
2	0–2.0
1	2.1–4.0
3	4.1–6.0
7	6.1–8.0
10	8.1–10.0
10	10.1–12.0
5	12.1–14.0
2	14.1–16.0

What is the capture rate for 78 attempts by this contractor? Because winning bids are openly revealed, the contractor is able to analyze his bidding strategy.

Discuss reasons for supporting the contractor's strategy if he desires no change in the future bidding environment. Now assume that the contractor wishes to improve his bidding performance. The average quotation is about $1 million and includes $50,000 for profit. Discuss the strategy in the light of this additional information. What adjustments to the bidding practice can you suggest?

9.12 A contractor defines its estimates in terms of the required accuracy. Determine the bid. Find the range of the estimate. In the worst possible case, is there sufficient profit to cover a higher-than-expected cost? Information is given as:

Work-Package Item	Company Practice	Bid, $	Required Accuracy
Direct labor		500,000	±2%
Direct material and subcontract		400,000	±%
Overhead	10% of direct cost		
Profit	5% markup on cost		

(*Hint*: The policy of a required accuracy is not uncommon, as a firm will have a range of acceptable errors, and it is presumed that the smaller error also means more estimating detail to achieve the policy.)

9.13 A contractor evaluates his bids against his costs for winning jobs. A quantity of C_e/C_a is determined for 60 won jobs. Lost jobs are not evaluated in this case. The following distribution is found:

No. of Projects	C_e/C_a
0	$C_e/C_a < 0.85$
3	$0.85 \leq C_e/C_a < 0.90$
12	$0.90 \leq C_e/C_a < 0.95$
19	$0.95 \leq C_e/C_a < 1.00$
16	$1.00 \leq C_e/C_a < 1.05$
8	$1.05 \leq C_e/C_a < 1.10$
2	$1.10 \leq C_e/C_a < 1.15$
0	$1.15 \leq C_e/C_a$

Sketch a histogram for this study. Find the expected bias for the estimates. What does $C_e/C_a = 1.0$, $C_e/C_a > 1$, and $C_e/C_a < 1$ mean? Discuss which category is ideal. Discuss whether the estimates should be nudged up or down, or left alone.

9.14 Find the cost of preparing an estimate if a fixed cost is $100, and the variable cost per element is $5/element. The empirically determined coefficient is 0.9. What is the cost for 10 elements? Discuss the general effect on cost if $a = 0, a = 1$, and $a > 1$.

9.15 Four individuals estimate the same job under the same conditions and business assumptions. Find the error-to-signal ratio. Their estimates are given as:

Person	1	2	3	4
Estimate, $	8,710	9,319	10,291	8,624

9.16 Find the material variances on (a) quantity first, and (b) material unit cost first. Prepare a sketch similar to Fig. 9.7.

	Units	$/Unit
Estimate information	200	17.38
Actual information	195	18.25

9.17 Estimates for a material are 650 lb at $17.15/lb. The record shows an actual usage of 665 lb at $17.10/lb. Find the variance for material cost first, quantity first, and total. Sketch a figure similar to Fig. 9.7.

9.18 A material take-off for a unit quantity estimate is made for a concrete wall, 25 ft 4 in. long by 9 ft 6 in. high. (See Fig. 7.2 and Table 7.2.) Concrete, 4000 psi quality, is estimated as 7.5 yd^3 for a unit cost of $78.50 delivered to the site. Crew unit man hours and unit cost are 3.5 and $95.00. Review of records for material shows that 7.7 yd^3 and $81.50 were charged against the job. Job tickets reveal 3.4 hr, and bookkeeper entries of crew labor rates of $92.50 are indicated.

Find the usage, labor, and total variance. Discuss the meaning of *favorable* and *unfavorable* variance in the context of this problem.

9.19 Approximate the overrun and slippage for the following 10^6^ data, and forecast variances. Sketch the *S*-curves. What is the final cost? Discuss the process of graphical forecasting. (*Hints*: Do the forecasting of future values using the sketch. Adopt the convention of using a dashed line for the forecast values. Label the curves, and show the cost and schedule variances.)

Period	1	2	3	4	5	6	7	8	9
BCWS	3	14	22	45	68	92	115	134	142
ACWP	4	10	19	60	83	113			
BCWP	1	4	15	28	52	82			

9.20 Determine the schedule and cost variance at period 4. Forecast BCWP and ACWP to project end using a sketch and find the overrun variance and slippage. (*Hint*: Money in the table is in units of $1000.)

Period	1	2	3	4	5	6
Period, %	20	40	60	80	100	120
BCWS, %	14.3	42.9	71.5	92.9	100	100
BCWS, $	915	2745	4575	5950	6400	6400
ACWP, $	250	1500	5000	6500		
BCWP, $	500	2000	3500	5000		

MORE DIFFICULT PROBLEMS

9.21 Two competing contractors compete against our firm. Information about both the contractors' and our performance in terms of their bids is known, as shown below:

B/C_e	P(our bid B < B_A)	P(our bid B < B_B)
0.98	1.00	1.00
1.00	0.98	1.00
1.02	0.94	0.98
1.04	0.85	0.80
1.06	0.65	0.70
1.08	0.35	0.60
1.10	0.13	0.25
1.12	0.05	0.09
1.14	0.02	0.05
1.16	0	0

Sketch the expected-profit function. Find the optimum bid location in terms of your cost, C_e, using your sketch. If your cost is $725,000, what is the bid to be laid on the table? What is the range of profit that may be anticipated?

9.22 An analysis of the construction-work estimates is made for the previous fiscal-year quarters. Dollar values are 10^3 units.

Quarter	Estimates Made	Estimates Won	C_p	C_d	C_a,	Business Value Lost	Estimates, $
First	216	112	$195	$191	$193	$380	$17.4
Second	237	118	201	198	202	426	17.2
Third	293	138	219	224	233	544	16.8
Fourth	338	149	221	227	244	625	10.2

Find quarterly and yearly estimating error on the basis of actual and preliminary estimates, capture rate, and productivity. What trends do you spot? Advise cost-estimating policy for the next quarter. Evaluate the performance of estimating over the last year. The contractor has a goal of bidding 250×10^3 from detail estimates. What can be expected?

9.23 A work and material take-off for a unit-quantity estimate is made for a concrete wall, 25 ft 4 in. long by 9 ft 6 in. high. A portion of the unit-quantity estimate is given below. Find the hours variance, labor-rate variance, and total variance. (*Hint*: Review Fig. 7.2 and Table 7.2 for the entire story and the notation of the construction-work estimate.)

Work and Material Take-off	Quantity, n_i	Unit Man Hours, H_{i2}	Unit Cost, R_{i2}, $	Actual Hours	Actual Wage, $
Lumber (carpenter crew)	481 ft^2	0.05	50.50	26.0	50.00
Reinforcing ties (steel worker crew)	395 lb	0.023	34.00	8.5	36.00
Concrete crew	7.5 yd^3	3.5	95.00	32.0	102.00

9.24 Determine schedule and cost variance and budget, project, and differences until completion. Sketch period number and period percent versus BCWS, BCWP, and ACWP, and BCWS%. Extend the data to completion and estimate cost at completion and project slippage. Discuss the connection from project trapezoid models to budgeted cost for work scheduled. (*Hint*: The $150,000 project is scheduled using a trapezoid model with $t_i = 40\%$ and $t_f = 70\%$. The contract time is 10 months with monthly payments. Review the material in Chapter 8.)

Month	Period, %	CBV, $	CBV, %	ACWP, $	BCWP, $
1	10	2,880	1.9	2,000	3,000
2	20	11,535	7.7	5,000	12,000
3	30	25,950	17.3	13,000	18,000
4	40	46,155	30.8	28,000	27,000
5	50	69,210	46.1	41,000	32,000
6	60	92,280	61.5	54,000	52,000
7	70	115,350	76.9	92,000	78,000
8	80	134,580	89.7	122,000	108,000
9	90	146,295	97.5		
10	100	150,000	100.0		
11	110	150,000	100.0		
12	120	150,000	100.0		

Interpret the early periods. What action do you think might have been taken on the basis of the early periods?

PRACTICAL APPLICATION

Form a team of classmates following guidelines that your instructor will provide. The goal is to interview several contractors and determine their bidding strategy. In preparation for the visits, the team will need to make a list of questions. Some of the questions may include:

How much does the contractor know about his/her competitors in advance of the bid opening?
What is the contractor's capture rate?
Do the contractors conduct any post estimate scrubdown that is analytical?
Is the contractor using any "analytical aids" for bidding.

Be business like in the handling of this interview and prepare a report that meets your instructor's requirements. As a courtesy to the contractors that are visited, the team may wish to send a thank-you letter along with a copy of the report.

CASE STUDY: COST-SCHEDULE PERFORMANCE REPORTING

When a contractor provides a proposal to an owner, the contractor promises to perform work at a specified cost within a time limit and to satisfy design requirements. For the contractor to fulfill a contract, specific work is scheduled at various times of the project. Some contracts require periodic preparation of cost-schedule performance reports that inform the owner what has been done and when, and additionally what needs to be done and when. A graph, such as Fig. C9.1, is often a significant part of the report. It differs from other CSPR figures in that an amount of ongoing project labor is considered fixed.

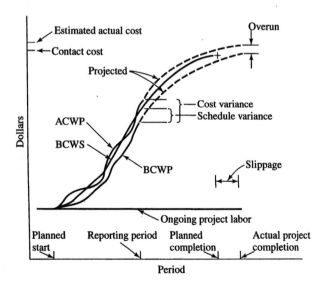

Figure C9.1 Cost/schedule performance reporting with ongoing project labor.

Create the specifications for a report system that builds on Fig. C9.1. Prepare your findings by word processing a detailed prospectus for management's consideration.

Consider what follows as necessary to your reporting procedure:

- Alert engineering to conditions that might affect either cost or schedule.
- Establish an orderly method of compiling supporting data.
- Permit timely and corrective action to minimize negative effects or maximize positive breakthroughs on project cost and schedule.

CHAPTER 10

Cost Analysis

In this chapter we present methods for applying cost analysis to construction designs. Procedures cannot be set out beforehand, unfortunately, in some recipe fashion. If that were possible, then by learning those procedures and strictly adhering to them, we would be absolutely confident of the result. The analysis, though, is special for each design. There are some principles that are helpful, and the student will find them worth learning.

Cost analysis is performed during construction "tradeoffs." These tradeoffs happen early in important construction projects–perhaps 60%–80% of the consequential cost/tradeoffs are made before and during design. Figure 10.1 describes the epochs that influence cost and scheduling.

10.1 FIRST PRINCIPLES FOR TRADEOFF STUDIES IN DESIGN AND CONSTRUCTION

An important work in engineering is the cost analysis for doing "tradeoffs," a slang term that describes the activity of understanding, collecting, computing, and making choices in design and construction. Engineering is the activity that determines the *comparative economy* of

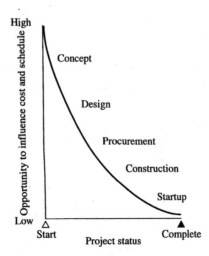

Figure 10.1 Illustration of the major events of a construction project and their influence on cost and schedule.

the construction work and project and makes the decisions in the tradeoff. As an aside, we employ the term "economy" in its original sense, meaning *thrift*. Software, company reference books, codes, design, and engineering skills are the foundations for these judgments.

Whether for the owner or contractor, this tradeoff activity affects the profitability of the enterprise. These are the elements in tradeoff studies:

- Search for design and construction alternatives
- Cost and estimates
- Analytical methods in comparative economy
- Constraints and conditions on the cost analysis
- Irreducible factors
- Selection

The beginning of cost analysis is the *search for design and construction alternatives*. The need for imagination in the forming of alternatives cannot be overstated. No matter how good an analysis and selection among two or more alternatives may be, if another and unidentified alternative is superior, then the selection is suboptimal.

An engineer is expected to consider a broad range of alternatives, while remaining within the realm of practicality. After a reasonably exhaustive list of alternatives is formed, there is a pruning of those that are less feasible within the constraints of time, effort, and money. Preliminary estimating methods may abridge the list to a few that are more promising. The converse is also true: as the analysis continues, opportunities may present themselves that were not known at the start.

This search for alternatives goes on during the estimating phases. For instance, should steel AISI A36 or 1040 be selected? What about the strength of the concrete, 3000 or 4000 psi? One or multiple stories for the commercial building? There are thousands of these questions—literally.

To choose between two energy alternatives for a building by listing the things we value, such as satisfaction, dependability, and lack of pollution, is not very helpful. Cost analysis cannot be conducted meaningfully on such lofty levels. More specific measures for selection between the alternatives are desirable, such as cost, heat loss, waste-heat recovery cost, and equal-marginal fuel cost. Using narrow measures of *design to cost*, there is a greater chance for successful analysis.

The word *cost* is meaningless when used alone. An adjective is needed in order to clarify the comparative economy of the construction alternatives. There are many modifiers that can be used, such as absolute, annual, current, direct labor, discounted, equipment, escalated, first, full, marginal, overhead, project, relative, salvage, short or long term—and the list continues. Refer to the index of this book for pages where these and other modifiers are used. Using precise language around the word *cost* aids the development of the constraints and conditions on the cost estimate.

These precise modifiers are indicated in the preparation of estimates. Once the cost facts (identified as facts because this information is the best that we have at this time) are available, various analysis methods are used for the cost tradeoffs. Cost-estimating encyclopedias, as described in Chapter 7, are applied for the simpler cases of construction-work tradeoff. For more complex levels, Chapter 8 gives the organization for the collection of the cost facts. These chapters describe the methods to find the estimates. Estimates are made for the alternative choices.

The principles for doing cost trades do not depend on the magnitude of the dollars. Whether the comparative economics are for several dollars or billions, the analytical principles are similar. Several analytical methods are presented in this chapter. These methods are appropriate for owners, contractors, and subcontractors.

An engineer may be unable to convert some aspects of a design into ordinary monetary units. Function, beauty, and quality of life are difficult to evaluate and are called *intangible* or *irreducible* factors. Although nonfundamental units for a scale of measurement or ranking might be forced on the intangibles, the engineer usually believes that it is not worth the effort and that the intangibles are truly inscrutable in terms of ordinary units. In that event, then there are other ways to consider them. If the stakes are not high, then it may be convenient to ignore them. Also, considering that "intangibles" advantageous to some may be disadvantageous to others, a simple listing of the intangibles, with descriptions both pro and con, may suffice. Executive engineering may be the court, judge, and jury on the merits of those intangible factors. Perhaps, too, the prominence of the intangible factors may be so high that they overshadow the economic comparison.

The step of *selection between the alternatives* is the last one in tradeoff analysis. Thus, for example, the recommendation is to use 3500-psi concrete, and this selection becomes a design feature on the working drawings, or a listing on the tabulated specifications. Consequently, all of the many selections are confirmed in this cost analysis, and they become an inseparable part of the business of construction.

With the better alternatives selected, and having gained an understanding of the importance of tradeoff study, we now consider several analytical methods for cost analysis.

10.2   CASH FLOW

Cash-flow analysis is compared to a reservoir receiving a stream of water. At certain times more water is received than at other times. Occasionally, there may be greater outflow than inflow. Money moving in and out of a company behaves like this metaphor and thus is frequently called a *cash stream*.

The meaningful considerations in cash-flow analysis are given as

- Income and expenses
- Depreciation
- Taxes
- Inflation and deflation

Even if the annual cost of the business venture is small in proportion to the inflow or accumulated cash surplus, the cash-flow document is necessary. A small contractor may require operating capital before construction revenue is received. It would find a cash-flow document necessary. Still, if the construction venture is a big one, then the company evaluates its cash position to meet its obligations. The cash-flow analysis is necessary. Bankers, venture capitalists, bonding and insurance companies, owners, and others who have a financial stake in the business want the reassurance that a cash-flow analysis gives.

The cash-flow statement can be prepared in advance of the spending and receiving of the money, in which case the project estimate and work package are the source for the information. On the other hand, if the cash flow is for past periods, accounting records supply most of the information.

PICTURE LESSON
Airport Terminal

The picture lesson, Denver International Airport, shows the central atrium of the terminal building. Dimensions of the architecturally exciting roof are 900 by 200 feet. The covering is a translucent tensile membrane roof. The outer waterproof shell is a Teflon-coated fiberglass material, while the inner membrane is made of uncoated woven fiberglass. Ten percent of visible light passes through the roof fabric for daytime illumination. With the fiberglass having little mass, it does not aid the storage of heat nor does it conduct heat. Its white-colored material reflects 90% of sunlight. The Teflon coating allows for easy washdown maintenance. The roof has 34 peaks, similar to those shown, and the cost was about $25 million for its construction.

For cash-flow analysis we define

$$F_c = (G - D_c - C)(1 - t) + D_c \tag{10.1}$$

or equivalently,

$$F_c = (G - C)(1 - t) + tD_c$$

where $\quad F_c$ = total cash flow of money, dollars per year

$\qquad G$ = gross income, dollars per year

$\qquad D_c$ = depreciation charge, dollars per year

$\qquad C$ = costs not estimated elsewhere, dollars per year

$\qquad t$ = effective income tax rate, decimal

This basic relationship shows the major factors for cash flow. While Eq. (10.1) is annual, a similar cash-flow equation can be written for a project having another reporting period. A tax-reporting period will follow the firm's fiscal-year reporting period, which may not correspond to the calendar year of January 1 through December 31. A taxable project period is the length of the project, which may be longer or shorter than a calendar year. This affects the taxable income.

The IRS gives special treatment for a long-term contract, such as a building, installation, and so on which is not completed in a tax year in which it is entered into. A method known as the percentage of completion, which is discussed in Chapter 4, allows gross income to be reported annually according to the percentage of the contract completed in that year. The completion percentage is determined by comparing costs allocated and incurred before the end of the tax year with the estimated total contract costs. All expenditures made during the tax year must be deducted, taking into account unused materials and supplies on hand at the beginning and end of the taxable period. Note that the IRS has clauses that distinguish home and residential construction contracts, which are not given the same exemptions as buildings, plants, and installations.

Once the owner accepts the bid, annual gross income is determined. In some situations, it is necessary to use the work packages as described in Chapter 8 and approximate project cash flow inside the various tasks. Smaller jobs that are completed within the year are a part of the fiscal revenue and expense.

Depreciation is introduced in Chapter 4, and those principles apply here. Depreciation is defined as a *noncash tax expense*. Depreciation is the recovery amount for the money that originally purchased the capital expenditure, which has been already spent. As a noncash tax expense, it acts to reduce the taxable income, but it is added back to the cash flow. Note this manipulation in Eq. (10.1).

The main types of taxes for project cost analysis that are important for an owner or contractor are summarized as

- Income taxes
- Property taxes
- Sales taxes

Income taxes are taxes on pretax income of the firm in the course of ordinary operations. Income taxes are also levied on gains on the disposal of capital property. Income taxes are usually the most significant type of tax to consider in construction-cost analysis.

Tax codes are complex, and their minutiae are beyond the needs of this text. But one needs to recognize that income taxes paid are just another type of expense, while income taxes saved (through business deductions, expense, or direct tax credit, and so forth) are similar to other kinds of reduced expenses, such as savings.

Property taxes are based on the valuation of property owned, such as land, equipment, buildings, inventory, and so forth, and the established tax rates. They do not vary with income and are usually much lower in amount. Still, they can be significant in terms of money or impact.

Sales taxes are taxes imposed on product or material sales, usually at the retail level. They are relevant in cost analysis only to the extent that they add to the cost of the purchased direct material or equipment.

At the end of each tax period (usually a year, but more frequently if necessary) a contractor or owner must calculate its taxable before-tax income or loss. An approximate relationship showing this maneuver is given as

$$
\begin{aligned}
\text{taxable income} \;=\; & \text{gross income} \\
& - \text{all expenses (except capital expenditures)} \\
& - \text{depreciation deductions} \qquad\qquad (10.2)
\end{aligned}
$$

Taxable income is often referred to as net income *before* taxes, and after income taxes are deducted, the remainder is called net income *after* taxes. There are two types of income taxes for computation purposes: ordinary income (and losses) and capital gains (and losses). We do not discuss capital gains or losses in this text.

Ordinary income is the net income before taxes that results from routine construction business performed by the contractor or owner. For federal income-tax purposes, all ordinary income adds to taxable income and is subjected to a graduated scale, which has higher rates for higher taxable income. Table 10.1 shows a corporate tax table.

TABLE 10.1   Illustrative corporate federal income tax table

If Taxable Income Is: Over	But Not Over	The Tax Is	of the Amount Over
$ 0	$ 50,000	15%	0
50,000	75,000	$ 7,500 + 25%	50,000
75,000	100,000	13,750 + 34%	75,000
100,000	335,000	22,250 + 39%	100,000
335,000	10,000,000	113,900 + 34%	335,000
10,000,000	15,000,000	3,400,000 + 35%	10,000,000
15,000,000	18,333,333	5,150,000 + 38%	15,000,000
18,333,333	...	6,416,667 + 35%	0

For example, suppose that in the last fiscal year a contractor had a gross income of $5,527,000, expenses (excluding capital expenses) of $3,290,000, and a depreciation of $1,650,000. Taxable income = 5,527,000 − 3,290,000 − 1,650,000 = $587,000. The federal income taxes on this amount are demonstrated in Table 10.2:

TABLE 10.2   Calculation of federal taxes on income of $587,000.

	Taxable Income	Income Taxes
Income taxes = 15% of first	$50,000	$7,500
+ 25% of next	25,000	6,250
+ 34% of next	25,000	8,750
+ 39% of next	235,000	91,650
+ 34% of the remaining	252,000	85,680
Total	$587,000	$199,580

More directly, the taxes $199,580 $(=113,900 + 0.34(587,000 - 335,000))$ can also be found using a single line in Table 10.1.

Income taxes are collected by the federal and state governments. While the state practices resemble those of the federal government, the tax rates vary. State tax rates are much lower than federal rates, but remain significant. The net effect of taxes is increase the risk and reduce the attractiveness of construction.[1] The rates for the states typically range from 4% to 9%. These two income-tax rates can be consolidated with the following approximation:

$$t = \text{federal rate} + \text{state rate} - (\text{federal rate})(\text{state rate}) \qquad (10.3)$$

where t = effective income tax rate, percent

For instance, if a corporation has a federal tax rate of 34% and a state rate of 6%, then the consolidated rate is 37.96%. Federal taxes allow a deduction for payment of state taxes from gross income, though the reverse is not true.

The assumption that costs and prices are relatively unchanged over extended periods of time is unrealistic. This error can lead to inconsistency and mistakes in choices between construction tradeoffs.

Price *inflation* or *deflation* is a general increase or decrease in the prices paid for materials, labor, subcontracts, and other goods and services and it affects the purchasing power of the monetary unit, the dollar. When inflation occurs, the purchasing power of the dollar decreases, and in the case of deflation it increases. The history of inflation is much more common. The last general decrease in the price of labor in the United States occurred in 1931. Contrariwise, deflation is seen with imported materials and with improved products. Inflation is charted with indexes, which were discussed in Chapter 5. The student will want to review that material.

The basic factors of inflation–deflation that influence the cash-flow model are given as

- Actual dollars
- Real dollars
- Inflation or deflation rate

Actual dollars are the actual amounts received or spent and not adjusted for inflationary factors. For instance, engineers anticipate their salaries two years hence in terms of actual dollars. The purchasing power of an actual dollar, at the time the cash flow occurs, includes the effect of price inflation or deflation. Sometimes actual dollars are referred to as *current dollars* or *escalated dollars*.

Real dollars are expressed in terms of the same purchasing power relative to a particular benchmark time, or base year. For example, if the future material, labor, and subcontract prices are changing rapidly, they are estimated in real dollars relative to some base year to provide a consistent floor for comparison. Sometimes real dollars are called *constant dollars* or *deflated dollars*.

The inflation or deflation rate is a metric of the change in the purchasing power of a dollar during a specified period of time. This metric is known as an *index* and is a ratio of the price of a market basket of goods and services over extended periods of time relative

---

[1] Taxes are important for a democratic society and necessary for its continuance to provide a broad range of needs and to give protection for its citizens. The tax rate is a lively subject for debate, but it is not discussed in this text.

to the price in a benchmark period. The Producers' Price Index is a popular index. Contractors will frequently chart special indexes for their business environment. These rates are dynamic and are influenced by many factors. Contractors and owners give significant attention to these rates.

The relationship between actual dollars and real dollars is given as

$$D_r = D_a \left(\frac{1}{1 + f} \right)^{n-k} \tag{10.4}$$

where D_r = real dollars at a point in time, dollars
           D_a = actual dollars as of the time it occurs, dollars
             f = inflation or deflation rate, decimal
             n = point in time, period, and usually years
             k = base time period used to define the purchasing power of the real dollar,
                 = usually 0.

This relationship may be used to convert actual dollars into real dollars or vice versa. While k is a general index of time and can be any period, for tradeoff studies between competing designs it is usually zero.

Consider the construction and operation of a chemical plant with cash flows given by Table 10.3. The estimated values are considered "real." Inflation is expected to be 5.3% per year for this period. The actual values also are shown in the table.

Now consider an example of a chemical plant. An owner will analyze these investments extensively, and the financial vision for the construction and operation is vital to the decision of whether to proceed with the design and build the plant. Assume that the chemical plant designs and estimates, which are expressed in real dollars and are indexed to the year 0, are summarized by the cash-flow analysis, as given by Table 10.4. The table consolidates capital expenditures, inventories, and accounts and allows receivables to build up.

A cost estimate involving bid and construction, schedules, marketing and sales rates, and capitalization costs for new equipment and plant enlargement is necessary to construct a cash-flow statement.

Construction is shown as a lump-sum amount at point-in-time 0. The analysis concludes with year 6. Capacity growth, expressed in terms of 10,000 tonnes/yr, reaches 100% in year 4, but design improvements will allow learning and greater capacity.

Project estimates provide the $700 million investment cost for the plant. Preoperating expenses and investment costs are first-year cash out, lumped at time zero for simplic-

TABLE 10.3   Conversion of real dollars into actual dollars for construction and operation of chemical plant

End of Year, n	Real Dollars, $	$(1.053)^n$	Actual Dollars, $
0	−$775,000	1.0	−$775,000
1	60,000	1.0530	63,180
2	125,000	1.1088	138,601
3	185,000	1.1676	216,000
4	250,000	1.2295	397,364

TABLE 10.4 Cash-flow statement for construction and operation of chemical plant with real cash values expressed as 10^3

Item	0	Year 1	Year 2	Year 3	Year 4	Year 5	Year 6
Capacity, percent		25	50	75	100	110	116
Production, 10,000 tonnes/yr		12	25	37	50	55	58
Construction cost of plant, $	700,000						
Startup expenses, $	75,000						
Sales revenue, $		60,000	125,000	185,000	250,000	269,500	282,500
Less expenses, $		15,000	31,250	46,250	62,500	68,750	72,500
Less depreciation, $		70,000	95,000	71,250	70,000	70,000	70,000
Net income before taxes, $		0	0	67,500	117,500	130,750	140,000
Less income taxes @ 40%		0	0	27,000	47,000	52,300	56,000
Net income after taxes (1–40%)		NOL*	NOL*	40,500	70,500	78,450	84,000
Add back tax depreciation, $		45,000	93,750	71,250	70,000	70,000	70,000
Net cash flow, annual, $	−775,000	45,000	93,750	111,750	140,500	148,450	154,000
Cumulative cash flow, $	−775,000	−730,000	−636,250	−524,500	−384,000	−235,550	−81,550

* Net operating loss.

ity. Production quantity and price are estimated, obviously with information from the marketing components of the business.

Expenses for the operation for the production quantity for the design flow sheet of the chemical plant are entered. These expenses involve working capital, which is made up of accounts receivable, raw material inventory, work in process, and finished-goods inventory. Direct labor and overhead are included. Increases in inventory require immediate cash outlays that delay cash flow from generating sales revenue. Ordinarily, a higher requirement for cash on hand occurs during periods when operations are increasing. Working capital is usually meant to be incremental capital (i.e., differential capital between present and prior year.)

Depreciation is a noncash tax expense, and a tax credit (a product of the depreciation times the firm's tax rate) helps to provide the inflow of funds. For this example, depreciation is straight-line over 10 years or $70,000 (=700,000/10). Typically, straight-line depreciation is not used, as the modified accelerated cost recovery system (MCARS) is the more common method. But notice in Table 10.4 that there is inadequate revenue to allow even the full depreciation until year 3, and this is a *carry-forward* process until sufficient revenue exists for allowable and full depreciation. The carry-forward provision of the tax code affects future tax computation.

Once expenses and depreciation are deducted from sales revenue, we have the net income before taxes. The effective tax rate of $t = 40\%$ is applied to this quantity. Net income percentage after taxes is $(1 - t) = 60\%$.

The arithmetic for net and cumulative cash flow is evident in Table 10.4. The payback time calculates the number of years required to regenerate, by means of profits, depreciation, and tax credits, the total investment of the fixed assets and preoperating expenses which are required to launch the product. It is seen that there is insufficient cash flow to recoup the capital investment and pretax operation expenses within 6 years. The results of this cash-flow analysis may give a negative decision to build the plant and produce the product. It may not be a good investment. Other designs will be examined to see if a more profitable chemical plant can be modeled that is satisfactory to the owner.

For our example, we consider the chemical plant as the only investment of the business, otherwise the losses of this operation are distributed to other parts of the business, and the total corporate income is taxed at another rate. Not knowing income from the other parts of the business, we assume that this plant is the only operating entity for this corporation. This is a good practice, because it requires the new opportunity to be a free-standing enterprise, thus showing its own warts or beauty marks.

The cash flow of Table 10.4 is based on a *nondiscounted* dollar (i.e., the face value of the real-dollar cash flows for each year). Nor do the cash flows deal with the concepts known as *time value of money*. Methods discussed later in this chapter show how these very important principles are applied to cash-flow analysis.

10.3   BREAK EVEN

Engineering provides estimates about designs to make future decisions. There is the premise that the action recommended by the estimate will add to the profit of the enterprise to make it worth the trouble. Typical situations include the following: An owner considers the construction of a chemical plant producing a soap liquid, or a consumer product is designed and will be marketed shortly, or a contractor is in business to haul construction materials. Not all future decisions assure that the owner or contractor is better off. Consider this case.

An electric utility serves several separated geographical regions. The utility employs transmission repair crews for high-voltage-line failures. The manager has permission to add another maintenance operator to a crew. In district A the average of the cost of repair per operator is estimated as $780, and in district B the estimate is $470. Our manager chooses to assign the operator to district B. But possibly the difference in the estimates occurred because the size of the repair crew in B was better adapted. If so, the new line mechanic may add less to cost reduction in B than he would in A. The reasoning that leads to the right choice is based on marginal cost and is important to engineering. Too often decisions are made on the basis of average cost or average revenue rather than on marginal cost or marginal revenue.

Marginal cost is the added cost because of adopting a change in construction, or making an engineering change order for a project. The change in operation usually implies increasing or decreasing construction work. The term *marginal cost* is interchangeable with *differential cost* or *incremental cost*. In a broad sense all estimates are marginal estimates, because they create changes from a current course of action. We explain marginal-cost analysis with simple arithmetic.

Activity Units	Estimated Total Cost	Analyzed Marginal Cost	Analyzed Average Cost
0	0	0	
1	$1000	$1000	$1000
2	1900	900	950
3	2700	800	900

This hypothetical illustration starts with the estimate of the total cost for the "activity units," which could be quantity, time, or some other measurable and known factor re-

TABLE 10.5 Total cost estimated for construction-materials hauling business for future annual period

Annual Truck Transport (Ton), n	Total Estimated Cost, C_T
5,000	$ 6,875
10,000	10,460
15,000	13,260
20,000	16,400
25,000	21,125
30,000	29,250

lated to the construction estimate. Once the estimating stage is concluded, then analysis on the estimated values begins, which leads to marginal and average cost.

The hypothetical cost is zero before any activity units. Marginal cost, by convention, is also zero, and average cost is undefined at this point. At two units, marginal cost is the amount added by the activity from one to two units, or $1900 - 1000 = \$900$. Average cost (arithmetic mean) is as usually determined. The brute-force way to explain marginal-cost analysis calculates extensive spreadsheets where total cost is enumerated for successive units (i.e., $4, 5, 6, \ldots$) in the region of interest. This is similar to the simple example above.

But a preferred approach to enumeration is to have a scattering of cost-estimated experiences and then fit linear and nonlinear regression lines. It is useful to remind ourselves of the assumptions for regression or least-squares: a cause-and-effect relationship is necessary, and there are no extraneous variables that make the regression of little value. The deviations of the dependent y values of "effect" are mutually independent, as the forcing variable x_j "causes" changes. The assumption is breached in the following example, as the dependent cost estimates for a controlled x_j factor, such as tons of construction material moved, are related to each other as the quantity increases. If a haulage firm uses trucks to move rock and rip-rap from one point to a second, then we assume that the fixed assets remain constant. Only the variable expenses of moving the quantity will change. Recall the assumptions for regression, given in Sec. 5.2.5.

There are many possible examples for regressing construction activity. As one example, a contractor has a truck fleet to haul rubble, rip-rap, and other bulk materials. Its central business is devoted to construction-material hauling. The trucks are of various sizes, ages, and capacities. Total annual costs for various tonnage on a specific transport route are estimated for the future year as shown in Table 10.5. Each activity value of n is separately estimated for total annual cost, C_T. To illustrate, we fit a first-, second-, and third-order polynomial regression line through the estimating data for a construction activity of tons of material moved.

These estimated data are scattered among arbitrary values for n tons. We then analyze the break-even properties of these data using regression models. It is simple to fit a polynomial regression equation of low order. Higher-order polynomials sometimes create problems. The number of data points for the polynomial equation should be three to eight times the order of the equation.

A model relating variable and fixed costs to an activity rate n is given as

$$C_T = nC_v + C_f \tag{10.5}$$

where C_T = total cost, dollars per period

        n = number of activity units per period

        C_v = variable cost, dollars per unit

        C_f = fixed cost, dollars per period

If at various values for n, C_v is the same, then we have the linear-cost model. If C_v varies, then we are concerned with nonlinear models. For the constant C_v, nC_v is a straight line increasing (or decreasing) at a constant rate per unit, and C_v is the slope of the variable-cost line. If the estimates are independently made for the activity variable, then there is no preknowledge of the slope, as we assume for this example.

The linear model, statistically fitted for the truck operational data, is $C_T = 0.84n + 1527$, where $C_v = 0.84$. The fixed cost is \$1527, which is the vertical-axis intercept for the plot of the linear model. The student may want to review the material in Chapter 5 about statistical fitting of data.

The average-cost model is given as

$$C_a = \frac{nC_v + C_f}{n} = \frac{C_T}{n} \tag{10.6}$$

where C_a = average cost, dollars per activity unit

If C_v is constant, we have the linear case again. Equation (10.6) is still valid if C_v is nonconstant; we have the nonlinear case, which is the usual fare.

For linear cost situations marginal cost equals C_v and is a constant. For the general case, we define *marginal cost* as

$$C_m = \frac{dC_T}{dn} \tag{10.7}$$

where C_m = marginal cost, dollars per activity unit

$\dfrac{dC_T}{dn}$ = derivative of total cost function with respect to activity variable n

By similar reasoning, if the output n is increased by an amount Δn from an established level n, and if the matching increase in cost is ΔC_T, then the increase in cost per unit increase in output is $\Delta C_T / \Delta n$. Marginal cost is the limiting value of this ratio as n gets smaller (i.e., marginal cost as the derivative of the total-cost function). It measures the rate of increase of total cost and is an approximation of the cost of a small additional unit of output from the given level. Sometimes marginal cost is called the slope or tangent of the total-cost curve at the point of interest.

Consider again the data of Table 10.5, which are regressed to a second-order polynomial, or $C_T = 6594 + 0.08n - 0.0000217n^2$, and marginal cost dC_T/dn becomes $C_m = 0.08 - 0.0000434n$. The marginal cost is a linear line inclining downward with a slope -0.0000434. Again, a third-order polynomial regression of the same data is $C_T = 840 + 1.527n - 7{,}418 \times 10^{-5}n^2 + 1.827 \times 10^{-9}n^3$ and the marginal-cost function becomes $C_m = 1.527 - 14.836 \times 10^{-5}n + 5.481 \times 10^{-9}n^2$, from which a marginal curve is plotted directly. The marginal-cost curve is a parabola, as shown in Fig. 10.2.

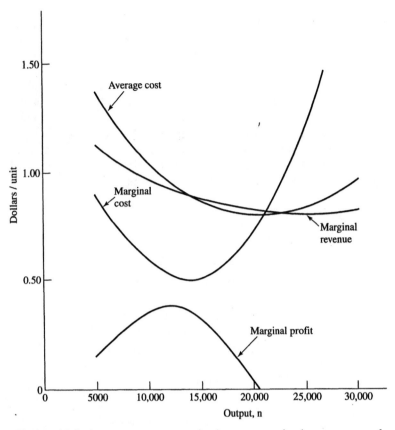

Figure 10.2 Average-cost, marginal-cost, marginal-revenue, and marginal-profit relationships.

After plotting the average-cost and marginal-cost curve, notice that the marginal-cost line intersects the lowest point of the average-cost curve. At this particular value of n, average cost is minimum and equal to marginal cost. This is shown in Fig. 10.2.

Reconsider the utility operation that is discussed above. The decision about the maintenance operator was made on the average-cost estimate. It should have been made on the marginal yield, which it promises. If district A has a greater marginal-cost reduction by the addition of the maintenance operator, it is the preferred location. The choice between the two districts should be based on the greatest negative C_m.

The parallel to marginal cost is marginal revenue or return. It involves many of the same concepts. In a construction-work design, marginal revenue is marginal savings; or in a project design, marginal revenue or marginal rate of return.

We assume that a specific economic measure can be chosen to reflect important differences among various values of the design variables. In a market situation, total return is the total money revenue of the producers supplying the demand, and the total money outlay of the consumers providing the demand. For our purpose a total-revenue function is estimated with knowledge about input price and demand, and knowledge that it increases or decreases with increasing output according to whether the demand is elastic or inelastic. With elastic demand and when price is decreased, increases in total revenue are found because quantity increases in greater proportion than the reduction in price, and vice versa.

In the opposite case of inelastic demand, when price is reduced, total revenue is reduced even though quantity may have increased.

To continue our analysis of the construction-material hauling business, total estimated revenue is given in Table 10.6.

Zero income for no tonnage moved is a trivial addition, but it aids the development of the data by a least-squares fitting. The total-revenue column contains no reference to the price for hauling a ton of material, but it is found by dividing total revenue by the units of output. We tacitly require that tonnage moved equals sales volume. In a similar fashion to the cost models above, we fit polynomial regression curves to those data and analyze those determined models on the basis of marginal principles.

We assume a model of the form

$$R_T = nR_v + R_f \tag{10.8}$$

where R_T = total revenue, dollars per period
        R_v = variable revenue, dollars per unit
        R_f = fixed revenue, dollars per period

This is a linear or nonlinear model if R_v is either a constant multiplier or a nonlinear term. If we consider the income of truck haulage, then revenue is not obtained until $nR_v > 0$. For a no-fixed-income condition, $R_f = 0$. If the total-revenue curve is linear, then the marginal revenue is a constant and equals R_v. For a contracting business this is the sales price. For average revenue, or average price, the model becomes

$$R_a = \frac{nR_v + R_f}{n} = \frac{R_T}{n} \tag{10.9}$$

where R_a = average revenue, dollars per activity unit

Marginal revenue is defined in a way similar to marginal cost, or

$$R_m = \frac{dR_T}{dn} \tag{10.10}$$

TABLE 10.6   Total estimated revenue for construction-material hauling business for future annual period

Annual Truck Transport (Ton), n	Total Estimated Revenue, R_T
0	0
5,000	$ 6,000
10,000	11,125
15,000	15,200
20,000	19,300
25,000	23,500
30,000	27,000

where R_m = marginal revenue, dollars per activity unit

$\dfrac{dR_T}{dn}$ = derivative of total-revenue function with respect to activity n

The data in Table 10.6, when regressed to a second-order polynomial, become $R_T = 90 + 1.138n - 8.13 \times 10^{-6}n^2$. Notice that this second-order polynomial equation does not pass through the origin $R_T = 0$ when $n = 0$. It might be assumed that at $n = 0$ there should be no revenue, which is correct. However, a displaced origin is not uncommon with a least-squares approximation. Additional methods beyond this text can force the equation to pass through the origin.

Practical constraints on the revenue are

$$R_T = \begin{cases} 0 & \text{if } n < 0 \\ R_v n + R_f & \text{if } n \geq 0 \end{cases} \qquad (10.11)$$

The marginal revenue of the second-order polynomial equation becomes $R_m = 1.138 - 16.26 \times 10^{-6}n$, and thus marginal revenue is a linear line inclining downward with a slope -16.26×10^{-6}.

For a third-order polynomial fit, $R_T = -48 - 1.23n - 1.64 \times 10^{-5}n^2 + 1.833 \times 10^{-10}n^3$. The marginal-revenue function would be $R_m = 1.23 - 3.28 \times 10^{-5}n + 5.499 \times 10^{-2}n^{-2}$, from which a marginal-return parabolic curve is plotted in Fig. 10.2. In the usual circumstance, the second-order marginal revenue model is considered more accurate than a first-order model.

Marginal cost and revenue estimating depend on the calculus. The marginal models are, for the most part, functions derived from regressed estimated and engineering data. The functions C_T or R_T are assumed continuous (despite the fact that they are routinely determined from arbitrary and selected activity levels) and differentiable. A good deal of information is uncovered from these basic models.

Returning to the maintenance-operator problem, the decision should be made on the marginal yield that it promises. Indeed, the addition of maintenance operators can proceed until the net yield is zero. The greatest negative marginal cost indicates the design with greatest cost reduction for least input resources (a negative cost change is a gain). Given adequate resources, both crews should be expanded until the yields are zero. If resources for expanding crews are not present, then the next strategy adopts the notion that marginal costs for districts A and B be the same. This may be done by reassigning operators from one district to another.

With curves and mathematical models, such as those described for the truck-transport example, it is also possible to optimize sales revenue and cost. The question is asked, "What shall we optimize?" Maximizing sales revenue may not guarantee maximum profit, nor does minimizing cost balance other factors for maximum profit. In some cases those actions may reduce profits. One of the traditional ways to study this interaction is by break-even analysis. A *break-even* point is defined as the point of operation at which there is neither a loss nor profit nor savings.

There may be more than one point of break-even operation. Several points of profit–loss neutrality are possible, and the location of those points is a straightforward exercise after the curves and models have been formulated from the data. When we use Eqs. (10.5) and (10.8), the linear approximation of a *point of indifference* occurs whenever $C_T = R_T$. Ignoring profits tax and solving for n, we have

$$n = \frac{C_f - R_f}{R_v - C_v} \qquad (10.12)$$

where n = break-even number of activity units per period

In large-scale construction projects, there may be revenue before work begins, because of earnest payments made for good faith and other contractual reasons. For a manufactured product that is commercially sold, income is not received until products are sold, and R_f as a fixed residual income does not exist, so $R_f = 0$. Model (10.12) can also include the class of problems where R_f is zero. In the case of models fitted to estimated data, R_f may exist to satisfy the best fit in a least-squares sense.

The denominator is unit profit without fixed cost or revenue. But, because fixed cost is absorbed in the business and sales volume, the net unit profit when divided into fixed dollars yields the break-even point in terms of dollars and units sold.

When using linear models $C_T = 0.84n + 1527$ and $R_T = 0.887n + 1284$, the break-even point is

$$n = \frac{1527 - 1284}{0.887 - 0.84} = 5170 \text{ tons}$$

This linear break-even point and the equations are pictured in Fig. 10.3.

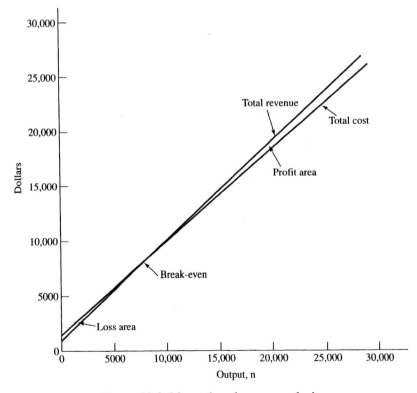

Figure 10.3 Linear break-even analysis.

Several aspects of linear break-even analysis are worth noting. These relationships are valid only for the short term, as defined by the estimating period. Discounting or time-value-of-money concepts are not used. For one firm, short term may be only weeks, and for another firm it may be an extended period of months and even years. To establish arbitrary rules about what is short or long term is dangerous unless specific cases are analyzed.

As demonstrated by the example, the mathematics and graphics involving linear (or nonlinear) costs and income provide information for break-even hauling of construction bulk materials. In the linear example of Fig. 10.3, a narrow wedge of profits begins at 5170 tons and continues indefinitely, so it would seem. However, the defect of pure linear models is obvious. Reduced revenue or unit price as an additional activity quantity is transported can occur from a variety of economic happenings. Increased unit costs of moving and hauling bulk materials beyond that which is normal are typical as activity increases indefinitely.

Although linear methods can be used for selecting the activity level above or below normal capacity and for cost-cutting tactics, the simplest way is to examine the plots of the original data. This is done by straightforward plotting of Tables 10.5 and 10.6, as shown in Fig. 10.4, where two break-even points, a lower and an upper, are indicated.

C_v is an increasing function and R_v a decreasing function. With this dual intersection of the revenue and cost lines, additional analysis of the meaning of optimum profit is called for. Thus, the indefinite profit above the profit break-even point of linear models can be misleading. Look at Fig. 10.2 again.

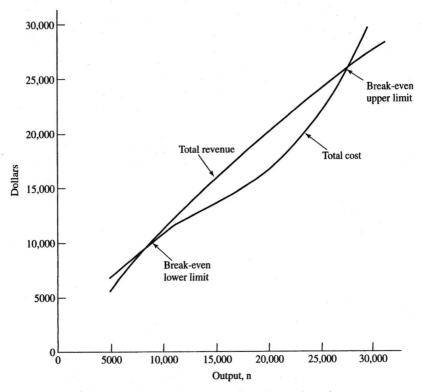

Figure 10.4 Total-revenue and total-cost functions.

The intersection of the slopes of the total-cost and total-revenue functions is given as

$$\frac{dR_T}{dn} = \frac{dC_T}{dn} \tag{10.13}$$

This intersection is the point at which marginal profit equals zero, or

$$\frac{dZ}{dn} = \alpha\frac{R_T}{dn} - \frac{dC_T}{dn} = 0 \tag{10.14}$$

where $\dfrac{dZ}{dn}$ = marginal profit, dollars per activity unit

If n increases beyond this point, then the cost of each unit exceeds the revenue from each unit, and total profits begin to decline, as they do beyond the upper break-even point in Fig. 10.4. Total profits do not become zero until the profit limit point, which is the beginning of a loss area. When the marginal profit is zero, we have the critical activity rate and maximum profit. This critical activity does not necessarily occur either at the rate corresponding to the minimum average unit cost or at the point of maximum profit per unit.

A plot of $dR_T/dn - dC_T/dn$ is given in Fig. 10.2. The point at which $dR_T/dn = dC_T/dn$ is also zero for marginal profit or dZ/dn. The critical production rate may be determined directly from differentiation of the profit function Z. This critical activity rate is zero at $n = 20{,}800$ tons, which is the point of maximum gross profit. But it is not the rate corresponding to the minimum average unit cost or the maximum profit per unit. The maximum profit per ton may be estimated from the level dome of the marginal-profit curve, or approximately 12,200 units. By using Fig. 10.2, we estimate minimum marginal cost as 14,500 units, and maximum marginal revenue is at the minimum sales projection of 5000 tons.

Although several optimal points of operation can be obtained from minimizing total or unit cost, maximizing total or unit revenue, or maximizing gross or unit profit, we should not conclude that all is lost if those exact objectives are never precisely realized. There is a redeeming feature if the actual and the ideal do not coincide. Fortunately, most optimums are relatively flat near the optimum. For those flat domes, operation on either side may be insensitive and not influence cost or profit very much. Naturally, a converse example can be found that demonstrates a sharp singularity as optimum. In this case, operation at this point is important.

Break-even analysis is customarily considered short term. In the next section, these concepts are extended with what is called engineering-economy analysis, where the important matters of discounting and time-value-of-money concepts are introduced.

This discounting allows a systematic method of comparing streams of costs and revenues that have differing moneys and times of payment. Generally, the discounting or present-worth model is used. If engineering desires an annual rate of profit as a 5% return, then a dollar received 1 year in the future is really worth $(1 + 0.05)^{-1}$ dollars right now, and received n years in the future is worth $(1 + 0.05)^{-n}$ dollars right now. Or if you presently had $(1 + 0.05)^{-n}$ dollars and lent it at 5% interest compounded, you would have back 1 dollar n years from today.

10.4 ENGINEERING ECONOMY

In the construction-business world, the purpose of converting money into buildings, plants, and equipment is to return an amount of money that exceeds the investment. This statement assumes that capital is productive and earns a profit for its owner. In efficiency terms, productive capital is related to a ratio of output to input. But unlike physical processes, the economic efficiency of capital, assuming long-term success and a capitalistic society, must exceed 1. The productivity of capital arises from the fact that money purchases more efficient procedures for supplying services and making goods than people can employ themselves. Those services and products are offered to the public at attractive prices that return cost and profit to the investor.

Both owners and contractors need to understand the principles of engineering economy. This is a brief encounter with the topic. Textbooks that enlarge upon it are listed in the References.

A central theme of this chapter is the importance of tradeoffs in the many phases of design to construction. Tradeoffs exist with capital investments, too. This is, perhaps, one of the earlier epochs for the tradeoff economics.

In earlier chapters, cost and price are emphasized as the metric of importance. For purposes of capital investment, however, the owner of the capital (which includes contractors, for example) uses a term called *return*. Like cost and price, return is expressed in several ways. Among those are total dollars, percent of sales, ratio of annual sales to investment, or return on investment. Return on investment is favored by project engineers.

The methods for calculation of return on investment are

- Average annual rate of return
- Payback period
- Engineering-economy rate of return

In this text we favor the compound-interest-based investment computation that considers the time value of money. However, as a background let us look at methods that are used because of their simplicity.

10.4.1 Average Annual Rate-of-Return Methods

In some economy studies, return on investment is expressed on an annual percentage basis. The yearly profit divided by the total initial investment represents a fractional return or its related percent return. This recognizes that a good investment not only pays for itself, but also provides a satisfactory return on the funds committed by the firm. There are several variations, of which this is one:

$$\text{return} = \frac{\text{earnings per year}}{\text{net bid value}} \times 100, \% \tag{10.15}$$

Earnings are after tax, and deductions for depreciation usually represent some average future expectation. For instance, consider the following example: The bid for new equipment is $175,000, salvage will provide $15,000, and an average earnings of $22,000 after taxes is expected.

$$\text{return} = \frac{\$22,000}{\$160,000} \times 100 = 13.75\%$$

Now, assume that an investment opportunity of $25,000 has come to your attention. This is broken down to $20,000 for the investment and $5000 for initial working capital (cash, accounts payable, and so forth). Annual operating and other expenses are estimated at $10,000, and income is estimated as $15,000 per year.

$$\text{return} = 5000/25{,}000 \times 100 = 20\%$$

Another variation is expressed as

$$\text{return} = \frac{\text{average earnings} - (\text{total investment} \div \text{economic life})}{\text{average investment}} \times 100, \text{percent} \quad (10.16)$$

The earnings in the formula are the average annual earnings after taxes, plus appropriate depreciation charges. The original investment is recovered over the economic life of the proposal by subtracting the factor of total investment divided by economic life from average earnings. This difference denotes the average annual economic profit on the investment.

The *average* investment is defined as the total investment times 0.5, acknowledging that the life of an investment for tax purposes and its true economic life are not the same.

The original investment is based on the normal physical life or as legally defined by the Internal Revenue Service. Average investment represents the profitable life of the investment, which is frequently a different period. If engineering desires, it may incorporate a risk element by further shortening economic life.

For example, an average after-tax earning of $22,000 is expected from an investment of $175,000 having an economic life of 10 years. Straight-line depreciation is assumed for a period of 12 years, and salvage is $15,000.

$$\text{return} = \frac{22{,}000 + (160{,}000/12) - (175{,}000/10)}{160{,}000 \times \frac{1}{2}} \times 100 = 22.3\%$$

10.4.2   Payback-Period Method

The payback method is easy to understand and is widely adopted. Essentially, the method determines how many years it takes to return the invested capital. The formula as normally given is

$$\text{payback} = \frac{\text{net investment}}{\text{annual after} - \text{tax earnings}}, \text{years} \quad (10.17)$$

The payback method recognizes liquidity as the important feature for the economic worth of capital expenditures. Payback dumps proposals of doubtful validity from those that call for additional economic analysis. Obviously, payback signals the immediate cash-return aspect of the investment, which may be desirable for corporations where a high-profit investment opportunity and limited cash resources exist. In some situations, the payback metric is used for those investment situations where the risk does not warrant earnings beyond the payback period.

Let us use an example to illustrate this. The installed cost for new equipment is $175,000, and old equipment will be sold for $15,000. Better productivity of the new equipment will earn $29,400 additionally. This assumes constancy of the earnings. For a composite 25% corporate tax rate, after-tax earnings amount to $22,000 (= 29,400 × 0.75).

$$\text{years payback} = \frac{175,000 - 15,000}{22,000} = 7.3$$

Thus, the investment requires about 7 years before the earnings have liquidated the first cost. Now, consider two investment opportunities:

	Equipment A, $60,000	Equipment B, $60,000
Revenue		
Year 1	20,000	30,000
Year 2	20,000	30,000
Year 3	20,000	30,000
Year 4	20,000	
Year 5	20,000	
Total annual after-tax earnings	$100,000	$90,000
Payback period	3 years	2 years

In this case, equipment B is preferred over A because of the shorter payback period. Notice that payback for equipment A is 3 years (=60,000/20,000). If sufficient resources to purchase investments were available, then an engineering fiat could approve for spending any investment where payback was under some arbitrary cutoff level, such as 5 years.

The average-annual-rate-of-return method is acknowledged to have faults. It assumes equal distribution of earnings throughout the economic life of the asset. Even if this assumption held, there is a significant difference between the value of the dollars earned in the first year and those earned in later years. The time value of money is ignored. A project yielding savings in early years of its life is more beneficial, because those funds become available for additional investment, or for alternative use. Those early funds are subject to less risk than savings projected many years ahead. Further, differences in salvage values and their relation to the time element are overlooked. Nor is interest on borrowed money in any way reflected in the above equations.

The payback method suffers similarly. The life pattern of earnings is ignored in payback formulas. In the example, equipment B had a shorter payback period than equipment A, yet A will return $10,000 more. New equipment may be profitable during the early part of the payback period. On the other hand, new equipment may be quite profitable in the future. Payback does not provide for a technique of ranking with other investment possibilities. Nor does it take into account depreciation or obsolescence or the earnings beyond the payback period. For example, the payback method does not recognize that one investment earning $10,000 the first year and $2,000 the second year is more desirable than another that earns $6,000 in each of the 2 years. The use for which payback is suited, and then only provisionally, is as a rough measure of evaluation.

The engineering-economy method of determining return overcomes those shortcomings. It is applicable to every possible type of prospective investment and it yields answers that permit valid comparisons between competing projects.

The time-value-of money approach applies compound-interest formulas to the additional cash flow produced by the investment. This concept enables engineering to place a value on the money that becomes available for productive use in the future as well as for the money available today. Fundamentally, the time value of money begins with *simple interest*, or

$$I = Pni \tag{10.18}$$

where I = interest earned, dollars

P = principal sum, dollars

n = number of periods

i = interest rate, decimal

This formula can be restated as the amount including principal and simple interest that must eventually be repaid, or

$$F = P + I = P(1 + ni) \tag{10.19}$$

where F = principal and interest sum at some future period, dollars

Payment of simple interest is made at the end of each time period, or the sum total amount of money is paid after a given length of time. Under the latter condition there is no incentive to pay the interest until the end of the contract time.

If interest is paid at the end of each time unit, then the lender could use the money to earn additional profits. Compound interest considers this point and requires that interest be paid regularly at the end of each interest period. If the payment is not made, then the amount due is added to the principal, and interest is charged on this converted principal during the following time unit.

An initial loan of $10,000 at an annual interest rate of 5% would require payment of $500 as interest at the end of the first year. If this payment were deferred, then the interest for the second year would be ($10,000 + $500)(0.05) = $525, and the total *compound* amount due after 2 years would be $10,000 + $500 + $525 = $11,025.

When interest is permitted to compound, the interest earned during each interest period is permitted to accumulate with the principal sum at the beginning of the next interest period. This compounding is shown in Table 10.7. The resulting factor, $(1 + i)^n$, is referred to as the single-payment compound-amount factor. Notice the arithmetic of the progression of the exponents.

TABLE 10.7 Derivation of basic compound-interest formula

Year, n	Principal at Start of Period	Interest Earned During Period	Compound Amount F At the End of Period
1	P	Pi	$P + Pi = P(1 + i)$
2	$P(1 + i)i$	$P(1 + i)i$	$P(1 + i) + P(1 + i)i = P(1 + i)^2$
3	$P(1 + i)^2$	$P(1 + i)^2i$	$P(1 + i)^2 + P(1 + i)^2i = P(1 + i)^3$
...			
n	$P(1 + i)^{n-1}$	$P(1 + i)^{n-1}i$	$P(1 + i)^{n-1} + P(1 + i)^{n-1}i = P(1 + i)^n$

Reconsider the quantity called F, as defined above, implying an interest and principal sum at a future period. Typically, the total amount of principal plus compound interest due after n periods is defined thus:

$$F = P(1 + i)^n \qquad (10.20)$$

$$\text{future sum} = (\text{present sum})(1 + i)^n$$

where F = future amount, dollars
  i = interest rate per interest period, decimal
  n = number of compounding periods

The single-payment compound-amount factor is used to solve for a future sum of money F, interest rate i, number of interest periods n, or a present sum of money P when given the other quantities.

Engineering-economy methods are preferred because they depend on time-value-of-money concepts. But do not conclude that all methods employing interest computations are useful for all occasions. Some have limited applicability. When those methods are given correct information and properly understood, their answers are equally valid. We present four distinct variations:

- Net present worth
- Net future worth
- Net equivalent annual worth
- Rate of return

Each of those methods measures a different factor of the investment. They give different evaluations. Nonetheless, they generally lead to the same recommendation for consistent decision making.

Each of the four methods is demonstrated with the same standard numbers. (Cents are dropped from calculations for ease of understanding.) The $1025 can represent the cost of construction equipment, for example. Notice that in this simple problem the revenues vary each year.

Year	Cost	Revenue
0	$1025	$0
1	0	450
2	0	425
3	0	400

Simplifying procedures are adopted for these engineering-economy methods. While the investment and revenues may be actually distributed throughout a year, we assume that nonuniform or annual revenues are instantaneously received at the end of year. Investment is scheduled at time 0, or the origin of the project.

10.4.3 Net-Present-Worth Method

The *net-present-worth* method is also known as the *net-present-value* or *venture-worth* method. It compares the present worth of future revenue with initial capital investment, assuming a continuing stream of opportunities for investment at a preassigned interest rate. The procedure compares the magnitude of present worth of all revenues with the investment at the datum time 0. A decision about the investment is made based on the magnitude of this comparison.

We may define net present worth as the added funds that will be required at the start of a proposed project, invested at a preassigned interest rate, to produce receipts equal to, and at the same time as, the prospective investment. For a given interest rate of 10%, the net present worth of the previously given problem is computed by discounting all revenues to year 0 at this rate and subtracting the proposed investment, or

Period, n	$1(1 + i)^n$ Present-Worth Factor at 10%	Amount, $
Year 1 to zero	$450 \times 0.9091 =$	$ 409
Year 2 to zero	$425 \times 0.8264 =$	351
Year 3 to zero	$400 \times 0.7513 =$	301
Total		$1061
Less proposed investment		1025
Net present worth		$36

The $36 is the amount that must be added to the $1025 to set up the amount that would have to be invested at 10% to achieve receipts equal to and at the same time as those predicted for the recommended investment, or

$$(\$1025 + \$36) \times 1.1 = \$1167$$
$$\text{less payment} \quad \underline{450}$$
$$\$ 717$$
$$717 \times 1.1 = \$789$$
$$\text{less payment} \quad \underline{425}$$
$$\$364$$
$$\$364 \times 1.1 = 400$$
$$\text{less payment} \quad \underline{400}$$
$$0$$

10.4.4 Net-Future-Worth Method

Assets and revenues can be invested at the preassigned interest rate where there is a continuous exposure of investment according to opportunities $[F = P(1 + i)^n]$. A comparison of investment of the original sum plus reinvestment of revenues at the preassigned interest is made against the standard alternative of investing only the original asset value. The calculation results in the added amount obtained at the end of the project's economic life. A common comparison uses the same stipulated interest rate i.

For the sample problem and 10%, the net future worth is computed by compounding future revenues to the terminal year, then subtracting from this the amount that would have resulted from the other alternative of investing the original asset at the same preassigned interest rate to the terminal year:

Period, n	$(1 + i)^n$ Compound Amount Factor at 10%	Amount
Year 1 to 3	$450 × 1.10^2 =$	$545
Year 2 to 3	425 × 1.10 =	468
Year 3	400 × 1.00 =	400
		$1413
Less disbursements compounded to terminal year at 10%		
Year 0 to 3	$1025 × 1.10^3 =$	1364
Net future worth		$49

The calculations point out that if the project is funded and if the revenues materialize as estimated, then a surplus of $49 will be expected over the simple alternative of investing only the asset of $1025. The same period of time and equal interest rates are parts of this comparison.

10.4.5 Net Equivalent Annual Worth Method

Management often wants a comparison of annual costs instead of, say, present worth of the costs. Here, we refer to net costs—that is, the net difference between any cost and revenues or credits. This method considers a supply of opportunities for investment of both assets and receipts at the predetermined interest rate plus a supply of capital at the same interest rate.

Now, the sample problem does not have uniform annual receipts, and the receipts first must be converted to total present worth and then to annual equivalents. The total present worth at time zero is $1061 (from Sec. 10.4.3). The annual equivalent is found by dividing by the sum of the present-worth factors, or

$$\text{annual equivalent amount} = \frac{1061}{(0.9091 + 0.8264 + 0.7513)}$$

$$= \frac{1061}{2.4868} = \$427$$

$$\text{net annual equivalent worth} = 427 - \frac{1025}{2.4868} = \$14$$

This $14 is the amount by which the anticipated revenues depart from the proposed investment, rated at 10% interest, and exceed the annual equivalent of the proposed investment:

		Amount
Anticipated equal annual receipts		$427
Less equal annual equivalent worth at 10%		14
Equal annual receipts to be generated by investing		$413
	$1025 \times 1.10 =$	$1128
Less payment		413
		715
	$715 \times 1.10 =$	$787
Less payment		413
		374
	$374 \times 1.10 =$	$413
Less payment		413
		0

Starting with the investment that earns interest and subsequently subtracting payments leads to a balance of zero dollars at the end.

10.4.6 Rate-of-Return Method

The rate-of-return method calculates the rate of interest for discounted values of the net revenues from a project to have the present worth of the discounted values equal to the present value of the investment. The rate-of-return method thus solves for an interest rate to bring about this equality.

Other titles also exist, such as *ROI* (return on investment), *true rate of return*, *profitability index*, and *internal rate of return*. The adjective "true" distinguishes true rate of return from other less valid methods that have been labeled rate of return [i.e., Eqs. (10.15) and (10.16)].

For this method there is no assumption of an alternative investment and no predetermined interest rate as was required for the previous methods. We define this interest rate at which a sum of money, equal to that invested in the proposed project, would have to be invested in an annuity fund in order for that fund to be able to make payments equal to, and at the same time as, the receipts from the proposed investment.

The solution for the interest rate is by repeated trials or by graphical or linear interpolation. Typically, two interest rates bound the value of the investment. For the sample problem, the two trial values are 5 and 15%, and the calculation is given as

Year, n	Revenue		$\dfrac{1}{(1+i)^n}$ PW Factor at 5%		Discounted Amount
1	$450	×	0.9524	=	$ 429
2	425	×	0.9070	=	385
3	400	×	0.8638	=	346
Total					$1160

Year, n	Revenue		$\dfrac{1}{(1+i)^n}$ PW Factor at 15%		Discounted Amount
1	$450	×	0.8696	=	$391
2	425	×	0.7561	=	321
3	400	×	0.6575	=	263
Total					$975

The two trial values bound the initial asset value of $1025. Note that the higher value, 15%, discounts the annual revenues into smaller values.

Interest Rate, %	Present Worth, $
5	$1160
To be determined	1025
15	975

Figure 10.5 shows a linear line between the two points, and the value of $i = 12.3\%$ is found. A better approximation occurs when the neighboring interest boundaries are narrower than that shown, as the line is nonlinear.

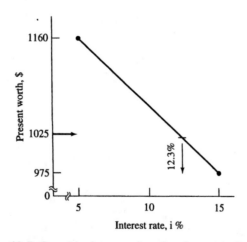

Figure 10.5 Graphical approximation for rate of return.

This rate of return is the interest rate at which the original sum of $1025 could be invested to provide returns equal to, and at the same time as, the three revenues of the prospective investment. The calculation below demonstrates this equivalence.

		Amount
	$1025 \times 1.123 =$	$1151
Less payment		450
		$700
	$700 \times 1.123 =$	$787
Less payment		425
		$362
	$362 \times 1.123 =$	$400
Less payment		400
		0

This calculation shows that the earning rate is true and is the actual return of the invested money. This method has the important advantage of being directly comparable to the cost of capital, which is the lending rate charged by banks and investors for the project.

10.4.7 Comparison of Methods

The results of the four different methods are summarized below.

Method	Amount
Net present worth at 10%	$36
Net future worth at 10%	$49
Net annual equivalent worth at 10%	$14
Rate of return	12.3%

We can demonstrate that the first three methods give commensurable answers. For instance, the net present worth of $36 can be compounded to the terminal year by using Eq. (10.20), or

$$F = 36(1.10)^3 = \$49$$

The preassigned interest rate, 10%, yields the net future worth or $49. The net annual equivalent worth of $14 can be found by dividing the net-present-worth value of $36 by a sum of present-worth values of 10%, or

$$\frac{36}{2.4867} = \$14$$

On the other hand, the rate of return cannot be calculated from the foregoing answers. It is only found directly from the data. Equivalency among the first three methods can be found for any arbitrary interest rate, and if those methods are equivalent to the rate of return, then they are equivalent at only one preassigned interest rate.

The differences in the first three methods are based on the choice of an interest rate. This arbitrary selection of an interest rate makes the three methods little more than decision tools for comparing projects. The rate of return is the only method that can provide a consistent measure of the extent of the economic productivity of prospective investments. The student needs to know that the four methods remain the most popular of all the engineering-economy methods, but no one single method or criterion of profitability analysis is preferred in all business situations.

10.4.8 Standard Approaches to Engineering-Economy Methods

The foregoing algebraic methods are instructive, but spreadsheets and software, calculators, equations, and factor tables are the techniques that are used for construction-cost tradeoff appraisal. Table 10.8 summarizes the equations and key symbols for engineering-economy calculations.

TABLE 10.8 Summary table of periodic compounding formulas for engineering-economy methods: Functional notation and formulas

Factor Name	Given	To Find	Functional Notation	Formula
Single payment				
Compound-amount factor	P	F	$(F/P, i\%, n)$	$F = P(1 + i)^n$
Present-worth factor	F	P	$(P/F, i\%, n)$	$P = F(1 + i)^{-n}$
Uniform payment series				
Sinking-fund factor	F	A	$(A/F, i\%, n)$	$A = F\left[\dfrac{i}{(1 + i)^n - 1}\right]$
Capital-recovery factor	P	A	$(A/P, i\%, n)$	$A = P\left[\dfrac{i(1 + i)^n}{(1 + i)^n - 1}\right]$
Compound-amount factor	A	F	$(F/A, i\%, n)$	$F = A\left[\dfrac{(1 + i)^n - 1}{i}\right]$
Present-worth factor	A	P	$(P/A, i\%, n)$	$P = A\left[\dfrac{(1 + i)^n - 1}{i(1 + i)^n}\right]$

where i = effective interest rate per interest period, decimal
n = number of interest periods, usually having units of years, but other units are allowed
P = present sum of money, dollars
F = future sum of money is an amount n interest periods from the present time that is equivalent to P at interest rate i, dollars
A = end of period cash receipt or disbursement in an equal payment series continuing for n periods, and the entire series is equivalent to P or F at interest rate i, dollars

The six relationships are divided into two groups, single- and equal-payment series. The notation consists of five symbols, P, F, A, i, and n, and their definitions are given. Cash flows are implicitly assumed to be at the start or the end of the year and are considered lump sum or discrete.

Functional notation is preferred over the writing of the equations. An example of functional notation is $(F/P, i\%, n)$. Rather than write $F = P(1 + i)^n$, we write $F = P(P/F, i\%, n)$, and a table factor is substituted. The meaning of functional notation, for example, F/P, says that you wish to find F given P.

There are other engineering-economy methods, such as continuous interest or continuous flow of funds, but for the purpose of early tradeoff of construction design and projects, our sense is that they are overkill when compared to the accuracy of preliminary estimates.

Now observe the abbreviated 5% sample factor table, Table 10.9. Clearly, spreadsheets and calculators do similar things, but published tables are helpful for instruction.

$(F/P, 5\%, 4)$ is the single payment, and given a value of P to find F and for $i = 5\%$ and $n = 4$ periods, examination of Table 10.9 gives a value of $(F/P, i\%, n) = 1.216$ as the factor. There are numerous tables, such as this one, but we only supply tables for 10% and 20%, which are given in the Appendix.

The solution to construction and engineering design alternatives, which have a simplified presentation in this book, involves several assumptions:

- Discrete cash flows occur at the year end
- Inflation effects are ignored, inasmuch as they affect the alternatives similarly (methods are provided earlier that allow inflation analysis)
- Analysis is before the effects of taxes
- The interest rate is constant during the lifetime of the project
- Nonquantifiable effects are ignored until the analysis is ended
  - Goodwill & prestige
  - Preferences
  - Political risks

TABLE 10.9   Abbreviated factor table for 5% interest rate for 1 to 5 years

	Single Payment		Equal Payment Series			
	Compound-amount factor	Present-worth factor	Compound-amount factor	Sinking-fund factor	Present-worth factor	Capital-recovery factor
n	To find F Given P F/P i, n	To find P Given F P/F i, n	To find F Given A F/A i, n	To find A Given F A/F i, n	To find P Given A P/A i, n	To find A Given P A/P i, n
1	1.050	0.9524	1.000	1.0000	0.9524	1.0500
2	1.103	0.9070	2.050	0.4878	1.8594	0.5378
3	1.158	0.8638	3.153	0.3172	2.7233	0.3672
4	1.216	0.8227	4.310	0.2320	3.5460	0.2820
5	1.276	0.7835	5.526	0.1810	4.3295	0.2310

A *minimum attractive rate of return* (MARR) i is predetermined before project analysis, and it meets or exceeds the benchmark of economic performance for the contractor or owner.

Those assumptions are employed because the information that is available at the early stages of finding the tradeoffs for a project are only dimly seen, and simple calculations expedite the decision making.

Construction projects have typically a long life. Buildings, bridges, plants, roads, and so forth may have a life of 50 or more years. Those long-lived projects are given a special analysis whenever the operating costs can overshadow the initial investment costs. This is known as life cycle-cost analysis.

10.5 LIFE CYCLE COST (OPTIONAL)

Life cycle cost is the summation of all estimated cash flows from concept, design, construction, operation, maintenance, and disposal of the system at the end of useful life. Intuitively, individuals have used LCC principles for economic evaluation of cars when they concern themselves not only with initial cost (sticker price), but with operating and maintenance expenses (gas mileage, worn parts, insurance, and license) and residual value (resale price).

Operating cash flows can easily be greater than the original R&D or the investment. Moreover, a construction system with higher engineering and investment costs but lower operation and maintenance costs may, depending on service life, be a least-LCC system. It has been shown that for military hardware systems approximately two-thirds of life-cycle costs are unalterably fixed during the design phase. Some states have legislation requiring LCC in the planning, design, and construction of state buildings. LCC encourages tradeoff analysis between one-time costs and recurring costs.

LCC has an important role because of the influence of total costs rather than initial costs of ownership. The selection is then focused on the total cost of ownership rather than the first cost of a building or equipment, for example. In the past, owners and contractors ignored those future costs, but the design decisions made today carry cost implications into the future.

LCC attempts to estimate all relevant costs, both present and future, in the decision-making process for the selection among various choices. General cost estimates that are required for LCC are the following:

- Design, development, and engineering
- Initial capital investment and financing
- Operation, maintenance, and functional use
- Replacement
- Alteration, refurbishing, and improvement
- Salvage and retirement

Figure 10.6 illustrates engineering, project, operation, and maintenance costs as separate cash flows. In LCC analysis, the estimates are scheduled as period cash flows, starting with $E = 0, +1, +2$, where E = moment of the estimate, N_c = end of costing period, and N_l = end of life-cycle. The periods are year designations.

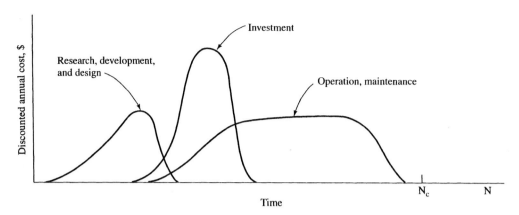

Figure 10.6 Cost-time phasing for a life cycle.

The length of the life cycle is sensitive to wear-out, casualty or destruction, obsolescence, economic, or technology factors. When obsolescence is a factor, qualified opinion is required, because the life of the equipment may be suddenly terminated by a change in company policy, buying habits, government legislation, competitive pressures, or new designs.

There are two major concerns in forecasting the life of an asset: annual deterioration and obsolescence. Where physical life or annual deterioration establishes the life, statistical data from past records become the basis for future prediction. If life is regarded as economic, then statistical methods find the probability that the economic life of a proposal will terminate during each year of its service life. In either of those two cases, deterioration or obsolescence, the life of a particular proposal terminates because it is worn out or because the equipment or service is no longer profitable.

In the first instance, the physical condition has deteriorated and does not produce the desired quality, or the cost of maintenance exceeds the cost of replacement. In the absence of statistical data, reliance is placed on the opinion of people having experience, such as engineers, operators, and the people producing the equipment. In the case of obsolescence, competition introduces substitute products, processes, or equipment with better prices, qualities, or services.

This raises the point: What is the life cycle? In public works 30 to 100 years can be used. In a weapons system the horizon may terminate within a period of months to 50 years. In commercial enterprises a system may logically extend from a fad period to an enduring life of 20 to 40 years. All kinds of priorities, political and social, affect the length of time. As the system life cycle becomes longer, the actual length becomes less crucial, because the discounting factor drops rapidly if either i or n or both increase.

Salvage value is another bewildering prediction. An experienced appraiser may give opinions of future land and building values. If the life is not expected to be great, then the engineer is in a position to trust this source of information. For the longer-lived equipment, information may be unavailable. Despite the uncertainty of distant predictions, errors in evaluating salvage value are not normally serious. Error in the present or the near future is given more study because the effects are greater.

Another prediction for equipment and buildings is the *utilization* or *efficiency*. Errors in the degree of utilization of equipment are considered serious, and predictions of this nature are studied closely. Efficiency is important because to some extent it controls labor costs.

PICTURE LESSON
Concrete

The cement used by the ancient Egyptians was calcined gypsum, and both the Greeks and Romans used a cement of calcined limestone. Roman concrete was made of broken brick embedded in a possolanic lime mortar. The cementitious matrix of the latter consisted of lime putty mixed with brick dust or volcanic ash. Hardening was produced by a prolonged chemical reaction between these components in the presence of moisture.

With the decline of the Roman Empire, concrete fell into disuse. A step was taken toward its reintroduction about 1790, when an Englishman, J. Smeaton, found that when lime containing a certain amount of clay was fired, it would set under water. This cement resembled that which had made by the Romans. This led to commercial production of natural hydraulic cement, which was used early in the nineteenth century. An important advance toward the manufacture of a dependable hydraulic cement was developed by an English mason, Joseph Aspdin, in 1824. His product was called Portland cement, because it resembled building stone that was quarried on the Isle of Portland. The first Portland cement plant opened in 1840 in England and in the United States about 1871.

Concrete, then, is a suitably proportioned mixture of coarse and fine aggregate, cement and water. The term "concrete" means to grow together.

The picture lesson shows a section of the Bonneville Dam spillway in stages of construction. Viewed in mixed sunlight and shade, the wooden forms are seen progressing higher as the framework. The Bonneville Dam straddles the states of Oregon and Washington and is a major source of power. It is one dam of several along the Columbia River.

Note that in Fig. 10.6 the cost of design precedes the investment of the project, which precedes the operating and maintenance costs. It is not necessary that the curves be symmetrically shaped. Also discrete spikes of lump-sum cash flows are possible—for example, salvage value. Costs conclude at the end of the life cycle.

In preparation for the LCC analysis, the following procedures and equations are required:

- Estimates
- Operating profile
- Maintenance schedule

The operating profile has a repetition time and contains all the operating and nonoperating modes of the equipment or building. It is sometimes possible to have operating profiles internal to other operating profiles. For tradeoff studies, candidates are evaluated with the same profile. The profile says when or in what way the equipment is operating.

Two parameters are Mean Time Between Failures (MTBF) and Mean Time To Repair (MTTR). Time between overhaul, power consumption rate, and preventive maintenance routines, such as cycle and the preventive maintenance rates, are required information.

Labor for preventive and corrective maintenance is calculated by using

$$PM \text{ actions} = \frac{SOH}{PM \text{ cycle time}} \qquad (10.21)$$

PM = preventive maintenance cycle time, units consistent with SOH
SOH = scheduled operating time, usually hours

$$C_{cm} = \frac{SOH}{MTBF}(MTTR)C_m \qquad (10.22)$$

where C_{cm} = cost for corrective maintenance, dollars
$MTBF$ = mean time between failures, usually hours
$MTTR$ = mean time to repair, usually hours
C_m = cost of maintenance labor, usually dollars per hour

Maintenance costs need to be framed into annual costs, if the dimensions deal with hours or maintenance cycles. Typically, most life-cycle costs are dimensioned into annual payments.

Here is an example for life cycle costing. We return to the concrete-pumping truck equipment, which distributes concrete into forms at the construction site. The student will

TABLE 10.10   Life cycle cost information for 195-yd³/hr capacity concrete-pumping truck equipment

Information Element	Amount
Equipment purchase cost	$1,200,000
Service life before first overhaul	5 yr
Overhaul cost	$125,000
Remaining life after overhaul	4 yr
Concrete travel and pumping operating hours, annual	2,600 hr
Standby hours, annual	3,000 hr
Salvage value, percentage of purchase cost	15%
Periodic and unscheduled maintenance cost, percentage of first cost, annual	2.5%
Spare parts and tire replacement, annual	$10,000
Gross hourly labor costs (available during all standby hours)	$21/hr
Diesel fuel, oil, and truck operating costs, annual for 3000 hours	$45,000
Cleanup material (sacks of Portland cement, etc.), annual	$7,800
Minimum attractive rate of return, $i\%$	20%

want to reread material in Chapter 2 and 7 about this equipment. In Chapter 7, discussion focused on fixed, semifixed, and variable costs and how we develop this equation to perform estimates. Life cycle cost analysis has a different slant than to give only estimates for work.

The portable concrete-pumping equipment has a maximum capacity of 195 yd³/hr and is purchased for $1,200,000 delivered cost. It has a service life of 5 years before the first overhaul, which restores the equipment nearly to its initial performance, and this major maintenance cost is $125,000. The refurbished equipment is eventually sold for 15% of the initial price.

Cost analysis is not rocket science. Variability and difference in approach are normal. But it is important to write down the assumptions that influence the collection and analysis of the data. For example, we use the start-of-year convention for expenses of labor, equipment operating costs, and investment costs as if they were prepaid. This means that the counting of the costs starts with zero time. The costs are considered discrete, inflation effects are ignored, and values are *constant dollar*. These and other facts are collected and are shown as Table 10.10. This example is without revenues from the equipment choices, although such additions are common for the general problem.

Table 10.11 shows the LCC analysis of the equipment estimates. An interest rate of 20% discounts all future cash flows.

Using the discounted annual and cumulative cost values from Table 10.11, the results are shown in Fig. 10.7. Unlike Fig. 10.6, which shows individual element graphs, the LCC curves in Fig. 10.7 are consolidated into annual and cumulative values. Two pumping capacities are compared, and it is seen that the smaller capacity, 175 yd³/hr, while it has a smaller first cost, is less attractive from a life-cycle-cost view.

In LCC, the notion is to experiment with several design configurations under various scenarios, which are exercised for future events in identical ways. Then the life-cycle-cost analysis allows for visibility of "tall poles," a jargon implying significant cost elements. In LCC it may not be necessary to find *total* or *absolute* or *full* cost, as *relative* costs may separate the better alternatives from those of dubious merit.

Much software is available to conduct LCC comparisons. There is software for concrete pipe, bridges, highways, equipment, building designs (with allowances for geographical location, materials of construction, maintenance requirements, and so forth), and other

TABLE 10.11 Life cycle cost analysis for concrete-pumping truck with 195-yd³/hr capacity

Cost Element and Year	0	1	2	3	4	5	6	7	8	9
Equipment	1,200,000									
Labor	63,000	63,000	63,000	63,000	63,000	63,000	63,000	63,000	63,000	
Equipment operation	45,000	45,000	45,000	45,000	45,000	45,000	45,000	45,000	45,000	
Maintenance	30,000	30,000	30,000	30,000	30,000	30,000	30,000	30,000	30,000	
Spare parts, tire replacement		10,000	10,000	10,000	10,000	10,000	10,000	10,000	10,000	
Cleanup material	7,800	7,800	7,800	7,800	7,800	7,800	7,800	7,800	7,800	
Overhaul						125,000				
Income from salvage sale										(180,000)
Total	1,345,800	155,800	155,800	155,800	155,800	280,800	155,800	155,800	155,800	(180,000)
$(P/F\ 20\%, n)$	1	0.833	0.695	0.579	0.482	0.402	0.335	0.280	0.233	0.194
Discounted annual value	1,345,800	129,781	108,281	90,2081	75,096	112,882	52,193	43,624	36,301	(34,920)
Cumulative value	1,345,800	1,475,581	1,583,862	1,674,071	1,749,166	1,862,048	1,914,241	1,957,865	1,994,166	1,959,246

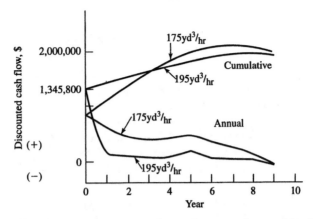

Figure 10.7 Life-cycle-cost comparison of two rated capacities for concrete-pumping equipment.

applications. Building analysis is even according to ASTM standards, for instance. The software handles estimates, inflation effects, taxes, and various discount rates, and many of the engineering design features.

The above cost-analysis methods are for entities incorporated as profit businesses. Various governmental agencies require special methods for cost analysis. For construction objectives, which are not funded by ordinary governmental budgeting, a technique known as benefit-cost analysis is available.

10.6 BENEFIT COST (OPTIONAL)

Benefit cost has a long and illustrious history, as it was initially applied to federal government improvement of navigation in 1902. Benefit-cost (B/C) analysis is practiced for water improvement, transportation, and related land use, and its application is controlled by federal statute. For the most part, benefit-cost analysis is adopted by governmental agencies. It is generally inappropriate for organizations that are incorporated for profit.

The important things to focus on are as follows:

- Public work projects
- Conventions in calculation
- Decision making

Benefit cost is a practical way of assessing the desirability of alternatives where it is important to take a long and broad view. It implies the enumeration and evaluation of all or nearly all of the selected costs and benefits.

We begin the discussion about B/C analysis with a small illustration on highway analysis. Costs have long been a factor in decisions made by highway engineers. From the many considerations of highway transportation, an economical road is achieved whenever the total cost is a minimum consistent with convenience, safety, transportation, and the ability to pay. The benefit-cost-ratio method compares the annual cost to highway users when vehicles are using an existing road in one case and an improved road in another. The equation is

$$\text{benefit cost ratio} = \frac{r_o - r_i}{(I_i - I_o) - (M_o - M_i)} \tag{10.23}$$

where $r_o - r_i$ = decrease in road user costs after improvements per year, dollars

$I_i - I_o$ = increase in investment costs per year, dollars

$M_o - M_i$ = decrease in maintenance cost per year, dollars

This would sim+plify to a ratio of decrease in user cost per year divided by net increase in investment costs per year. A similar approach is found for irrigation, recreational development, dams, and so forth. Benefit-cost ratios are calculated for the logical alternatives and are compared with the basic condition.

Criticism of the B/C method usually deals with the methods of analysis and overlooks the schemes of estimating, their definition, logic, and so forth. One of the confusing choices in benefit-cost analysis is that the benefit to some is a disbenefit to others, or a cost to a governmental agency serves as income to other firms. For purpose of this discussion, we rule that relevant consequences to the governmental units are classified as costs.

The benefits and costs are analyzed by present worth or equivalent annual cost. For simplicity, we adopt a present-worth approach and confine discussion to that method. The interest rate, i is preestablished, often by legislative statute.

Two methods of calculating the benefit-cost ratio exist. The first involves subtracting annual benefits from annual costs to establish a net annual benefit. Annual net benefits are discounted back to the date of the program's inception and summed to establish a present value of discounted net benefits. The benefit-cost ratio is then formed by relating this figure to the capital cost of the program. This approach is vaguely similar to a business calculation of the rate of profit that can be earned by capital. The second approach is to establish the gross benefits and costs for a typical year. The costs include annual operating costs and amortization of investment. No discounting is used.

Of the several ways to estimate a benefit-cost ratio, the one chosen here uses discounting. Principally, the B/C ratio has in the numerator the present worth of all benefits, and in the denominator the present worth of all costs. General notation is as follows:

B_n = benefits in year n, dollars

C_n = costs in year n, dollars

i = interest rate, decimal

n = year, $n = 0, 1, \cdots, N$

We adopt an end-of-the-year convention, meaning that benefits and costs, however they may occur, are assumed to be instantaneously received or paid at the end of the year. An exception is an investment or first cost which occurs at time $= 0$. The excess of the present worth of benefits over the present worth of costs is

$$\sum_{n=0}^{N} \frac{B_n}{(1 + i)^n} - \sum_{n=0}^{N} \frac{C_n}{(1 + i)^n} > 0 \tag{10.24}$$

Investment costs are already at time 0 and do not require discounting. If the difference in Eq. (10.24) is negative, then the contemplated action is unfavorable. A present-worth approach to benefit-cost ratios is determined by

$$P_b = \sum_{n=0}^{N} \frac{B_n}{(1 + i)^n}$$

$$P_c = \sum_{n=0}^{N} \frac{C_n}{(1 + i)^n}$$

$$\frac{B}{C} = \frac{P_b}{P_c} \qquad\qquad (10.25)$$

Consider the stream of cash flows for project X,

Year	0	1	2	3
Costs, C_n,	6	3	4	5
Benefits, B_n	0	10	12	15

The analysis for project X, where the interest rate $i = 8\%$, is given as

Year	$(1 + i)^n$	$\dfrac{B_n}{(1 + n)^n}$	$\dfrac{C_n}{(1 + n)^n}$
0	1	—	6.
1	1.080	9.529	2.778
2	1.166	10.292	3.431
3	1.260	11.905	3.968
Total		31.726	16.177

The value of the discounted cash flows of benefits and costs is $31.726 and $16.177. The difference between the present-worth benefits over costs is 15.549 (= 31.726 − 16.177), and the B/C ratio is 1.96 (= 31.726/16.177). In this example, we assume that the benefits were amenities, such as might develop from increases in irrigation, recreation, and fish and wildlife revenues, and we assume that they were nonexistent before.

In another situation, such as the straightening of a road, the benefits that accrue are reductions in road-user costs and are a favorable consequence to road users. We then look on the difference between the improved and the original road as a benefit to the road user.

Consider now projects X and Y, and determine their benefit-cost difference and ratio. Project X is in bad condition, and reconditioning is planned. Project Y is new and involves higher investment but lower costs to the user. Operation and maintenance costs of the project also differ.

	Project X	Project Y
Investment	6.000	9.000
PW of maintenance	10.177	13.115
PW of user costs	31.726	25.267

By using present-worth calculations and using $P_b = P_{bx} - P_{by}$ as the difference in benefits, $P_b = 31.726 - 25.267 = 6.459$, and the difference in costs is $P_c = P_{cy} - P_{cx}$. Then $P_c = (9.000 + 13.115) - (6.000 + 10.177) = 5.938$. $P_b - P_c > 0$ can also be expressed as the ratio 1.09 (= 6.459/5.938). The conclusion is to adopt project Y, but barely, on the basis of the marginal B/C ratio.

TABLE 10.12   Benefit-cost ratios for four mutually exclusive system choices

Design	Present-Worth Benefits	Present-Worth Costs	B/C
A	$120,000	$60,000	2.00
B	112,000	53,000	2.11
C	75,000	58,000	1.29
D	64,000	34,000	1.87

As with other cost-analysis measures, the B/C ratio has limitations that we cannot cover fully. It is a ratio—that is, simply benefits divided by costs. At that point, however, complications set in. Which costs and whose benefits are to be included? How are they to be valued? At what interest rate are they to be discounted? What are the relevant constraints? There is bound to be arbitrariness in answering those questions.

In most cases the scope and nature of the system to be analyzed are clear. A wide class of costs and benefits accrue to organizations other than the one sponsoring the system, and an equally wide issue arises of how the parent agency should consider them. For instance, a hydroelectric dam can be costed and benefits determined, but what about the relative costs and benefits from recreational amenities, water for farming, and improvements in scenery? The net rise in rents and land values is a result of the benefits of hydropower. Those secondary benefits may be more important to one governmental agency than to another, and calculations can impute those financial spillovers in a more or less favorable way.

If there are several designs requiring a decision for one system, invariably the cost and benefits are different, and a marginal-cost analysis method must be used to find the best system. Several guidelines are required. First, the same interest rate is used to figure costs and benefits. The same period of life, N, is necessary for the system alternatives. We initially calculate the B/C ratio for each system, such as that given in Table 10.12. In this table there are four mutually exclusive choices. System choices that are less than 1 are rejected.

Inspection of those B/C ratios suggests that alternative B is chosen because its ratio is the greatest. This is an improper selection.

Table 10.13 shows the required calculations. The choices are considered in order of increasing cost. System D is used as the initial base, since it requires the minimum present-worth cost. The first row of this table repeats row D from the previous table. Insofar as this first calculation is concerned, the decision is to accept D. Next, the marginal increase in cost and benefits is determined by using the next alternative above the least costly alternative. This would be C, and the C-minus-D values are indicated for the row. The marginal B/C ratio is 0.46, which is less than 1, and C is rejected.

TABLE 10.13   Marginal benefit-cost ratios

Design	Marginal Present-Worth Benefits	Marginal Present-Worth Costs	Marginal B/C Rates	Decision
D	$64,000	$34,000	1.87	Accept D
C–D	11,000	24,000	0.46	Reject C
B–D	48,000	19,000	2.52	Accept B
A–B	8,000	7,000	1.14	Accept A

We proceed by considering the alternatives in order of increasing costs. Now choice B is compared to D, the B/C ratio exceeds 1, and design B is preferred to alternative D. System design B is the current best choice. Last, the marginal gain and loss of design A minus design B are computed. The ratio of marginal present-worth benefits to marginal present-worth costs is greater than 1, indicating that design A is preferable to B. The final choice is A, and it assures that the equivalent present-worth benefits will be less than the equivalent present-worth costs and the system return is maximized. Choice A is contrary to the initial selection B.

Thus, B/C analysis deals with the first step of estimating the costs and benefits. If this is successfully achieved, then the next step is to rank alternatives in order of increasing investment, such as the present worth, and then check to see if the marginal investment is effective. This is tested by checking the differences between successive pairs of B/C alternatives against a status-quo, or do-nothing, condition. A conclusion is reached whenever the last alternative is compared with the last acceptable alternative.

Consider another example: A project is planned to provide water for industrial, municipal, and domestic use in connection with anticipated development of coal and oil reserves. It would increase irrigation supplies for production of livestock feeds and it would also benefit recreation, fish and wildlife, and flood control.

Refer to Table 10.14, which is a summary of project and operation estimates and benefits. Project estimates are determined by using guidelines given in Chapter 8, and they are listed for each of the separate projects. In turn, operation estimates that use the advice in Chapter 7 are averaged for the span of life, 100 years in this example. Project costs are nonrecurring. Operation estimate provides for the labor, materials, and overhead to operate and

TABLE 10.14 Summary of benefit-cost estimates for industrial and domestic water development in connection with exploitation of gas and oil reserves

Project Items	
Costs	
Dam 1	$11,850,000
Dam 2	6,400,000
Dam 3	5,100,000
Canals and diversion dams	23,870,000
Laterals and drains	5,350,000
Operating equipment	330,000
Fish and wildlife equipment	610,000
Recreational facilities	1,246,000
Construction total	$54,756,000
Operational costs	
Operation, maintenance, and replacement for 100 years, annual	$147,200
Benefits	
Average annual values for 100 years	
Industrial, municipal, and domestic water use	$4,590,900
Irrigation	869,800
Fish and wildlife	245,800
Flood control	11,000
Less benefits lost	9,900
Average yearly net benefits	$5,845,300

TABLE 10.15   Cost allocations and repayment summary

	Project Costs	Annual Operation, Maintenance, and Replacement Costs
Reimbursable costs*		
Industrial, municipal, and domestic	$39,330,000	$61,200
Irrigation	10,160,000	40,100
Recreation	166,500	17,800
Fish and wildlife	375,000	
Subtotal	$50,031,500	$119,100
Nonreimbursable		
Recreation	$2,765,500	$27,500
Fish and wildlife	1,823,000	400
Flood control	136,000	200
Subtotal	$ 4,724,500	$147,200
Total	$54,756,000	
Repayment		
Industrial, municipal, and domestic use*		
Prepayment**	$474,000	
Water conservancy district	38,856,000	$61,200
	$39,330,000	$61,200
Irrigation		
Prepayment	$126,000	
Water conservancy district	6,720,000	$40,100
Apportioned to others	3,314,000	
	$10,160,000	$40,100
Recreation, fish and wildlife***		
Nonfederal interests	$541,500	$17,800
Total	$50,031,500	$119,100

* Reimbursed over 50 years.

** Repayment rate at 3.5% annually.

*** Nonreimbursable expenses for project investigations.

maintain the investment. Occasional replacements of minor capital requirements are, for convenience, included in the operation estimates. Operation estimates are frequently recurring as annual costs.

The benefits are determined by using price schedules of surrogate opportunities. For example, irrigation water has a market value, which is found by using comparison estimating methods. In circumstances where there is no market advice, opinion estimates lead to imputed values. It may be possible that some benefits reduce financial gain. For instance, the impounding of water into a reservoir will reduce the grazing pasture for cattle, and those efforts cause *disbenefits*, or benefits lost. It is a practice that benefits lost are charged against the benefits and not added to the denominator cost term.

With the benefits and cost posted to the pro-forma summary (Table 10.15), it is possible to determine the net present worth of the cash flow and the B/C ratio. We use an interest rate of 6%.

$$P_b = \sum_{n=0}^{100} \frac{B_n}{(1 + i)^n} = \$5,845,300(16.618) = \$97,137,195$$

$$P_c = \sum_{n=0}^{100} \frac{C_n}{(1 + i)^n} = \$54,756,000 + 147,200(16.618) = \$57,202,170$$

$$P_b - P_c > 0$$

$$\frac{B}{C} = 1.7$$

Once the tangible benefits and costs are recognized, classes of *reimbursable* and *non-reimbursable cost allocations* are made. In a B/C situation, though, the problem is in the proper distribution of the costs of the features that serve several purposes. This problem does not arise in the cost of a single-purpose project or when national policy has determined in advance that the purpose to be served outweighs all costs. The costs of a multiple-purpose project are composed of the costs of individual project features, such as irrigation canals, power houses, or navigational works, which serve only a single purpose. A dam, of course, serves several single purposes, and its cost must be allocated to both reimbursable and nonreimbursable services.

Broad principles of cost allocation are possible. Each purpose should share equitably in the savings resulting from multiple-purpose construction within the limits of maximum and minimum allocations. The maximum allocation to each purpose is its benefits or alternative single-purpose cost, whichever is less. The minimum allocation to each purpose is its specific or its separable cost. *Joint costs* are apportioned without regard to the ability of any particular purpose to pay.

In an oil-and-gas water project as major as that shown in Table 10.14, it is necessary to recognize that many of the costs can be repaid by various public or private entities that will benefit. Land rents, payment for irrigation water, and recreational amenities by the general public are typical opportunities. Laws require repayment, and practices have established the categories of reimbursable and nonreimbursable costs.

Reimbursable costs for Table 10.15 have a period of 50 years for repayment. Often, interest of a few percent may be added to those costs. Some expenses are considered nonreimbursable and are borne by the federal government, such as the expenses for project investigations. Note, however, that the total of both costs is used in determining the B/C ratio.

Some critics suggest that a B/C analysis sidesteps the social issues by requiring public agencies to operate on a commercial basis, leaving resource allocation to be resolved through an artifice of the pricing system. But welfare economics, the well-being of people, income redistribution, market imperfections, and the like, make a reasonable demonstration of B/C in a commercial environment difficult. Thus, a B/C philosophy has to have a comprehensive public viewpoint. When constrained by laws and appropriations, the B/C ratio is best used as a means of ranking various systems. Those higher B/Cs are considered better from the B/C viewpoint.

SUMMARY

This chapter presented several fundamental techniques for cost analysis. Those techniques are applied to the important "tradeoff," where competing candidates in design and construction are compared side by side under similar rules of calculation and engagement. The eventual objective of this work is selection of the optimal design.

The techniques presented in this chapter are the following:

Tradeoff principles	Engineering economy
Cash flow	Life cycle cost
Break even	Benefit cost

As the design and construction project is unique, so must the technique be adapted to the circumstances. Estimates are prepared that give a plethora of costs for those alternatives. There are intangible factors that must be logically considered and evaluated, for they too are significant. Codes, specifications, design, practice, and legal and professional obligations are involved in selection of the design. In the next chapter we consider contracts and their effects upon construction-cost estimating.

QUESTIONS

1. Give an explanation of the following terms:

Tradeoffs	End-of-year convention
Cash stream	NEAW
Noncash expense	Salvage value
Real dollars	Operating profile
Escalated dollars	MTBF
Carry-forward depreciation	MARR
Marginal cost	Benefit cost
Average revenue	Disbenefits
Break-even point	Life cycle cost
Upper break-even point	Taxable income
Return	Net future worth
Payback	Time value of money
Simple interest	

2. Is there a difference between the interest calculated for project funding and the interest included in time-value-of-money concepts?

3. Describe two pseudo return methods and indicate why they are faulty. What advantages do they serve?

4. What is meant by "equivalent or equality" with methods of net present worth, net future worth, and net annual equivalent worth?

5. What criteria do you suggest for making project decisions? Will a shortage of capital influence your decision? Do you believe that a successful company is cash poor and bank-mortgage rich?

6. Describe some tradeoffs for a construction project that you are familiar with.

7. List some irreducible factors for a construction project that you are familiar with.

8. What is the current rate of inflation or deflation for construction in your area?

9. Do you believe that construction-cost analysis should use actual or real dollars? Support your decision.

10. How does a scattering of estimated points aid break-even analysis?

11. Discuss how marginal revenue and marginal cost parallel each other.

12. Several points of optimum location are possible with break-even analysis. Describe the nature of these points.

13. What are the problems with methods such as average annual rate of return?

14. Why is the present-worth method considered a good one for capital investments?

15. Give the nature of problems that lend themselves to life cycle costing.

16. What sort of problems lend themselves to benefit-cost analysis? Find a B/C example in your area.

17. Study the US corp of Engineer's WEB site, www.USACE.mil and determine differences in cost analysis that arises from dealing with a federal agency.

PROBLEMS

10.1 A contractor earns $1,050,000 of gross income and incurs operating expenses of $825,000. There are interest payments on borrowed capital for equipment, which amount to $85,000. Depreciation for the year is $115,000. Find the taxable income for this firm. If interest had been $120,000, what would be the taxable income? What are the consequences of operating losses?

10.2 A contractor has a gross income for the year of $3,125,000, expenses of $2,100,000 (excluding capital expenditures), and depreciation of $510,000. Find the taxable income and federal income taxes.

10.3 A contractor has a gross income of $5,250,000, expenses (excluding capital expenditures) of $2,900,000, and a depreciation of $1,870,000. Find the taxable income and federal income taxes. Determine the average tax rate.

10.4 Suppose that the federal tax rate is 34% and the state tax rate is 9% on ordinary business operations. Find the effective income tax rate.

10.5 A contractor has the following yearly facts for its business operations: sales revenue, $200,000; depreciation, $40,000; and other expenses, $130,000. The income-tax rate for this firm is 20%. Find the net income before taxes, net income after taxes, and net cash flow.

10.6 An owner forecasts the following cash flow for a commercial development:

Year	Cost	Revenue	Amount for Depreciation
0	$60,000,000	0	0
1	5,000,000	$20,000,000	$9,000,000
2	5,000,000	20,000,000	9,000,000
3	3,000,000	15,000,000	9,000,000
4	4,000,000	15,000,000	9,000,000
5	5,000,000	10,000,000	9,000,000

The first-year cost includes $45,000,000 as fixed assets and $15,000,000 as initial operating expenses. If the company has an effective tax rate of 40%, determine the annual net cash flow. Find the period when positive cumulative cash flow starts. (*Hints*: Assume that this investment is the only asset for the company. Use undiscounted cash flows.)

10.7 The estimated after-tax cash flow in actual dollars for a construction design alternative is given:

End of Year, n	Cash Flow, $
0	−$72,500
1	−20,500
2	75,500
3	82,00
4	110,000

If the general price inflation rate is expected to average 5.2% per year during this period, what is the real dollar equivalent of these actual amounts? (*Hints*: Use a convention called end of year, which says that the dollar amounts are seen as lump sums that occur precisely at this point in time. The base or reference time period is year 0.)

10.8 A bid contract is dimensioned as a unit-price and the following table is estimated. The owner requires a bid that is expressed in current dollars, but allows *escalation* of the unit prices based on evidence of inflation as demonstrated by a reliable index.

End of Year, n	Unit-Price Estimate, $
0	$538
1	538
2	538
3	538
4	538

Inflation is expected to average 4.7% per year for the next several years. Find the actual dollars per year for the unit-price contract.

10.9 A municipal plant considers two methods to treat gray water, A and B. Information is as follows:

	A	B
Labor, $/yr	140,000	190,000
Depreciation, $/yr	100,000	$56,000
Chemical & treating materials, $/m³	0.15	0.10

Find the annual amount of water to have the two methods equal. (*Hint*: This problem is evaluated as short term.)

10.10 A contractor has net sales of $600,000 annually. The annual fixed costs are $350,000 and the direct costs are 31.5% of the net sales dollars for this level of business. Find the gross profits. What is the break-even point in terms of the sales dollars? Find the annual sales to have a profit of $80,000.

10.11 A business is evaluating the short-term economics of chemical plant construction and operation. Fixed costs for operation are $10,000,000 per year. The variable cost per unit is $3.00 up to 100% of capacity, and $3.30 above 100%. This project can produce 12,000,000 units per year at 100% of capacity. All units are sold at a net selling price of $5. The composite tax rate for all taxing entities is 31.5%. Find the break-even point per year. Find the net profit per year at 80% and 120% of production capacity.

10.12 An opportunity sale of an additional 200 units is received above an existing construction activity. The buyer is prepared to pay $35,000 for 200 units. Cost for the current 500 units is $95,000. Total cost for installing the 700 units is $127,700. What is the total marginal cost? Find the unit marginal cost? What is the profit or loss? What is the minimum opportunity sale price to break even?

10.13 A sphere has the following weight and cost information (lb x, cost y): $(2, 2), (3, 3), (4, 5)$, and $(5, 8)$. Sketch the cost curve, average-cost curve, and marginal-cost curve. Do those data comply with the *economy-of-scale* law? (*Hint*: The dimensional units are $/lb.)

10.14 The total cost for constructing a project varies according to the number of units and is given as follows:

Project Units	Total Cost, $	Project Units	Total Cost, $
0	$0	5	$19,200
1	10,000	6	20,500
2	15,000	7	23,000
3	17,500	8	28,000
4	18,700	9	38,000

Sketch the total-cost curve, average-cost curve, and marginal-cost curve. (*Hint*: Use one graph for the three curves.)

10.15 (a) New construction equipment costs $225,000, salvage is $25,000, and an average annual earning of $20,000 after taxes is expected. Find the average annual rate of return. If earnings are doubled, then what is the rate?

(b) Consider part (a). The investment has an economic life of 10 years. Straight-line depreciation for 8 years is used. Find the percent return.

10.16 Investment for new equipment is $100,000, and salvage will be $10,000 eight years hence. Average yearly earnings from this equipment are $15,000 after taxes. What is the non-time-value-of-money return? When is the payback?

10.17 Two investment opportunities are proposed:

	Equipment A	Equipment B
Total investment	$60,000	$60,000
Revenue (after tax)		
Year 1	20,000	30,000
Year 2	20,000	30,000
Year 3	20,000	30,000
Year 4	20,000	
Year 5	20,000	

For an interest rate of 10%, which equipment has the least period of time before investment recovery is complete? (*Hint*: Use the cumulative amount of present worth to determine a payback number.)

10.18 Two bids are compared for performing the same work:

	Bid A	Bid B
Total bid	$100,000	$150,000
Revenue (after tax)		
Year 1	25,000	25,000
Year 2	30,000	40,000
Year 3	35,000	55,000
Year 4	30,000	40,000

   (a) On the basis of payback, which bid is preferred?

   (b) For an interest rate of 5%, which bid has the least period of time before the bid value is returned? (*Hint*: Use the cumulative amount of present worth to determine a payback number.)

   (c) Discuss the pros and cons of the payback measure of evaluating an investment option.

10.19 (a) What is the amount of interest at the end of 2 years on $450 principal for a simple interest rate of 10% per year?

   (b) If $1600 earns $48 in 9 months, what is the nominal annual rate of interest?

   (c) An investment of $50,000 is proposed at an interest rate of 8%. What is the future amount in 10 years?

   (d) What is the present worth of $1000 for 6 years hence if money is compounded 10% annually?

   (e) What is the compound amount of $3000 for 15 years with interest at 7.25%?

   (f) Find the annual equivalent value of $1050 for the next 3 years with an interest rate of 10%.

   (g) How many years will it take for an investment to triple itself if interest is 5%?

   (h) An interest amount of $500 is earned from an investment of $7500. What is the interest for 1 year? For 2 years?

10.20 (a) Find the principal if the interest amount at the end of 2.5 years is $450 for a simple interest rate of 10% per year.

   (b) A loan of $5000 earns $750 interest in 1.5 years. Find the nominal annual rate of interest.

   (c) In 1626 Native Americans bartered Manhattan Island for $24 worth of trade goods. Had they been able to deposit $24 into a savings account paying 6% interest per year, how much would they have in year 2000? At 7% interest?

   (d) What payment is acceptable now in place of future payments of $1000 at the end of 5, 10, and 15 years if the interest rate is 5%?

   (e) What is the compound amount of $500 for 25 years with interest at 10%?

   (f) Calculate the annual equivalent value of $1000 for the next 4 years with an interest rate of 10%.

   (g) How many years will it take for an investment to double itself if the interest rate is 10%?

   (h) An investment of $10,000 earns an interest amount of $750. Find the rate of interest if the amount is earned over 1, 2, or 3 years.

10.21 A prospective venture is described by the following receipts and disbursements:

Year End	Receipts	Costs
0	$0	$800
1	200	0
2	1000	200
3	600	100

For $i = 15\%$, find the desirability of the venture on the basis of net present worth.

10.22 (a) The cash flows of costs and revenues for a project are given as:

Year	Cost	Revenue
0	$1025	$0
1	0	450
2	0	425
3	0	400

Calculate the net present worth, net future worth, and net equivalent annual worth if the interest rate is 11%.

(b) Repeat (a) for an interest rate of 9%.

10.23 A project estimate has the following cost and revenue cash flows. The cash flows are assumed to occur at the end-of-year.

Year	Cost	Revenue
0	$800	$0
1		450
2		425
3		400

(a) If interest is 10%, find the net present worth, net future worth, and net annual equivalent worth.

(b) Find the rate of return.

(c) Present a summary of the four methods.

10.24 Construction-equipment cash flows of two competing choices, A and B, are given below. Assume that an interest rate of 20% is used. Project life is over after 5 years. Present a summary of the four methods of engineering-economy evaluation. Is A or B preferred? Discuss the advantages of the time-value-of-money concepts for a situation like this. (*Hint*: Use the 20% interest table for the computations.)

	Equipment A	Equipment B
Investment	$60,000	$60,000
Annual after-tax earnings		
Year		
1	20,000	30,000
2	20,000	30,000
3	20,000	30,000
4	20,000	
5	20,000	
Total	$100,000	$90,000

10.25 (a) Evaluate the cash-flow diagram given by Fig. P10.25. Determine a present sum, equivalent annual payment, and a future sum. Use an interest rate of $i = 10\%$. (*Hint*: Arrows are end-of-the-year cash flows. Arrows pointing downward represent costs and those pointing upward are revenues.)

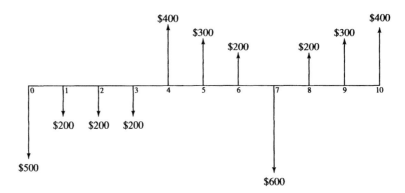

Figure P10.25

(b) Repeat (a) with an interest rate of $i = 20\%$.

10.26 (a) Evaluate the cash-flow diagram, Fig. P10.26. Find present value, annual value, and a future value. Use an interest rate of $i = 10\%$. (*Hint*: Arrows are end-of-the-year cash flows. Downward-pointing arrows are costs and upward-pointing arrows are revenues. The present value is timed to occur at $n = 0$, and the future value is timed to occur at $n = 10$.)

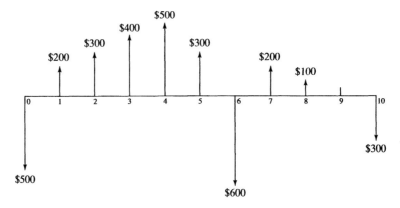

Figure P10.26

(b) Repeat (a) with $i = 25\%$. Use a spreadsheet to find your solutions.

10.27 HVAC designers are evaluating two systems for heating and cooling the air in a new building. Life of the building is 40 years, and the alternatives are expected to have a twenty-year life. Evaluate the alternatives and find the preferred design using an interest rate of 10%. (*Hint*: Find the preferred choice using the annual-cost method.)

	Natural-Gas Engine Powered Heat Pump	Refrigeration & Domestic Heating
Investment	$900,000	$600,000
Annual energy costs	$15,000	$35,000
Annual maintenance costs	$22,000	$30,000
Incremental annual taxes	$4,000	

10.28 The U.S. Army Corps of Engineers issues purchase instructions informing potential suppliers that roadway graders will be evaluated according to the following life cycle cost model:

$$LCC = \text{unit operating cost, dollars}$$

$$= (\text{unit price} + \text{logistic cost}) \div \text{service life}$$

The instructions require that the selected equipment supplier demonstrates service life in a post award reliability acceptance test. If the reliability test does not meet the level guaranteed by the contract, then a penalty function deducts from the unit price as

$$\text{penalty cost} = (1 - \text{test value } MTBF/\text{quoted } MTBF) \times (\text{unit price} + \text{logistic cost})$$

where $MTBF$ = mean time between failures, hours

logistic cost = value supplied by U.S. Army Corps of Engineers, dollars

The following bids are received:

Contractor	Unit Price	MTBF, Hr
A	$30,000	1000
B	35,000	1200
C	60,000	1635

The buyer determines that the logistic cost is $25,000 and supplies this information to qualified suppliers. Find the winning supplier for the road equipment. Now suppose that the supplier fails to meet the quoted $MTBF$ by 10%. What are the penalty cost and the final price for the equipment?

10.29 The U.S. Army Corps of Engineers establishes a life-cycle-cost model to evaluate tires used in construction equipment. The model is given as

$$LCC = (\text{number of tires})(\text{unit price} + \text{shipping cost} + \text{maintenance cost}), \text{dollars}$$

Three tire manufacturers are invited to bid. Each supplier provides a sample for simulated working tests to determine the number of miles per tire. Then the Corps of Engineers determines the number of tires for each company's quote based on a tire-mile index, which is given by number of required miles divided by best performance. The number of required miles is 12,000,000. Shipping costs are evaluated from the supplier to a central inventory. Maintenance cost to change a tire is $95. Make a bid evaluation to determine the LCC price and recommend the source of purchase.

Supplier	Miles per Tire	Shipping Cost per Tire	Bid Price per Tire
A	1100	$24	$1380
B	1050	18	1280
C	950	23	1360

10.30 A contractor announces that construction equipment is to be evaluated according to the following LCC model:

$$C_u = \left(\frac{SOH}{MTBF}\right)(MTTR)C_m + \text{Unit Price, dollars per unit}$$

where SOH = scheduled operating hours

$MTBF$ = mean time between failures, hours

$MTTR$ = mean time to repair, hours

C_m = cost of maintenance labor, dollars/hour

The following bids are received. Which supplier do you choose?

Company	Unit Price	SOH	MTBF	MTTR	C_m
A	$15,000	50,000	10,000	150	$75
B	28,000	45,000	15,000	140	75
C	39,000	40,000	20,000	130	75

10.31 **(a)** Cash flows for a project A are given as follows:

Year	0	1	2	3	4
Costs, C_n	20	5	5	6	8
Benefits, B_n	0	25	20	15	10

Find the present worth of costs and benefits for an interest rate of i = 5%. Calculate the net-present-worth benefit-cost ratio.

(b) Cash flows for project B are given as

Year	0	1	2	3	4
Costs, C_n	20	8	6	5	5
Benefits, B_n	0	10	15	20	25

By using an interest rate of i = 5% find the present worth of costs, benefits, their difference, and the benefit-cost ratio. Repeat for an interest rate of i = 10%. Discuss the effect of the interest rate on the benefit-cost ratio and the decision.

10.32 Two projects, Endicott and Columbia, have the following cash flows:

Year	1	2	3	4
Project Endicott				
Maintenance	5	5	6	8
User Cost	25	20	15	10
Project Columbia				
Maintenance	4	4	4	4
User Cost	20	15	10	5

(a) Investment costs for projects Endicott and Columbia are $20 and $30. The interest rate used in the analysis is 5%. Find the present worth of cash flows. Find the marginal B/C between the two projects. Is the more expensive project Columbia worth the additional investment? (*Hint*: Develop benefits on the basis of the differences between the projects.)

(b) Repeat part (a) for an interest rate of 10%.

10.33 A governmental agency desires to select one of three projects. U.S. regulations stipulate that the benefit-cost ratio be determined at an interest rate of 5%.

Project Values	Project A	Project B	Project C
Initial cost ($ $\times$ 10^6)	40	50	60
Annual benefits ($ $\times$ 10^6)	15	20	25
Life, years	4	4	4

(a) For each project, find the net present worth. Determine the B/C ratios. Rank the projects according to the B/C ratio. Determine the marginal benefit-cost ratios. Re-rank the projects. Which project do you recommend? (*Hint*: Marginal B/C ratios are differences between the benefits and costs of the ranked projects.)

(b) Reconsider part (a) with an interest rate of 10%.

10.34 A welfare program initially costs $650,000 with a $20,000 annual operating cost. Benefits are anticipated to be $40,000 in the first year and to increase $30,000 each year for 4 years. They will decline to no benefits in 2 years. The interest rate for this type of project is 10%. Find the benefit-cost ratio with and without discounting.

10.35 A river reclamation and flood control project by the U.S. Army Corps of Engineers costs $350,000 with a $20,000 annual operating cost. Benefits are estimated to be $40,000 in the first year and to increase by $30,000 each year for 4 years. The benefits drop to zero in 2 years. At that time no additional costs or benefits accrue. Sketch the cash-flow diagram. Using engineering-economy functional notation, write separate equations for the benefits and for the costs. Find the benefit-cost ratio with and without discounting. (*Hints*: Have the benefit arrows point upward for the sketch. Use an interest rate of 5%. Assume the construction cost occurs at time = 0, and otherwise let the costs and benefits be end-of-year.)

MORE DIFFICULT PROBLEMS

10.36 A cost at the end of this year is estimated as $100,000. This year end is the base point for considering inflation. Find the equivalent worth 5 years from this base point for the following conditions:

$$i_r = \text{real interest rate} = 4\%$$

$$f = \text{inflation rate} = 6\%.$$

(*Hints*: This problem indicates two rates, interest and inflation, and asks for a future equivalent value. The real and the inflation rate have a compounding effect, as given by the equation below.)

$$i_c = (1 + i_r)(1 + f) - 1$$

where i_c = combined interest rate, percent per period

10.37 There was an epoch of time (early 1900s to mid 1920s) when horse-drawn wagons and internal-combustion-engine trucks competed for economical haulage of construction materials. Consider the following choices:

Comparison Item	Single-Horse Wagon	700-lb-Capacity Light Delivery Truck
Initial cost	$430	$910
Estimated life, year	10	2.5
Total operation cost per working day (300 days per year)	$3.40	$7.10
Total cost per idle day	$1.74	$1.21
Average mileage per day	20	60

Construct a break-even analysis and make suggestions for operation during this period. (*Hint*: Assume that the load-carrying capacities are identical between the competing methods (which incidentally was nearly true), except that distance does vary.) The use 10% for a NAEW comparison and find the number of miles for break even.

10.38 A concrete plant is being considered for a new booming market. Total fixed cost for design, land, and plant construction is 10^5. The variable cost of producing concrete is a function $C(n) = 100n - 3 \times 10^{-3}n^2 + 10^{-7}n^3$, where n = cubic yards of concrete produced.

(a) Determine the total-cost function $C(T)$. (*Hint*: It includes the cost of producing the concrete and fixed cost.)

(b) Find the marginal-cost function C_m for producing the concrete.

(c) Find the cubic yards for minimum marginal cost. What is the marginal cost at this point? What is the average cost per cubic yard at this level of production?

(d) Total sales is given by $S(n) = 250n$. Find the marginal price. Determine the functions for total profit, and marginal profit.

(e) Find the point to maximize profit.

(f) Graph these cost equations. Your instructor will give additional instructions.

10.39 The marginal revenue for a construction activity n is given as $100 - 0.02n$, where n is the number of activity units per period. Total cost, including fixed costs, is expressed as $2 \times 10^{-4}n^2 + 10,000$. Find the break-even point, maximum profit and the construction rate at which n occurs; and the minimum average cost and the construction rate at which n occurs.

10.40 Engineering studies of three plant sites, located at mile markers 103, 111, and 123 along a to-be-constructed highway, show a daily demand of 20,000, 7,000, and 6,000 yd³ of concrete in each location for 6 days per week. Two plans, A and B, meet this demand.

In plan A, one central and large plant having a daily capacity of 40,000 yd³ is located equidistant between the locations and produces concrete with a fixed cost of $20,000 per day and a variable cost of $1.50 per yd³.

Alternatively, plan B has portable and distributed operations in each location with capacity of 24,000, 8000, and 8000 yd³. Fixed costs are $15,000, $7000, and $7000 per day. Because of lower transportation costs, variable costs per yd³ are $1.00, $1.10, and $1.20 per yd³.

(a) Which alternative, A or B, is more desirable based on demand requirements? (*Hint*: Consider this tradeoff study as "short term.")

(b) If sales were to increase to production capacity, then does the best choice change?

(c) To what variable cost per cubic yard may plan B increase for a break-even to A at demand capacity? At full capacity?

10.41 Two alternate bridge designs are available for a river crossing. Estimates are given as follows:

Design	Project Estimate, $	Life, years
Wood	80,000	10
Steel	100,000	15

Any salvage value of the replaced bridge is offset by its removal at the end of its life. Maintenance costs are equal for both bridges.

(a) Sketch the cash-flow diagrams. Find the preferred design for an interest rate of 10% for a required life of 30 years. Find the annual costs of both designs. Use functional notation to solve this problem. (*Hints*: In the sketch, have the project estimate cost arrows point downward and the equivalent cash flows upward. Assume two replacement of the wooden bridge and one replacement of the steel bridge. There is no replacement at the 30th year period. Why? Use the 10% table in the appendix.)

(b) Repeat part (a) for 20%.

10.42 A construction job is to last 2 years. Significant quantities of bulk materials must be moved. Two solutions are proposed: A fixed belt conveyor costing $750,000 is estimated to have an annual operating cost of $200,000, and a $150,000 salvage value at the end of the second year. An alternate choice is mobile equipment costing $300,000, and its estimated operating cost is $700,000 annually. Salvage value of the mobile equipment is $100,000 at the end of two years. Both of these alternatives have a longer physical life than the project life.

(a) Sketch the cash-flow diagrams. Find the preferred construction method for an interest rate of 10%. Find the annual cost of both methods. Use functional notation and the 10% table in the appendix to solve this problem. (*Hint*: The operating costs are applied at $n = 0$ and 1 end-of-year periods.)

(b) Repeat part (a) for 20%.

10.43 Two materials are evaluated for best choice for a commercial building. Evaluate the proposals at 10% interest using the annual-cost model. Estimates are given in the following table:

Alternative	Brick	Wood
Initial cost	$180,000	$75,000
Annual maintenance	$500	$7000
Expected life, years	30	10

10.44 The U.S. Army Corps of Engineers announces that road equipment is up for bid and an LCC approach is deemed mandatory. The Corps advertises that an LCC model to select the winning bid is based on the model

$$\text{cost per unit} = \frac{\text{unit price} + \text{unit stocking cost}}{\text{bid } MTBF}$$

The stocking cost of $11,000 per unit is the total cost for storage, transportation from central inventory, mobilization at the job site, setup and dismantling. Each bidder supplies this information:

Bidder	Unit Price	Bid MTBF
1	$100,000	800
2	125,000	615
3	117,500	917

Which bidder wins the contract? Discuss the merits of this sort of analysis. Propose another scheme to evaluate construction equipment.

10.44 A federal agency is considering three projects for an identical requirement. Each project has a 5-year life. Possible choices and estimates of costs and benefits are shown in the table.

Project Values	Project A	Project B	Project C
Investment, $(\$10^5)$	$144	$50	$380
Annual net benefits, $(\$10^5)$, 5 years	40	16	100
Residual value $(\$10^5)$	0	0	10

(a) The interest rate is 5%. Find the net present worth for each project. Which project should be chosen? Calculate the benefit-cost ratios. Determine the marginal benefit-cost ratios. Which project do you now recommend? Discuss your approach.

(b) Reconsider part (a) with an interest rate of 10%.

10.45 The B/C test is applied to welfare and job training for underprivileged youth. In this case an undiscounted B/C ratio is given as

$$\text{benefit cost ratio} = \frac{B_p - B}{C_a - T_n}$$

where B_p = graduate earnings after training, dollars

B = original earnings of student before training, dollars

C_a = annual amortization payment, dollars

T_n = taxes on net increased earning of student, dollars

The following estimated items are given:

	Amount
Direct program training cost	$2,175,000
Allocation of center overhead based on planned enrollment	2,850,000
Subtotal	$5,025,000
Capital investment cost at 5%	251,250
Job Corps cost at 25%	1,256,250
Total cost	$6,532,500
Number of graduates	400
Total cost/graduate	$16,329
5-year amortization cost (C_a), annual	$3,267
Average starting salary (B_p), annual	$12,666
5-year average taxes (T_n), annual	$1,686
Original earning power of students (B), annual	$3,120

Let the interest rate be 5%. Find the undiscounted and discounted benefit-cost ratio. What happens to the B/C ratio as the interest rate increases above 5%? Decreases?

10.46 An existing highway, ABC, originally constructed in 1954, is 8.5 miles long. Average daily traffic is 10,000 vehicles with 10% trucks. Now requiring reconstruction, the unit estimated cost of improvements to the existing highway is $2 million per mile. The right-of-way will cost $550,000 per mile.

A 7-mile supplemental location, ADC, can be constructed for $3 million per mile. The right-of-way costs $220,000 per mile. ADC requires two viaducts for $500,000 each. Long-term main-tenance cost of either road is an average $10,000 per mile per year. This maintenance includes occasional resurfacing. Trucks cost $1 per mile, and cars are evaluated to cost $0.25 per mile. If ADC were constructed, then it is estimated that 40% of the traffic between points AC would move to ADC and the remainder would use ABC.

Describe the two alternative plans and determine their capital cost. Based on a project estimate of capital cost only, which plan do you advise? Find the undiscounted and discount-ed traffic costs for 20 years for the alternative designs. Use an interest rate of $i = 5\%$. Construct a B/C analysis for this problem and advise a solution.

10.47 Four mutually-exclusive construction designs have their benefits and costs estimated.

Design	P.W. Benefit	P.W. Cost
A	$48,000	$38,000
B	35,000	24,000
C	37,000	31,000
D	45,000	34,000

Determine the individual B/C ratios, and analyze the four choices on the basis of the margin-al benefit-cost ratio to find the best design. (*Hint: Marginal analysis* implies an arithmetic dif-ference between the individual first best choice and the next one of increasing benefits.)

10.48 A U.S. contractor will design, fabricate, and supervise construction of a project in country X. Designing and fabricating are performed in the United States, while erection, assembly, and con-struction are done in country X by using the national labor of that country. The deal is agreed to at base time zero, and reimbursed costs are adjusted for inflation within country X and for the exchange rate. The contract at time zero for the adjustment of the final cost is as follows:

Time	U.S. Material	U.S. Cost for Construction in Country X
0	$60 million	$20 million

The engineer believes that the rate of inflation will be 10% and 25% annually in the United States and country X. The exchange rate of country X to the U.S. base is $1:1, 2:1$, and $3:1$ for years 1, 2, and 3. U.S. material cash flow is spread evenly over the first 2 years, and erection oc-curs during year 3 in country X. The contract requires that the lump cash sum be paid at year-end 3, adjusted for inflation and exchange rate.

If the U.S. material costs are spread uniformly over the first 2 years, then roughly what total amount is spent because of inflation? (*Hint*: Assume that the inflation and exchange rates are independent of each other. Consider that inflation increases at a compound amount, and that money spent at the start of the year will inflate for the entire year while money spent at the end of the year will not inflate for that year. Thus, consider average inflation for the time period.)

If the contractor will pay the national labor in that country's currency during year 3, then how many country-X value units are roughly expected? How many U.S. dollars? What ap-proximate sum of U. S. money is due to the contractor at year-end 3 from country X?

PRACTICAL APPLICATION

There are many opportunities for a practical experience in cost analysis for design and construction. Your instructor may suggest several.

1. An interesting possibility is finding a B/C ratio for your own education along the lines of the one in Prob. 10.45. You have two choices: one that considers the point at which you started post-high-school education and concludes with your degree, and another that forecasts the B/C for an additional program and degree.

2. Contact and visit a construction design firm and inquire about the cost analysis they do. Prepare a list of questions in anticipation of the visit. Be professional in your arrangements and appearance, and afterward write a report detailing the lessons you have learned. Send them a courtesy letter expressing your thanks after the visit. Include the report.

CASE STUDY: LIFE CYCLE COST ANALYSIS OF PUMP FOR MUNICIPAL PLANT

A new pump for a pumping station is specified by Mr. Clayton Lyell, a municipal planning engineer. Mr. Lyell collects the following estimates for the new pump:

	Amount
Construction and installation cost (1 year from design)	$220,000
Preventive maintenance cycle per year	2190 hr
Pump operating hours per year	6570 hr
Preventive maintenance action	40 hr
$MTBF$	3285 hr
$MTTR$	120 hr
Power demand	60 kW
Power cost	$0.10/kW

Mr. Lyell estimates design cost as $40,000. Maintenance labor for a three-person crew costs $125 per hour. In addition to the consumption charge for power, there is a monthly demand charge of $690 for power.

Find the undiscounted cash-flow total over a ten-year span of pump operation. What is the discounted total LCC assuming a 10% interest rate? Sketch the discounted cash flow for the lifetime of the pump.

(*Hints*: The LCC analysis begins at year zero, where design costs are assumed to be instantaneously bunched. Assume that all maintenance costs are collected at the start of the year. There are no labor costs for operation of the pump. There is no effect of inflation for this analysis. For the sketch, have the *x* axis as years and the *y* axis as discounted dollars, and smooth the values as if continuous cash flow were expected. Do not include any consideration for pump replacement.)

C H A P T E R   1 1

Contracts and Ethics

If the customer favors the construction bid, then the owner and contractor proceed to contract discussions. Because legal requirements are important, the engineer needs an understanding of contracts. Simply, owners and contractors deal with contracts.[1] Stipulations regarding contract changes, claims, disputes, and many other clauses become important. An intermediate step to a signed contract may be negotiation.

Two major families of contracts are *fixed price* and *cost reimbursable*. Construction requires insurance and bonds. Auditing and ethics are other considerations. This chapter gives an introduction to those practices.

11.1   IMPORTANCE

Though the engineer is not a lawyer, he or she needs a fundamental understanding of contract law, because preparation and submission of the estimate and its formalized offer may involve legal considerations. Construction work and project estimates have differing legal requirements. The engineer needs to understand his or her responsibilities to the employer, client, or customers as well as professional obligations to society at large and public law. In particular, knowledge about the following practices is important:

- Contracts
- Insurance and bonds
- Negotiation, award and audit
- Ethics

Self interest leads the parties to consent to an exchange—an agreement which may be recognized by a handshake or by a formal contract. Because the estimate is a measure of the economic want, it may lead to a statement of price and an amount of money in a legal contract, which is the usual means of agreement. This requires a buyer and seller, or an owner and contractor. Negotiation may or may not be involved.

---

[1] We do not provide copies of various contracts in this text. The student can find them on the web or in books listed in the references. For example, the following sites may be browsed for information on contracts: American Society of Civil Engineers (*www.ASCE.org*); National Society of Professional Engineers (*www.NSPE.org*); American Institute of Architects (*www.AIA.org*); Associated General Contractors of America, Inc. (*www.AGC.org*); and U.S. Army Corps of Engineers (*www.USACE.mil*).

PICTURE LESSON
Turbine Rotor

The picture lesson shows a boilermaker sitting in the entry section of a steam turbine for construction detailing and final adjustments. The turbine is temporarily located in a shed near a power plant before its move to the turbine room.

With the cast steel shroud removed, the rotor is resting on the shaft and bearings and is exposed for work. Boiler steam has single entry to the high-pressure stages but is divided into a double exit, which gives efficiency advantages of obtaining more available work from the expanding steam.

Notice the rotor. The axial turbine stage comprises a row of fixed guide vanes or nozzles (often called a stator row) and a row of moving blades or buckets (a rotor row). Each rotor side has five stages. The circular row of nozzles that are nonrotating are fixed to the shroud housing (called the stator), which is not shown. The blades or buckets of each stage have different entry angles for the steam as it expands and does work. The final stage receives the steam at its greatest expansion and lowest pressure. This final stage does not have the supporting ring at the largest diameter, as do the higher-pressure and smaller-diameter stages. In two-dimensional theory of turbomachines, it is usually assumed that the axial velocity remains constant. The fluid motion assumes that radial velocities do not exist.

We use the term *contract* to describe a variety of agreements or orders for the procurement of design, construction, services, or materials. Modification to contracts may alter the design, specification, delivery point, rate of delivery, contract period, price, quantity, or other provisions. Contracts discuss administrative procedures for payment.

Significant construction projects have simultaneous contractual arrangements in force between the owner, engineer, architect, contractor, management services, subcontractors, and other prime contractors. Construed as a whole, these contracts establish a complicated structure for responsibilities of many kinds.

11.2 BASIC CONTRACT TYPES

There are two fundamental contract families to scale the value of the estimate, or

- Fixed price
- Cost reimbursable

Contracts are known by many titles and there are numerous variations. Table 11.1 is a simple list.

The element of risk, for instance, or the willingness of the parties, or the competition, advanced or ordinary technology, complexity of construction, urgency, and life cycle may influence the general type of contract selected. Cost-reimbursable contracts transfer the economic risk to the buyer or owner, while fixed-price contracts place the economic risk on the contractor, supplier, or vendor. It is as necessary to estimate accurately for cost-reimbursable contracts as it is for fixed-price contracts.

Firm fixed-price arrangements have in common that one party—a contractor, subcontractor or a supplier—is to deliver construction work in accordance with the terms and conditions of the contract, and another party—an owner, buyer, firm, agency, or government—is to to pay a price equal to that specified by the contract for money spent, subject to restrictions and any special negotiated understanding. Almost all construction and well-developed production consumer products adopt firm fixed-price contracts.

The fixed-price contract requires that the design and construction work be delivered as described on a predetermined schedule. The price is fixed for the life of the contract, precluding changes allowed by the contract. However, the terms and conditions may allow for

TABLE 11.1 Illustrative types of contracts

Contract Family	Variation
Fixed price	
	Firm fixed price
	Unit price
	Productive labor rates
	Time and material
	Fixed price incentive
	Fixed price with redetermination
	Quote or price in effect
Cost reimbursable	
	Cost plus incentive fee
	Cost plus award fee
	Cost plus fixed fee

adjustment. Though the fixed-price contract provides the greatest risk to the contractor, it also offers incentive and opportunity to realize the greatest profit. The contractor fully re-covers savings due to cost reductions. Thus, if the actual costs are less than estimated costs, then greater profit is realized. Too, the fixed-price or lump-sum contract is popular from an owner or buyer's viewpoint, because the total project cost is known in advance. Almost all public project contracts, as well as a large portion of private construction projects, are se-lected by fixed-price competitive bidding.

Competitive-bid contracts for projects that are fixed price have interesting variations. A *unit-price construction contract* is based on the placement of certain well-defined items of work and the costs per unit amount of each. For example, a price per linear foot of pile or cubic yard of excavation allow a reasonable variation to be made in the driven length of the individual piles or quantity of excavation. Thus, a contractor submits a bid made by the number and depth of the piles, which are verified by the owner's engineer working in the field. Some metal-machining shops do much the same thing by providing a "time" ser-vice—that is, by charging a productive-hour cost rate that will include a portion for profit.

If the project cannot be accurately estimated, then the fixed-price contract may not be suitable. Large complex construction is a candidate for a cost-reimbursable type of con-tract. The *cost-reimbursable contract* places the risk on the owner and is adopted whenev-er research, construction, design, or urgency are entailed. The owner has to be assured that the contractor is reputable in quality, delivery, and design. Inherent in a cost-reimbursable contract is the best effort by the contractor to complete the work. Even so, the costs are passed on from the contractor to the owner.

Cost-reimbursement contracts have provisions for payment of allowable, allocable, and reasonable costs incurred in the performance of the contract. Certain procedures, to be described shortly, permit adjustments for fees, incentives, and penalties.

Often the raw and standard commercial materials are listed in a catalog, price list, or schedule regularly maintained by a manufacturer or vendor which may be published or made available for inspection by customers. The information may state prices of current interest or give prices to buyers. Though those values may not be formal contracts, they are established in the usual course of business between buyers and sellers, who are either free or-not free to bargain and agree. Competition establishes the prices, and thus is a sufficient standard for a contract. E-commerce is a popular way to gather those prices. Those cata-log prices, or the value of the estimate, say from $0.05 to $50 million, do not diminish the importance of the estimating, negotiating, or contractual procedure.

Construction operations involve direct labor and direct material in the estimating process. Labor costs may be contracted between the company and the union or between the company and an individual willing to work for the conditions as specified. Negotia-tions may be a factor. At the lower wage level, federal laws may dictate the amount of the wage. (Labor laws are discussed in Chapter 2.)

Some subcontractors bid for work on the basis of a gross-hour cost rate. For exam-ple, a specialty subcontractor specifies a labor-rate schedule and then performs the work. Markup rates depend on the skill, market demand, and competitive pressures. This is a form of *fixed pricing*. Refer to Table 11.2 and Eq. (8.10), which adjust the cost on the basis of profit markup.

Obviously, the buyer needs to be satisfied that a fair and reasonable time is charged against the contract. Some companies maintain an open-purchase contract allowing this arrangement between the owner and a subcontractor.

A *time and material contract* is between an owner and subcontractor for work at a fixed and specified rate (hourly, daily, etc.) that includes direct labor, indirect costs, profit,

TABLE 11.2 Illustrative labor rates construed as a fixed-price contract

Construction Craft	Estimated Gross-Hour Cost Rate, $/hr	Markup (%)	Quoted Hourly Cost Rate, $/hr
Lather	$39.16	10	$43.08
Mosaic & terrazzo	90.98	20	109.18
Plasterer	32.19	15	37.02
Painter	42.33	5	44.45

and materials. The materials may be at cost or cost plus profit. This contract is suitable for construction where the amount or duration of work is unpredictable or insignificant. *Repair work* is often handled on a time and material contract. The labor is a fixed part, as the owner agrees to a preestablished gross-hour wage schedule, and a timekeeper verifies the numbers of hours. The material is cost reimbursable, since the nature of repair materials is unknown at the time of the contract.

11.3 FIXED-PRICE ARRANGEMENTS

The parties agree to the price before a *firm fixed-price* (FFP) contract is awarded. Standard forms of agreement between an owner and contractor are available from several sources. The student is encouraged to check the references or various web sites.

The price is firm for the life of the contract unless revised according to the change clauses given in the contract. An example for a $1 million contract under varying consequences is given below.

Contract price	$1,000,000	$1,000,000	$1,000,000
Actual cost	900,000	$1,000,000	1,100,000
Realized profit	$100,000	0	$(100,000)

The contractor either gains or loses based on performance.

In another fixed-price contract, a negotiated pricing formula motivates and rewards the contractor for performance. In those *fixed-price incentive* (FPI) contracts the process involves an estimated cost, target profit, target price, ceiling price, and profit sharing for costs incurred above or below the estimated cost. The example above with the $1 million contract price is a 0/100 sharing arrangement. This means that the owner does not share in gains or losses, and the contractor accepts 100% of the difference between price and actual cost. Figure 11.1 shows the basic fixed price with 0/100% sharing.

The fixed price incentive has a target cost, and a target profit percentage along with a contract ceiling price and a price-adjustment formula.

Estimated cost	$900,000
Target profit (11.1%)	100,000
Target price	$1,000,000
Price ceiling (125%)	$1,250,000
Cost sharing above estimated cost	80/20
Cost sharing below estimated cost	90/10

The owner does not pay any money above the price ceiling.

PICTURE LESSON
United States Capitol Dome Under Construction

President George Washington laid the cornerstone of the United States Capitol building in 1793 and construction proceeded steadily. Built, burned by the British in 1814, rebuilt, redesigned and refurbished several times, it has been refreshed with American spirit and idealism and inhabited over the years by lively, colorful, and dedicated people. The U.S. Capitol inspires awe and appreciation. As the capitol building expanded its wings to accommodate the increasing number of elected senators and representatives, and as the country grew, it became apparent during the 1850s that the dome was not attractive, as it was dwarfed by the magnitude of the enlarged capitol. The Bulfinch (architect during the period 1818–1829) dome was built of masonry and wood with copper sheathing. Later the architect Thomas Walter used design elements of the great masterpieces of that time—St. Peter's in Rome and St. Paul's in London—to produce the enlarged dome that is instantly recognizable. The picture shows the construction in May 1861, less than a month after President Abraham Lincoln was inaugurated the first time. Troops were garrisoned to protect the Capitol during the Civil War. Notice the troops and resting rifles in the foreground of the picture.

Cast iron frames were constructed over the existing dome. Statuary for the Senate pediment is in front of the steps. Construction on the great cast iron dome was halted by the Army Corps of Engineers during the summer of 1861, but President Lincoln's personal determination prevailed that construction forge ahead. The Statute of Freedom atop the Capitol dome was installed amid a proud cannon salute of 35 guns (one for each state) on December 2, 1863, during Lincoln's first term.

My country, 'tis of Thee, Sweet Land of liberty, of Thee I sing:
Land where my fathers died, Land of the pilgrims' pride,
From ev'ry mountainside Let freedom ring.

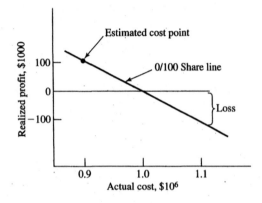

Figure 11.1 FFP arrangement with 0/100 sharing line.

Assume that a contractor experiences an overrun of 15% of estimated cost, or $135,000 (= 0.15 × 900,000). The reimbursement is as follows:

Target cost	$900,000
Plus 80% of $135,000 (owner's share)	108,000
Total reimbursement cost	$1,008,000
Target profit	$100,000
Less 20% of profit (contractor's share)	20,000
Total profit	$80,000
Cost plus profit reimbursement	$1,088,000

Thus, the owner covered 80% of the increased cost, but the contractor lost 20% of profit. Instead of being 11.1% (= 100,000/900,000), profit declines to 7.9% (= 80,000/1,008,000). The profit return can be computed another way. On the basis of actual and received costs, the profit = 5.1% ((= (1,088,000 − 1,035,000)/1,035,000), where $1,035,000 (= 900,000 + 135,000) is the final job cost.

Suppose that the contractor completed the job, but costs were $820,000, or $80,000 less than estimated. Under the terms of cost sharing of 90/10 below estimated cost, we find the following:

Target cost	$900,000
Less 90% of $80,000 (owner's share)	72,000
Total reimbursement cost	$828,000
Target profit	$100,000
Plus 10% of $80,000 (contractor's share)	8,000
Total profit	$108,000
Cost plus profit reimbursement	$936,000

Instead of being 11.1%, the profit increases to 13% (= 108,000/828,000). Note that profit is a percentage of the total reimbursement cost of the contract.

The FPI represents joint responsibility for ultimate cost. This arrangement shares in any dollar difference between the estimate and final cost. In an 80/20 example, the contractor is responsible for 20% of the difference either as an addition to or a deduction from target profit. Though shares are always a total of 100%, the proportions vary because of uncertainty, amount of target profit, and the spread between estimated cost and ceiling price. Various expressions of the owner/contractor or government/contractor shares are 60/40, 75/25, 50/50, and so on. These proportional shares are determined though negotiation.

The share line can be linear both above and below estimated cost. An 80/20 above estimated costs can be matched with a 90/10 below estimated costs. Slope changes such as those depend on the risk, design and negotiation.

Assume a 70/30 sharing proposition for the following cost facts:

Price ceiling	$11,500,000
Target price	$10,850,000
Estimated cost	$10,000,000
Final cost	$9,600,000
Difference (underrun)	$400,000

The contractor receives $120,000 (= 0.30 × 400,000) as an increase in profit. This adds to a target profit.

Target profit	$850,000
Contractor's share	120,000
	$970,000

The owner or government receives 70%, or $280,000 (= 0.70 × 400,000) difference, as a reduction in price.

Final cost	$9,600,000
Final profit	970,000
Final price	$10,570,000
Target price	10,850,000
Price reduction	$280,000

Now assume a final cost of $10,500,000.

Estimated cost	$10,000,000
Final cost	10,500,000
Difference (overrun)	$500,000

The contractor receives 30%, or \$150,000 (= 0.30 × 500,000) as a decrease in profit.

Target profit	\$850,000
Less 30% of overrun	150,000
Final profit	\$700,000

The owner receives 70%, or \$350,000 (= 0.70 × 500,000) as an increase in price.

Final cost	\$10,500,000
Final profit	700,000
Final price	\$11,200,000
Target price	10,850,000
Price increase	\$350,000

If the final cost is \$12,000,000, or \$500,000 in excess of the price ceiling, then the ceiling of \$11,500,000 is the final price.

A possible contract arrangement could include fixed-price incentives, with successive sharing ratios for early development, design, and construction.

Fixed price with redetermination (FPR) is another possible arrangement. The redetermination may be prospective or retroactive. In prospective, negotiation is undertaken for a fixed price in a future period, and then successive fixed prices are renegotiated periodically. Data available are past costs and performance.

The retroactive type of FPR contracts provides for adjusting contract price after the work is completed. A ceiling price is initially determined and actual audited costs are the starting point for negotiation. The terms of the contract do not provide for mutual sharing. The retroactive type requires an opinion of the contractor's performance by the owner or government.

Quote or price-in-effect (QPE) types of contracts (introduced in Section 3.4) are also known as *escalation* contracts. They protect the contractor against unanticipated inflationary increases in labor, material, overhead, and other costs in the performance of the contract. Long-term uncertainties that result from inflation or deflation effects are the reason for their use. The contractor will estimate costs and then establish a mutually agreed on benchmark or index. The escalation ties to some official index such as labor rates for the area or basic raw material prices. Adjustment is mostly up, but downward adjustment is possible. The price-adjustment clause must identify a base or benchmark period and one or more indexes to measure changes in price level in relation to the reference cost and reference period.

11.4   COST-REIMBURSEMENT ARRANGEMENTS

The *cost-plus-incentive-fee* (CPIF) contract develops an incentive sharing formula and is used in lieu of cost reimbursement with a 0/100 share. CPIF states a ceiling price, and costs that comply with contract terms are reimbursed. The total of reimbursed costs is the final cost of the contract. In FPI plans the reasonableness and necessity of costs are established, and cost is finally found after negotiation. However, under a CPIF, the maximum and minimum fees are limited by negotiation. A cost level above and below estimated cost is negotiated for minimum and maximum levels. Contract sharing ceases, and the contract reverts, in effect, to a cost-plus-fixed-fee 100/0 sharing plan. In contrast, the ceiling price in a fixed-price contract establishes a point over target cost where the owner ceases to share and the contract becomes an FFP with a 0/100 share model.

The parties to a CPIF contract believe the cost risk is too great for a realistic price ceiling. CPIF arrangements encourage incentive over a greater variation from estimated cost than would be expected with an FPI contract. What follows is an example of a CPIF plan.

Estimated cost	$10,000,000
Target fee	750,000
Maximum fee	1,350,000
Minimum fee	300,000
Share formula	85/15

Assume a trial cost of $9,000,000.

Estimated cost	$10,000,000
Final cost	9,000,000
Difference (underrun)	$1,000,000

The contractor receives 15%, or $150,000 (= 0.15 × 1,000,000), as an increase in fee.

Target fee	$750,000
Share of underrun	150,000
Total fee	$900,000

The owner receives 85%, or $850,000 between estimated and final cost, as a reduction in price.

Final cost	$9,000,000
Final fee	900,000
Final cost plus fee	9,900,000
Estimated cost plus fee	10,750,000
Reduction in price (underrun)	$850,000

In the above example the incentive is an effective basket of $7,000,000, including an underrun of 40% and an overrun of 30%. The contractor share of a $4,000,000 underrun is 15%, or $600,000. The share of a $3,000,000 overrun is 15%, or $450,000. Those adjustments add to or subtract from a target fee of $750,000. The maximum and minimum fee ranges from $1,350,000 to $300,000. If the actual cost is greater than +$3 million or less than −$4 million from the estimated cost, the plan fixes the fee at either the maximum or minimum level.

Another cost-reimbursement plan, the *cost plus award fee*, provides a base fee and an additional fee that may depend entirely or partially on performance. The cost-plus-award-fee plan does not include targets or automatic fee adjustment. The amount of the fee is subject to unilateral judgment by the owner or government.

The *cost-plus-a-fixed-fee* (CPFF) plan pays the contractor a fixed fee above reimbursable costs. The fee may change when the work breakdown structure changes. Conceptually, a CPFF is the opposite of a FFP arrangement. The FFP has a 0/100 sharing and CPFF a 100/0 sharing.

Estimated cost	$15,000,000
Fixed fee	900,000
Estimated CPFF	$15,900,000

The fee is unalterable concerning actual costs. CPFF is used when the risk is great and when effort is inconsistent with performance.

11.5 CONTRACT CLAUSES

Organizations have procurement policies expressed as standard or boilerplate contract clauses. These printed statements are usually an attachment to the main body and should not be dismissed as unimportant.

There are many contract clauses. Patent rights, value engineering, excusable delays, retainage, progress payments, conduct of the work, interpretation, excess material, shipping papers, insurance bonding, hiring of women and minorities, sex discrimination, and overtime requirements are typical ones. See Table 11.3 for a listing of contract clauses. Many clauses exist and only a few are discussed here.

The following points are stressed:

- Importance
- Supplemental contracts
- Parties to the clauses

The existence of real or potential competition before selection and award ends when the contract is signed. A competing offeror becomes a sole-source contractor or supplier. Contract clauses provide for circumstances that may alter some parts of the terms and conditions of an agreement. Contract changes are actions before the estimated cost and price and directly affect those documents.

TABLE 11.3 Illustrative titles for contract clauses

Party to the Contract	Contract Clause
Owner	Financial obligations
	Surveys, legal limitations, easements
	Right to stop work
	Right to carry out work
Contractor	Supervision and construction procedures
	Labor and materials
	Warranty
	Taxes, permits, fees, and notices
	Allowances
	Progress schedule
	Shop drawings, product data, and samples
	Cleaning up
Engineer/architect	Owner's representative
	Site visitation
	Certificate of payment
	Interpretation of design and work
	Rejection and acceptance of work

Construction contracts give the owner the right to make changes in the design or work after the contract has been signed. Similarly, the U.S. government has the authority to initiate changes.

A general commercial practice accepts what is known as the change clause, which is applicable to fixed-price and cost-reimbursable contracts. The customer may, without notice, make certain changes in the contract requirements if such changes are declared by the customer as not constituting a change in scope. If the change affects, for instance, cost, delivery, schedule, or performance, then the contractor must serve written notice to initiate a review, negotiation, and an equitable resolution of the claim. If the customer authorizes a change that may increase cost, then it is assumed that there is adequate money to pay for the change. Construction clauses require that the contractor is not to continue unless there is written authorization.

When those change orders are issued, a supplement to the contract is prepared. The supplement is made on the lump-sum or cost-reimbursable arrangement.

A change order is a modification of the contract. The parties understand that modification has considered subsequent negotiations, terms, designs, and conditions.

A disputes clause of the contract provides procedures where resolution may be initially thwarted for a variety of reasons. The dispute clause provides that the customer will initially decide any dispute concerning a question of fact arising under the contract—for example, failure to agree to a price adjustment resulting from a change.

Details differ between governmental and commercial activities. Construction contractors will first attempt to settle with the owner. There are levels of appeal from an adverse decision, including arbitration and courts, depending on whether the dispute involves questions of fact or law. In a similar way, the government will arbitrarily send a written notice to the contractor. After the decision is rendered by the court, the contractor may appeal.

The addendum statement for cost-reimbursable contracts provides for control and regulation. Costs caused by the contract are allowed and reimbursed. Criteria for determining allowable costs are defined in various documents that are referenced by the contract, or the terms of the contract may define the kinds of allowable costs.

Reasonableness and prudence are governing practices in accepting allowable costs. For example, a subcontract material that was not competitively bid may not be allowable. Thus, this material may be subject to negotiation to determine value.

The inspection and correction of defects, warranty, and full or conditional protection, for example, are clauses that obligate the contractor, vendor, or seller to correct defects, deficiencies, and inability to meet specification in various ways. Those terms are so broad that we are unable to offer specifics, except that the cost estimate needs to include reasonable costs based on the wording of the clause. Mean time to failure and mean time to repair are ideas originally described in Chapter 10. Percentage returns, conditional service contracts, customer assistance, field correction, recalls, and bulletins alerting the customer are features of a seller's costs. They are the contractual terms that provide specificity. Costs are generally reimbursed in a cost-reimbursable contract.

A subcontract is an agreement between a general contractor and a subcontractor or a contractor and supplier. The subcontractor agrees to perform specialized work at a construction site or provide certain materials. A supplier agrees to supply materials, subassemblies or assemblies, or subcontract material.

A subcontract binds only the parties to the agreement. Many of the same clauses that are required of the general contractor are applied similarly to a subcontractor, although the value of the contract will dictate its complexity. A $1-million subcontract has greater specificity than one with a $1000 value. Provision of the general contract, including changes in work, minimum wage laws, warranty clauses, and other laws, may extend to the subcontractor.

In fixed-price agreements, competitive bidding is encouraged before subcontracts are awarded. The subcontract clauses require the contractor to meet some stipulations in cost-reimbursable contracts. The contractor may be required to advise the owner or government of the anticipated subcontracts. Approval by the owner may be necessary, depending on the size of the award. Consent by an owner to use a specific subcontractor does not relieve the major contractor of failure to perform.

Contracts conclude in a variety of ways. The usual way is full and satisfactory performance by both parties. Another way is breach of contract. Failure of progress payments and unreasonable delays of the project are the most common breaches by the owner. The contractor is entitled to damages caused by the owner's inability to discharge responsibilities required by the contract. Default or failure to perform as required by the contract are the more common breaches by the contractor. Nonperformance, poor quality, failure to show progress, disregard of laws or instructions are actions that may allow an owner to invoke the termination clause.

A convenience termination clause allows the customer an opportunity to decide that the material under procurement is no longer required and that the customer is prepared to assume losses associated with termination. The contractor ceases work and issues cancellations on purchase orders and subcontractors. Eventually, the contractor provides a termination claim that includes incurred costs of work performed and special expenses associated with the termination effort. In cost-reimbursable contracts, the claimed costs are verified by an audit, and the fee is negotiated. In a fixed-price contract, the legitimate costs and fees are determined both by submittal of evidence and by negotiation.

11.6 INSURANCE AND BONDS

The United States is a country that requires a risk management and a comprehensive approach to handling exposure to loss of many kinds. In devising an overall policy to handling risk, commercial insurance is a cornerstone to financial protection. Certainly insurance does not eliminate the risks involved in construction contracting, but it does shift most of the financial loss to a professional risk-bearer. There is, of course, self insurance by the contractor, but that may be done in conjunction with a state-approved reinsurer.

As construction is a hazardous sector of business, accidents can be frequent and severe. This fact requires the contractor to protect itself with a variety of complex and expensive insurance coverage. An *insurance policy* is a contract under which the insurer promises to assume financial responsibility for a specified loss or liability. Table 11.4 gives a sample list of various kinds of insurance policies.

An insurance contract contains many provisions affecting the loss against which it affords protection. While the law of insurance contracts is similar to the general law of contracts, it is tied to public welfare, and the insurance field is closely controlled by federal and state statute. Each state has a regulatory agency that administers the state's insurance codes and imposes requirements on the insurer as to investments, reserves, annual financial statements, periodic audits, and contract responsibility.

Typically, the contractor will prepay the premium for the protection that the insurance gives. If there is a loss, the contractor cannot recover more than the loss. A profit cannot be made at the expense of the insurance company. The law dictates mandatory

TABLE 11.4 Illustrative types of insurance for contractor

Typical types of insurance
Property insurance on project
All-risk builder's risk
Named-peril builder's risk
Property insurance on contractor's own property
Fire
Motor vehicle
Liability
Employer's liability
Contractor's public liability and property damage
Liability insurance to protect owner/contractor/engineer/architect
Comprehensive general liability
Malpractice
Employee
Workmen's compensation
Federal Income Contribution Act
Unemployment
Disability
Business, accident, and life
Owners, landlords, and tenants premise exposure
Medical: mandatory/voluntary
Special risks or conditions and personal injury
Contingent liability for acts of engineer/owner/contractor/subcontractor

insurance of various kinds, but the contractor does have a prerogative to decide which insurance it shall carry, and from which company to buy that protection. The elective insurance is for its own property mostly. If the contractor bought insurance for every conceivable risk, the burden of the cost for the premiums could cause financial exigency in its operations. That balances against ruinous legal costs to cover the costs of perils. At times, meticulous engineering and construction procedures and safety can minimize a risk and may preclude the necessity for insurance. Skill and extraordinary precautions balance against the cost of insurance.

Construction bonds and construction insurance are different. A bond offers guarantees and protection to losses by the owners against default of the contractors. Insurance, by contrast, provides protection against losses due to damage or injury. Bonds are a popular form of contract within the construction business. Table 11.5 gives a simple listing.

A performance bond or a *surety bond* guarantees performance of a contract or an obligation. In law a surety is a party that assumes liability for the default, debt, or failure of another. A surety bond, then, is the contract that describes the conditions and obligations pertaining to the agreement. Those surety bonds extend the credit by the surety, not in the sense of a financial loan, but as an endorsement. The original definition of the word *surety* implies the sense of security.

In the terms of the surety bond, the surety agrees to indemnify the owner against default or failure in duty of the contractor. Contract bonds are three-party agreements that guarantee the work is completed according to the contract documents and that all construction costs will be paid. After the contractor finishes the job, and after the warranty period concludes, the bond agreement expires. If the contractor fails to fulfill its contractual obligations, the surety must complete the contract and pay all amounts up to the face value of the bond.

TABLE 11.5   Illustrative types of bonds

Typical types of bonds
Performance
Payments for labor and materials
Bid
Miscellaneous surety bonds
Discharge of liens or claims
Indemnify owner against liens
Protect owners of rented equipment and leased property
Fidelity
License or permit
Termite/soil expansion protection
Union wage

A surety bond does not further describe the construction work but accepts the construction contract in its provisions, and the obligations of the bond are identical to the owner-contractor agreement. The contractor is unable to extend the agreement with the owner as a result of the bond.

Contract bonds are not required on all jobs, although they are required by law on public jobs. Much private work is not bonded in this fashion.

It is usually the practice for construction contracts to require separate contract bonds[2]—performance and payment for labor and materials. A performance bond acts primarily to protect the owner. It guarantees that the contract will be performed and that the owner will receive his design built in accord with the terms of the contract. This bond covers any warranty period that is stated by the contract. The bond value is limited to the face value, which is related to the total contract price.

A payment bond protects any third parties to the contract and guarantees payment for labor and materials. The owner is protected against liens that can be filed on his property by unpaid parties to the work.

There are national surety companies specializing in this business. They are subject to public regulation similar to insurance companies. Because the true worth of the bond is no better than the ability of the surety bonding company to pay upon default of the contractor, the owner is allowed an opportunity for negotiating the selection of the company. Indeed, in some cases the owner requires that the surety company be approved by the owner before the final signing of the contract. The owner does not desire default of the surety company, of course.

A contractor will have an upper limit of bonding capacity. The surety will examine the past records of the contractor, seeing cash-flow statements and credit reports and determining the contractor's net worth or net working capital. The balance sheet and income statement are conveyed to the surety company for this examination.

Bid bonds, also called *proposal bonds*, are supplied with a bid to the owner. They provide bid security, guaranteeing that the contractor on being declared the successful bidder will enter into contract with the owner for the amount of the bid and will provide contract bonds as required. Rather than a bid bond, some owners require a certified check or

[2] The American Society of Civil Engineers list several documents on its web site that give omnibus description of contract documents and lists various types. The American Institute of Architects supplies three standard forms: Performance Bond Agreement, Labor and Material Payment Bond, and Bid Bond. The student may wish to check the contents of these agreements at *www.ASCE.org* or *www.AIA.org*.

some other form of negotiable security. The owners will return the security of unsuccessful bidders.

If a contractor is awarded the contract, and then does not sign the contract or is unable to secure a performance bond, the bid bond assures that the owner is reimbursed, sometimes for the difference between the bid and that of the next lowest responsible bidder. The responsibility to pay this difference is that of the contractor or the surety company issuing the bond.

11.7 NEGOTIATION, AWARD, AND AUDIT

The owner and contractor pay attention to the phases of negotiation, award, and audit. Our summary focuses on some principles that relate to cost analysis and estimating.

Negotiation may begin once the price and technical proposal are conveyed to an owner or buyer. Consider the situation where several bidders are responding to an RFQ. If the RFQ is for standard commercial materials and ordinary designs, the lowest (or another responsible bidder) may be selected. Contrariwise, if the design is complicated and significant in terms of money, then negotiation may be required.

Negotiation is a broadly used term and has come to mean tactics and maneuvers by both parties in an effort to reach a decision on whether to contract together. "Horse trading" in negotiation is a simplistic picture when dealing with complicated designs and estimates. Negotiation requires complex maneuvers.

Not all procurements are negotiated. For technical and complex designs, the various bidders make exceptions or claims regarding the RFQ and their technical response.

Negotiation involves engineers, owners, contractors, and contract administrators. The discussion relates to engineering and design, costs, schedules, and so on, for which the team approach is useful. Representatives are informed about price, schedule, contract type, and the design before negotiation. For big contracts, there may a "day in court" during which the submitter discusses its offering and explores the possibility of reaching contractual agreement. There are occasions where performance and price are considered separately, and the best selection is made on one or the other factor. The object of technical negotiation is to explore and clarify points to establish merit of various engineering approaches, and to overcome objections or demonstrate that the bid satisfies or does not satisfy minimum requirements of the designs and specifications, and therefore is responsive or nonresponsive.

It is an inviolate principle to this author that cost estimates be factual. However, profit and pricing are an executive decision. Pricing refers to the fair and reasonable values and is negotiable.

In small purchases, negotiation consists of letters or telephone calls. In more significant purchases, negotiation is face to face and lasts many days. A plan or list of discussion points is necessary for effective negotiation. Issues regarding the design, RFQ, contract, terms, schedule, estimate, and performance are open to discussion. Each bidder may have exceptions or additional claims for the RFQ. For instance, two bidders on a high-voltage electrical transmission line may have a low cost but high voltage-line losses, and vice versa.

The owner's engineers in negotiation must be prepared to ask questions such as: Are there issues that can be traded off if necessary? How realistic is the construction schedule? If the construction schedule lengthens, is a lower price possible?

If a price reduction is desired, then a vague statement that the "price is too high" represents a weak approach. A price-reduction request must be plausible and businesslike. In certain situations the cost estimate is privileged information. In others, such as government or public construction work, it may be open to examination. The contractor's engineer defends cost-estimating relations, wage rates, overhead rates, and so on, on a factual basis. The owner's engineers may perform a technical analysis on the estimate.

A variety of contract terms and contract types exist. Usually, standard contract terms are nonnegotiable, although an able negotiator takes nothing for granted. Terms of special clauses are another matter. Those terms need to be carefully examined, because many of the contract terms and specifications may have financial and serious implications. Specifications mean additional practices, such as a particular type of inspection, quality control, and so on. They may place a burden on an owner or contractor who is unaware of the consequences.

Negotiation may deal with penalty clauses, retainage amounts, or progress payments or omissions, nebulous requirements, inaccuracies, inconsistencies, and impossible or very expensive requirements. In disputes about engineering changes, patents, warranty claims, and so on, litigation may follow, and the estimate may be a document of evidence.

A competitive negotiation will provide an opportunity for discussion by the offeror and will conclude with the award of a contract to the offeror whose price and design, for instance, are most beneficial to the owner, buyer, prime contractor, or government.

Once the contract is ongoing, terms of the contract may allow audits to verify allowable costs, possibility of fraud, compliance, and documentation. Auditing is a common occurrence for government prime contractors. Audits may be by the customer or may be internal by the firm or by a consultant hired by the organization. We are not referring to an accounting-firm auditor, who examines the balance sheet and profit-and-loss statement and issues a public notice. We are referring to an audit that deals with the cost-analysis and estimating function, although it may be difficult to uncouple that from engineering, buying, accounting, or management. Internal audits infer the monitoring of cash flow, accounting, estimating, contracting, and the general business conduct of the firm's operations. Internal audits of this type are commonplace in business. General contractors may also audit subcontractors.

Audit may follow the work breakdown structure, bill of material, Master Format scheme, construction check-off sheets, and so on, and it must have provision for field compliance inspection. These activities are extensive, and discussion is beyond the needs of this book. Here we provide overall ideas to aid the learning of cost analysis and estimating.

Administrative audits deal with several cost factors for a period of time that include overhead cost rates, labor-hour and equipment rates, efficiency factors, and so on. They determine whether present or future conditions negate the appropriateness of the cost factors.

Estimating audits are concerned with mistakes and omissions, procedures, and contingency assessment. Mistakes are $2 + $1 = $4 and are unavoidable. Despite the popularity of mistakes, avoidance begins with attention to detail and checks by others and faultless arithmetic. Though there is justifiable allegiance to software and computers, their reliability and application are fair game for auditors. Nor should the engineer always conclude that the computer and software are perfect. They are not.

Procedures are a significant concern for audit. Consistency of estimate with the accounting system and to other estimates is important. The audit path of verifiable facts is conducive to reassessing consistency.

Separation of direct and indirect costs is audited closely. Direct costs are those identified as having been incurred for a particular product, work order, job, or contract. Indirect costs are material, labor, and expenses that affect two or more products, work orders, jobs, or contracts, where the amount of cost charged to a specific one cannot be precisely determined. No universal rule exists that under every estimating and accounting system there is an assurance that items of cost be treated fairly as direct or as indirect cost. It is essential that within estimating, each item of cost be treated consistently. Some material costs may be confused as either direct or indirect, even though they are clearly incurred to the final project work order. Paint, for example, which appears on an end item, is an awkward estimate and may be called an indirect charge. Minor hardware is another optional choice. Because of these typical alternate choices, auditors check estimates for inconsistent treatment of direct and indirect cost.

Significant direct material and subcontractor cost estimates are audited carefully. Much of what is included as estimated direct material will have been purchased from outside sources. The auditor examines the principal items within each material cost category and will check, for example, sources, quantities, unit prices, and losses as shown on the direct material and subcontract estimate. Whenever engineering is complete and working drawings, specifications, work-breakdown structure or a bill of material are available, key WBS or part numbers are traced. For most design efforts, estimates are prepared from less data than appear on a bill. The WBS provides the skeleton for tracing, or the engineer may prepare a tentative bill of material from preliminary drawings.

Estimating forms may serve as the preliminary *bill* or, if the Master Format is used, then that is a possible document of examination. Another means is the engineer's project manual, a bound record, where costs for the material and labor are estimated and tied to the design. For developmental work where costs are uncertain, planning quotations obtained from potential suppliers are verifiable.

Other options are open to auditing material costs and range from routine supply problems to uncertain engineering efforts. Follow-on procurement provides realistic costs by using earlier projects, and data can apply to projects even though they may be developmental. In follow-on procurement, reasonable projections of historical costs account for price reductions caused by removal of original design, equipment, rearrangement, excess spoilage, and other startup costs. Economic factors, normal increases or decreases in price, and changes in mobilization rates and quantities are considerations for follow-up estimating. Auditors are aware of those options.

A priced bill of material allows for scrutiny, especially in the amount of materials used. Once the design is fixed, auditors compare the quantity of material to that specified by the bill. This comparison is random, because an entire bill is seldom checked. Historical citations of scrap, waste, obsolescence, and spare-parts percentages are helpful in supporting estimates. A priced work-breakdown structure can be audited. If a contract work-breakdown structure is developed, then estimates, subcontractor bids, purchase offers, vouchers, and bills of lading can be tracked from paper to physical hardware.

11.8   ETHICS AND ENGINEERING

It is fitting that a book on cost analysis and estimating close with a discussion of ethics. It is no less important than the very first sections of the book, which deal with the necessity of profit and wise stewardship. Consider the *American Heritage Dictionary's* definition: Ethics (eth'iks). 1. A principle of good conduct. 2. A system of moral principles or values.

This book focuses on an important role of the engineer, but we have not yet defined engineering. "Engineering is an important and learned profession. As members of this profession, engineers are expected to exhibit the highest standards of honesty and integrity. Engineering has a direct and vital impact on the quality of life for all people. Accordingly, the services provided by the engineers require honesty, impartiality, fairness and equity, and must be dedicated to the protection of the public heath, safety, and welfare. Engineers must perform under a standard of professional behavior that requires adherence to the highest principles of ethical conduct."[3]

The task facing the engineer is providing a measure of the *economic want* for the design. During the planning period it is not uncommon that political or business pressures are applied on the people who are estimating. It is natural that a design engineer will believe that the new design is cheaper, or that the sales staff will promote an opportunity that gives encouragement to marketing. Though those motives are understandable, engineers need to maintain objectivity in fact finding and cost analysis. Subjectivity that may inappropriately influence estimating out-of-pocket future costs seems to be unprofessional.

The estimate deals with elements of material, work, and money. Because competition is the nature of business, there are occasions in which the propriety of some trade practices is questionable. Revealing confidential quotations to subcontractors or vendors with the hope that a new bidder will submit an even lower bid is improper. "Bid shopping" and "bid peddling" are the terms applied to those practices. On the other hand, some contracts are required to be public knowledge, and, on those occasions, integrity would require that the same value be disclosed equally to all candidate bidders.

Firms known to be unqualified to perform work or supply the product should not be invited to bid. Unless it is understood as a clause in the contract or is mandated by public law, the price and cost estimates of one competitor should not be made known to another competitor.

Acts of collusion or conspiracy with the implied or express purpose of defrauding clients, suppliers, or subcontractors, and business practices that are not fair and honest, are condemned and may be illegal as well as unconscionable to professional engineering practice. There are ethical engineering standards that comment on those practices. See the NSPE Code of Ethics on the web that adds specifics to these statements.

SUMMARY

Professionals who work in construction need to have knowledge of contract fundamentals. We have discussed practices of

Fixed-price contracts	Auditing and negotiation
Cost-reimbursable contracts	Ethics and engineering
Insurance and bonds	

QUESTIONS

1. Define the following terms:

Contractor insurance	Shares
Surety bond	FPI

[3] National Society of Professional Engineers web statement, *www.NSPE.org.*

Bid bond   QPE
Unit-price contract   Target cost
FFP   Contract clauses
Time-and-material contract   Bid peddling

2. In the sharing arrangements for FPI, what are the pros and cons of a sharing plan such as 90/10 versus 50/50 to the contractor and owner for cost reductions below estimated cost?

3. Contrast fixed-price versus cost-reimbursable types of contracts for (a) very high voltage transmission line over rugged terrain and (b) prototype manufacture of an ultrahigh-vacuum-chamber environmental test unit.

4. When would an owner prefer an FPI contract? Why would a contractor desire one?

5. List the advantages and disadvantages of CPIF from the owner's and contractor's viewpoint.

6. Prepare an outline of a negotiation strategy for any project estimate for which you have knowledge.

7. Assume that the high-voltage transmission-line project estimated in Chapter 9 is awarded. Itemize a list of documents that will be useful for an auditor on this project.

8. What is the purpose of a surety company?

9. What are the circumstances that require a bid bond?

10. Look at the web site, www.ASCE.org and list the titles of construction contract documents.

PROBLEMS

11.1 A lump-sum contract is to be procured by an owner. The contractor, in preparation for the fixed-price bid, estimates the following work packages:

Work Package	Amount, $
Direct labor	$3,500,000
Direct material and subcontract	4,000,000
Facilities and equipment	700,000
Engineering	55,000
Estimated interest	120,000

Overhead cost for the office and the job is 4% and 5% of direct cost, respectively. Profit is 7.5% markup on total cost. Determine the firm fixed-price bid. After the completion of the job, the actual cost is found to be $9 million. What is the final profit amount? What is the profit as a percentage of cost? Repeat for $10 million. Discuss the consequences of profit and loss for the contractor. (*Hint:* Assume that interest is not direct cost or overhead and is a listed line item.)

11.2 A contractor provides a unit-price contract:

Pile Diameter (in.)	Price, $/ft
12	$1250
18	2576
20	2685

The owner's engineer observes that 18 12-in. piles were driven to 60 ft, 14 18-in. piles were driven to 52 ft, and 5 20-in. piles were driven to 41 ft. What is the net realized contract value?

11.3 A subcontractor is considering a fixed-price contract where the gross-hour cost rates are quoted. The subcontractor uses a *markup-of-cost-value* method of pricing. The subcontractor varies the markup rates depending on the perception of negotiation advantage of the craft work.

    (a) What are fixed-price quotations? How does markup rates of cost value differ from markup rates on sales?

    (b) The customer estimates the potential number of hours. What total fixed price is quoted by the subcontractor?

11.4 Information for a fixed-price incentive contract is given as

Construction Craft	Gross-Hour Cost Rate	Markup (%)	Job Estimate (Hours)
Lather	$39.16	10	80
Mosaic & terrazzo	90.98	20	40
Plasterer	32.19	15	20
Painter	42.33	5	15

What is the *cost plus profit* paid to the subcontractor where an overrun cost is $135,000?

Contract Item	Quantity
Estimated cost	$900,000
Target profit (11.1%)	100,000
Target price	$1,000,000
Price ceiling	$1,125,000
Cost sharing above estimated cost	75/25
Cost sharing below estimated cost	80/20

Now assume that a cost underrun is $150,000, and find the cost plus profit that is paid to the subcontractor.

11.5 A FPI project is bid in this way:
The cost-sharing arrangement is 80-20 above and 90-10 below the target cost.

Contract Line Item	Amount, $
Target cost	100,000
Target profit (10%)	10,000
Ceiling cost (125%)	125,000

    (a) The contractor experiences an overrun of 20% in the contract cost. Find the reimbursement that is paid to the contractor and the percent profit. (*Hints:* In no case is the contractor reimbursed above 125%, or $125,000. The ceiling includes all costs, including profit. Profit is expressed as a percent of the cost of the contract.)

    (b) Now assume that the contractor completes the contract and shows costs of $80,000. Find the total reimbursement and the percent profit.

11.6 A contractor estimates and negotiates the following fixed-price incentive contract:

Sketch the contractor's FPI profit chart, where profit is the y axis and cost dollars are the x axis.

Contract Item	Quantity
Estimated cost	$10,000,000
Target profit	$ 850,000
Target price	$10,850,000
Price ceiling	$11,500,000
Cost sharing below estimated cost, %	70/30
Cost sharing above price ceiling, %	0/100

11.7 A contract is negotiated as CPIF with the following plan:
If the job cost is $28 million, find the estimated cost and fee. What is the reduction in the price?

Contract Item	Amount, $
Estimated cost	$30,000,000
Target fee	2,000,000
Maximum fee	2,400,000
Minimum fee	500,000
Share formula, %	75/25

11.8 A contract is negotiated as CPIF with the following plan:

Contract Item	Amount, $
Estimated cost	$6,000,000
Target fee	400,000
Maximum fee	500,000
Minimum fee	200,000
Share formula, %	70/30

If the actual project cost is $7 million, find the final price. Determine the variance. Is the variance favorable or unfavorable?

11.9 A bid has an estimated cost of $5 million, target fee of $600,000, and maximum and minimum fee of $800,000 and $450,000. The share proportion is 80/20. Find the contractor's CPIF and the owner's cost for a final cost of 15% above estimate. Repeat for 10% below estimated cost. Sketch a chart for 80/20 sharing. (*Hint*: The x axis is actual cost while the y axis is fee.)

11.10 Hot-rolled alloy steel bar, $\frac{1}{2}$ in. O.D. $\times$ 20 ft, AISI 4140 oil hardening annealed grade, machine straightened, is quoted on a QPE contract. Three quantities are shown, along with their price, and the quotation is tendered on the index benchmark period of 0 time.

Weight (lb)	Price ($/100 lb)
120 (= 1 bar)	$115.00
2000	71.00
6000	66.50

AISI 4140 steel material is indexed matched to high-carbon steel scrap that has shown a 5%, 10%, and 8% increase per period since the benchmark period. Find the escalated schedule of prices expressed in $/100 lb for the last period.

11.11 The High-Voltage Transmission-Line Construction Company is under contract to build the transmission line as estimated in Sec. 8.4.4. The price is given by Table 8.12. Please review that information. The owner invokes the right to two changes:

(1) The owner increases the length over which the transmission line will travel by 0.6 mile.

(2) The owner stipulates that construction roads are to be "pioneer" style, thus being less and more environmentally pleasing.

What plan of action do you advise for this company? What contract change clauses will you base your price adjustments on? What is an approximate increase in cost? Give your recommendations in a word-processed single page.

MORE DIFFICULT PROBLEMS

11.12 A construction estimate gives the following costs:

Work-Package Item	
Direct labor	$10,000,000
Direct material	20,000,000
Subcontract	1,400,000
Job and office overhead	600,000

Find the target cost and bid if the markup of full cost is 5%. Now the contract is a fixed-price incentive with a 120 % ceiling and a split of 75 − 25 above target cost and a 80 − 20 split below target cost.

(a) If the contractor incurs an actual cost of $33 million, what is the reimbursement (including profit) to the contractor? What is the percent profit?

(b) Find the reimbursement if actual cost is $31,000,000, and find the percent profit.

(c) Find the reimbursement if actual cost is $40,000,000, and find the percent profit.

11.13 The following data are determined:

Work-Package Item	Amount
Construction labor	40,000 hours
Construction gross hourly rate	$28 per hour
Overhead rate on basis of direct labor costs	20%
Material	$50,000
Material overhead rate	15%
Engineering labor	1500 hours
Engineering hourly rate	$40 per hour
Overhead rate	25%
Profit rate as a markup of full cost	10%

The contract that is based on those figures is awarded with a 120% ceiling and a split of 75/25 above estimated costs and 80/20 below estimated costs.

    (a) If the contractor actually achieves the figures, then what is the reimbursement to the contractor for costs and profit?

    (b) If the incurred costs amount to an overrun of 10% above the estimated costs, then what is the reimbursement?

    (c) If the incurred costs amount to 5% under the contract, then find the reimbursement.

PRACTICAL APPLICATION

Many possibilities exist for a practical application using the objectives of this chapter. For example, consider the following:

    1. Obtain and examine a contract. Use directed reading for the contract, and outline the contract, essential parties to the contract, provisions that protect one or the other party, and so forth. Determine points of neutrality between the parties and how that aspect is handled. Would you want to be the contractor or the owner with the contract? Your instructor will provide other questions that you are to answer. Prepare your thoughts in a report.

    2. Arrange for a discussion with a contractor or owner who has experience with construction contracts. Before the visit prepare a list of questions and know your objectives. Make this a focused learning experience. Perhaps, you can encourage the owner or the contractor to visit the class and discuss his or her experiences.

CASE STUDY: FUNDAMENTAL CANONS OF ETHICS

As an important profession, engineering deals with the public, employer, clients, and colleagues in a work that affects the quality of life for all people. There are fundamental canons of ethics for exemplary professional behavior that include:

- Hold paramount the safety, health, and welfare of the public
- Perform engineering services in the area of competence
- Issue public statements in an objective and truthful manner
- Act for the employer or client as faithful trustee
- Avoid deceptive acts
- Conduct yourself honorably, responsibly, ethically, and lawfully

These NSPE fundamental canons are the driving force for expanded rules of practice that give detail to the weave and substance of the fabric of masterful, learned, and skilled work.

    The essence of this case study is to expand on these rules using a number of approaches and sources. You are to write a list of sustainable professional objectives for construction-cost analysis and estimating that is meaningful. Among the many slants you may wish to think about are wisdom, instruction, understanding, discretion, learning, equity, courtesy, and counsels.

    There are several sources to consider:

- Examine the NSPE web address, *www.NSPE.org*, which provides rules of practice.
- Consider the Book of Proverbs in the Bible, which is given to the teaching of many ethical and moral principles. The peculiarity of this book is that it largely espouses the principles of behavior in personal and business life by contrasts.
- Invite a representative of NSPE to speak on ethics for the class.
- Interview a professor who teaches ethics on your campus

Standard Normal and t Distributions; 10% and 20% Tables of Interest

APPENDIX 1　Values of the Standard Normal Distribution Function

		Areas under the Normal Curve $$F(z) = \int_0^z \frac{1}{\sqrt{2\pi}} e^{-z^2/2} \, dz$$								
z	0.00	0.01	0.02	0.03	0.04	0.05	0.06	0.07	0.08	0.09
0.0	0.0000	0.0040	0.0080	0.0120	0.0159	0.0199	0.0239	0.0279	0.0319	0.0359
0.1	0.0398	0.0438	0.0478	0.0517	0.0557	0.0596	0.0636	0.0675	0.0714	0.0753
0.2	0.0793	0.0832	0.0871	0.0910	0.0948	0.0987	0.1026	0.1064	0.1103	0.1141
0.3	0.1179	0.1217	0.1255	0.1293	0.1331	0.1368	0.1406	0.1443	0.1480	0.1517
0.4	0.1554	0.1591	0.1628	0.1664	0.1700	0.1736	0.1772	0.1808	0.1844	0.1879
0.5	0.1915	0.1950	0.1985	0.2019	0.2054	0.2088	0.2123	0.2157	0.2190	0.2224
0.6	0.2257	0.2291	0.2324	0.2357	0.2389	0.2422	0.2454	0.2486	0.2518	0.2549
0.7	0.2580	0.2611	0.2642	0.2673	0.2704	0.2734	0.2764	0.2794	0.2823	0.2852
0.8	0.2881	0.2910	0.2939	0.2967	0.2995	0.3023	0.3051	0.3078	0.3106	0.3133
0.9	0.3159	0.3186	0.3212	0.3238	0.3264	0.3289	0.3315	0.3340	0.3365	0.3389
1.0	0.3413	0.3438	0.3461	0.3485	0.3508	0.3531	0.3554	0.3577	0.3599	0.3621
1.1	0.3643	0.3665	0.3686	0.3708	0.3729	0.3749	0.3770	0.3790	0.3810	0.3830
1.2	0.3849	0.3869	0.3888	0.3907	0.3925	0.3944	0.3962	0.3980	0.3997	0.4015
1.3	0.4032	0.4049	0.4066	0.4082	0.4099	0.4115	0.4131	0.4147	0.4162	0.4177
1.4	0.4192	0.4207	0.4222	0.4236	0.4251	0.4265	0.4279	0.4292	0.4306	0.4319
1.5	0.4332	0.4345	0.4357	0.4370	0.4382	0.4394	0.4406	0.4418	0.4430	0.4441
1.6	0.4452	0.4463	0.4474	0.4485	0.4495	0.4505	0.4515	0.4525	0.4535	0.4545
1.7	0.4554	0.4564	0.4573	0.4582	0.4591	0.4599	0.4608	0.4616	0.4625	0.4633
1.8	0.4641	0.4649	0.4656	0.4664	0.4671	0.4678	0.4686	0.4693	0.4699	0.4706
1.9	0.4713	0.4719	0.4726	0.4732	0.4738	0.4744	0.4750	0.4756	0.4762	0.4767
2.0	0.4772	0.4778	0.4783	0.4788	0.4793	0.4798	0.4803	0.4808	0.4812	0.4817
2.1	0.4821	0.4826	0.4830	0.4834	0.4838	0.4842	0.4846	0.4850	0.4854	0.4857
2.2	0.4861	0.4865	0.4868	0.4871	0.4875	0.4878	0.4881	0.4884	0.4887	0.4890
2.3	0.4893	0.4896	0.4898	0.4901	0.4904	0.4906	0.4909	0.4911	0.4913	0.4916
2.4	0.4918	0.4920	0.4922	0.4925	0.4727	0.4929	0.4931	0.4932	0.4934	0.4936
2.5	0.4938	0.4940	0.4941	0.4943	0.4945	0.4946	0.4948	0.4949	0.4951	0.4952
2.6	0.4953	0.4955	0.4956	0.4957	0.4959	0.4960	0.4961	0.4962	0.4963	0.4964
2.7	0.4965	0.4966	0.4967	0.4968	0.4969	0.4970	0.4971	0.4972	0.4973	0.4974
2.8	0.4974	0.4975	0.4976	0.4977	0.4977	0.4978	0.4979	0.4980	0.4980	0.4981
2.9	0.4981	0.4982	0.4983	0.4983	0.4984	0.4984	0.4985	0.4985	0.4986	0.4986
3.0	0.4987	0.4987	0.4987	0.4988	0.4988	0.4989	0.4989	0.4989	0.4990	0.4990
3.1	0.4990	0.4991	0.4991	0.4991	0.4992	0.4992	0.4992	0.4992	0.4993	0.4993

*This table gives the probability of a random value of a normal variate falling in the range $z = 0$ to $z = z$ (in the *shaded area in figure*). The probability of the same variate having a deviation grater than z is given by $0.5 - $ probability from the table for the given z. The table refers to a single tail of the distribution; therefore the probability of a variate falling in the range is $\pm z = 2 \times$ probability from the table for the given z. The probability of a variate falling outside the range $\pm z$ is $1 - 2 \times$ probability from the table for the given z.

The values in this table were obtained by permission of author and publishers from C. E. Weatherburn, *Mathematical Statistics*. Cambridge University Press, London, 1946.

Source: From Ostwald, *Engineering Cost Estimating*, Prentice Hall, 1992.

APPENDIX 2   Values of the Student *t* Distribution

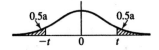

| Degrees of Freedom | Probability α | | | |
ν	0.10	0.05	0.01	0.001
1	6.314	12.706	63.657	636.619
2	2.920	4.303	9.925	31.598
3	2.353	3.182	5.841	12.941
4	2.132	2.776	4.604	8.610
5	2.015	2.571	4.032	6.859
6	1.943	2.447	3.707	5.959
7	1.895	2.365	3.499	5.405
8	1.860	2.306	3.355	5.041
9	1.833	2.262	3.250	4.781
10	1.812	2.228	3.169	4.587
11	1.796	2.201	3.106	4.437
12	1.782	2.179	3.055	4.318
13	1.771	2.160	3.012	4.221
14	1.761	2.145	2.977	4.140
15	1.753	2.131	2.947	4.073
16	1.746	2.120	2.921	4.015
17	1.740	2.110	2.898	3.965
18	1.734	2.101	2.878	3.922
19	1.729	2.093	2.861	3.883
20	1.725	2.086	2.845	3.850
21	1.721	2.080	2.831	3.819
22	1.717	2.074	2.819	3.792
23	1.714	2.069	2.807	3.767
24	1.711	2.064	2.797	3.745
25	1.708	2.060	2.787	3.725
26	1.706	2.056	2.779	3.707
27	1.703	2.052	2.771	3.690
28	1.701	2.048	2.763	3.674
29	1.699	2.045	2.756	3.659
30	1.697	2.042	2.750	3.646
40	1.684	2.021	2.704	3.551
60	1.671	2.000	2.660	3.460
120	1.658	1.980	2.617	3.373
∞	1.645	1.960	2.576	3.291

*This table gives the values of *t* corresponding to various values of the probability α (level of significance) of a random variable falling inside the shaded area in the figure, for a given number of degrees of freedom ν available for the estimation of error. For a one-sided test the confidence limits are obtained for α/2.

This table is taken from Table III of Fisher and Yates, *Statistical Tables for Biological, Agricultural, and Medical Research*, Oliver & Boyd Ltd., Edinburgh, 1963.

Source: From Ostwald, *Engineering Cost Estimating*; Prentice Hall, 1992.

APPENDIX 3 10% Interest Factors for Annual Compounding Interest

	Single Payment		Equal Payment Series			
	Compound-amount factor	Present-worth factor	Compound-amount factor	Sinking-fund factor	Present-worth factor	Capital-recovery factor
n	To find F Given P F/P i, n	To find P Given F P/F i, n	To find F Given A F/A i, n	To find A Given F A/F i, n	To find P Given A P/A i, n	To find A Given P A/P i, n
1	1.100	0.9091	1.000	1.0000	0.9091	1.1000
2	1.210	0.8265	2.100	0.4762	1.7355	0.5762
3	1.331	0.7513	3.310	0.3021	2.4869	0.4021
4	1.464	0.6830	4.641	0.2155	3.1699	0.3155
5	1.611	0.6209	6.105	0.1638	3.7908	0.2638
6	1.772	0.5645	7.716	0.1296	4.3553	0.2296
7	1.949	0.5132	9.487	0.1054	4.8684	0.2054
8	2.144	0.4665	11.436	0.0875	5.3349	0.1875
9	2.358	0.4241	13.579	0.0737	5.7590	0.1737
10	2.594	0.3856	15.937	0.0628	6.1446	0.1628
11	2.853	0.3505	18.531	0.0540	6.4951	0.1540
12	3.138	0.3186	21.384	0.0468	6.8137	0.1468
13	3.452	0.2897	24.523	0.0408	7.1034	0.1408
14	3.798	0.2633	27.975	0.0358	7.3667	0.1358
15	4.177	0.2394	31.772	0.0315	7.6061	0.1315
16	4.595	0.2176	35.950	0.0278	7.8237	0.1278
17	5.054	0.1979	40.545	0.0247	8.0216	0.1247
18	5.560	0.1799	45.599	0.0219	8.2014	0.1219
19	6.116	0.1635	51.159	0.0196	8.3649	0.1196
20	6.728	0.1487	57.275	0.0175	8.5136	0.1175
21	7.400	0.1351	64.003	0.0156	8.6487	0.1156
22	8.140	0.1229	71.403	0.0140	8.7716	0.1140
23	8.954	0.1117	79.543	0.0126	8.8832	0.1126
24	9.850	0.1015	88.497	0.0113	8.9848	0.1113
25	10.835	0.0923	98.347	0.0102	9.0771	0.1102
26	11.918	0.0839	109.182	0.0092	9.1610	0.1092
27	13.110	0.0763	121.100	0.0083	9.2372	0.1083
28	14.421	0.0694	134.210	0.0075	9.3066	0.1075
29	15.863	0.0630	148.631	0.0067	9.3696	0.1067
30	17.449	0.0573	164.494	0.0061	9.4269	0.1061
31	19.194	0.0521	181.943	0.00.55	9.4790	0.1055
32	21.114	0.0474	201.138	0.0050	9.5264	0.1050
33	23.225	0.0431	222.252	0.0045	9.5694	0.1045
34	25.548	0.0392	245.477	0.0041	9.6086	0.1041
35	28.102	0.0356	271.024	0.0037	9.6442	0.1037
40	45.259	0.0221	442.593	0.0023	9.7791	0.1023
45	72.890	0.0137	718.905	0.0014	9.8628	0.1014
50	117.391	0.0086	1163.909	0.0009	9.9148	0.1009

APPENDIX 4   20% Interest Factors for Annual Compounding Interest

	Single Payment		Equal Payment Series			
	Compound-amount factor	Present-worth factor	Compound-amount factor	Sinking-fund factor	Present-worth factor	Capital-recovery factor
n	To find F Given P F/P i, n	To find P Given F P/F i, n	To find F Given A F/A i, n	To find A Given F A/F i, n	To find P Given A P/A i, n	To find A Given P A/P i, n
1	1.200	0.8333	1.000	1.0000	0.8333	1.2000
2	1.440	0.6945	2.200	0.4546	1.5278	0.6546
3	1.728	0.5787	3.640	0.2747	2.1065	0.4747
4	2.074	0.4823	5.368	0.1863	2.5887	0.3863
5	2.488	0.4019	7.442	0.1344	2.9906	0.3344
6	2.986	0.3349	9.930	0.1007	3.3255	0.3007
7	3.583	0.2791	12.916	0.0774	3.6046	0.2774
8	4.300	0.2326	16.499	0.0606	3.8372	0.2606
9	5.160	0.1938	20.799	0.0481	4.0310	0.2481
10	6.192	0.1615	25.959	0.0385	4.1925	0.2385
11	7.430	0.1346	32.150	0.0311	4.3271	0.2311
12	8.916	0.1122	39.581	0.0253	4.4392	0.2253
13	10.699	0.0935	48.497	0.0206	4.5327	0.2206
14	12.839	0.0779	59.196	0.0169	4.6106	0.2169
15	15.407	0.0649	72.035	0.0139	4.6755	0.2139
16	18.488	0.0541	87.442	0.0114	4.7296	0.2114
17	22.186	0.0451	105.931	0.0095	4.7746	0.2095
18	26.623	0.0376	128.117	0.0078	4.8122	0.2078
19	31.948	0.0313	154.740	0.0065	4.8435	0.2065
20	38.338	0.0261	186.688	0.0054	4.8696	0.2054
21	46.005	0.0217	225.026	0.0045	4.8913	0.2045
22	55.206	0.0181	271.031	0.0037	4.9094	0.2037
23	66.247	0.0151	326.237	0.0031	4.9245	0.2031
24	79.497	0.0126	392.484	0.0026	4.9371	0.2026
25	95.396	0.0105	471.981	0.0021	4.9476	0.2021
26	114.475	0.0087	567.377	0.0018	4.9563	0.2018
27	137.371	0.0073	681.853	0.0015	4.9636	0.2015
28	164.845	0.0061	819.223	0.0012	4.9697	0.2012
29	197.814	0.0051	984.068	0.0010	4.9747	0.2010
30	237.376	0.0042	1181.882	0.0009	4.9789	0.2009
31	284.852	0.0035	1419.258	0.0007	4.9825	0.2007
32	341.822	0.0029	1704.109	0.0006	4.9854	0.2006
33	410.186	0.0024	2045.931	0.0005	4.9878	0.2005
34	492.224	0.0020	2456.118	0.0004	4.9899	0.2004
35	590.668	0.0017	2948.341	0.0003	4.9915	0.2003
40	1469.772	0.0007	7343.858	0.0002	4.9966	0.2001
45	3657.262	0.0003	18281.310	0.0001	4.9986	0.2001
50	9100.438	0.0001	45497.191	0.0000	4.9995	0.2000

References[1]

JOURNALS

Construction Industry—Management, E. & F. N. Spon, London, 1983. Journal.

Journal of Construction Engineering and Management, American Society of Civil Engineers, New York, 1983. Quarterly Journal.

ESTIMATING DATA[2]

Process Plant Construction Standards, Richardson Engineering Services, Mesa, AZ, 1997, Four Volumes.

Sweet's Catalog Files, Sweet's Division, McGraw-Hill Information Systems Co., New York, NY.

Kiley, Martin D., *National Construction Estimator*, Craftsman Book Company, Carlsbad, CA, 1997, 557.

Waier, Phillip R. Sr., ed., *Means Unit Price Estimating Methods: Standards & Procedures For Using Unit Price Cost Data,* R. S. Means Co., Kingston, MA, 1999, 416.

–, *Walkers Building Estimator's Reference Book*, 25th Ed., Frank R. Walker Co., Lisle, IL, 1996.

BOOKS

Adrian, James J., *Construction Accounting: Financial, Managerial, Auditing, and Tax, 2d ed.,* Prentice Hall, Englewood Cliffs, NJ, 1986, 366.

Adrian, James J., *Construction Estimating: An Accounting and Productivity Approach*, Reston Pub. Co., Reston, VA., 1982, 528.

Adrian, James J., *Quantitative Methods In Construction Management*, American Elsevier Pub. Co., New York, N.Y, 1973, 491.

Ahuja, Hira N., and Walter J. Campbell, *Estimating from Concept to Completion*, Prentice-Hall, Englewood Cliffs, NJ, 1988, 166.

Banafa, Ahmed Mohammed, *Design Reliability for Estimating Costs of Pile Foundations: From Theory to Application of a Probabilistic-Fuzzy Approach*, University of Colorado at Boulder, Boulder, CO, 1991, 392.

---

[1] This list of references can be augmented by checking a web site, such as *www.AMAZON.com* or *www.BARNESANDNOBLE.com*. Additionally, many college libraries have web-based search engines able to help the student find material and books.

[2] Partial list. There are over 25 estimating encyclopedias that are routinely published annually. There are quarterly N regional editions that are published too.

Beeston, Derek T., *Statistical Methods for Building Price Data*, E. & F.N. Spon, London, 1983 175.

Bledsoe, John D., *From Concept to Bid: Successful Estimating Methods*, R. S. Means Company, Inc., Kingston, MA, 1992, 288.

Brooks, Frederick P., *The Mythical Man Month*, Addison-Wesley Publishing Co, Reading, MA, 1975, 194.

Canada, John R., William G. Sullivan, and John A. White, *Capital Investment Analysis for Engineering and Management*, Prentice Hall, Upper Saddle River, NJ, 1996, 566.

Clark, John E., *Structural Concrete Cost Estimating*, McGraw-Hill, New York, N.Y., 1983, 298.

Collier, Keith, *Construction Contracts*, 2d ed., Prentice Hall, Englewood Cliffs, NJ, 1986, 352.

Collier, Keith, *Estimating Construction Costs, A Conceptual Approach*, Prentice-Hall, Englewood Cliffs, NJ, 1984, 315.

Collier, Keith, *Fundamentals of Construction Estimating and Cost Accounting with Computer Applications*, 2d ed., Prentice Hall, Englewood Cliffs, NJ, 1993, 352.

Cook, Paul J., *Estimating For The General Contractor*, R.S. Means Co., Kingston, MA, 1982, 224.

Cooper, George Henderson, *Building Construction Estimating*, McGraw-Hill, New York, NY: London, England, 1945, 282.

Dagostion, Frank R., *Estimating for the Building Construction*, 4th ed., Prentice-Hall, Englewood Cliffs, NJ, 1993, 418.

Fatzinger, James A. S. *Basic Estimating for Construction*, Prentice Hall, Englewood Cliffs, NJ, 1996 352.

Geddes, Spence, *Estimating for Building and Civil Engineering Works*, Butterworths, London; Boston, 1985, 424.

Gillette, Halbert Powers, *Handbook of Construction Cost*, McGraw-Hill, New York, NY, 1922, 1734.

Gillette, Halbert Powers, *Handbook of Cost Data for Contractors and Engineers*, Myron C. Clark, New York, NY, 1905, 610.

Gould, Frederick E., *Managing the Construction Process: Estimating, Scheduling and Project Control*, Prentice-Hall, Englewood Cliffs, NJ, 1997, 338.

Geer, Willis R., ed., and Daniel A. Nussbaum, *Cost Analysis and Estimating, Tools and Techniques*, Springer-Verlag, New York, NY, 1990.

Helton, Joseph E., *Simplified Estimating for Builders and Engineers*, 2d ed., Prentice-Hall, Englewood Cliffs, NJ, 1991, 224.

Hornung, William J., *Estimating Building Construction, Quantity Surveying*, Prentice-Hall, Englewood Cliffs, NJ, 1970, 210.

Kharbanda, O.P., *Process Plant and Equipment Cost Estimation*, Craftsman Book Company, Solana Beach, CA, 1979, 235.

McNeill, Thomas F., and Donald S. Clark, *Cost Estimating and Contract Pricing*, American Elsevier Pub. Co., New York, NY, 1966, 514.

Navarrete, Pablo F., *Planning, Estimating, and Control of Chemical Construction Projects*, Marcel Dekker, New York, NY, 1995, 324.

Nisbet, James, *Estimating and Cost Control*, B.T. Batsford, London, England, 1961, 227.

O'Brien, James J., *Preconstruction Estimating: Budget Through Bid*, McGraw-Hall, New York, NY, 1994, 596.

Ostwald, Phillip F., *AM Cost Estimator*, 4th ed., Penton Publishing, Cleveland, OH, 1987, 560.

Ostwald, Phillip F., *Engineering Cost Estimating*, 3d ed., Prentice-Hall, Englewood Cliffs, NJ, 1992, 510.

Ostwald, Phillip F. and Jairo Muñoz, *Manufacturing Processes and Systems*, 9th ed., John Wiley & Sons, New York, NY, 1997, 782.

Patrascu, Angel, *Construction Cost Engineering*, Craftsman Book Company, Solana Beach, CA., 1978, 302.

Peurifoy, Robert Leroy, W. B. Ledbetter, and C. J. Schexnayder, *Construction Planning, Equipment, and Methods*, 5th ed., McGraw-Hill, New York, NY, 1996, 633.

Peurifoy, Robert Leroy, and G. D. Oberlender, *Estimating Construction,* 4th ed., McGraw-Hill, New York, NY, 1989, 473.

Pulver, Harry E., *Construction Estimates and Costs*, McGraw-Hill, New York, NY, 1969, 644.

Ross, Timothy J., *Fuzzy Logic With Engineering Applications*, McGraw-Hill, Inc., New York, NY, 1995, 600.

Schuette, Stephen D., and Roger W. Liska, *Building Construction Estimating*, McGraw-Hill, New York, NY, 1994, 346.

Steinberg, Joseph, and Martin Stempel, *Estimating For The Building Trades*, American Technical Society, Chicago, IL, 1965, 374.

Shirley-Smith, H., *The World's Great Bridges,* Phoenix House, London, 1953, 250.

Stevens, Joseph E., *Hoover Dam, An American Adventure*, University of Oklahoma Press, Norman, OK, 1988, 326.

Tucker, Spencer A., *Cost-Estimating and Pricing With Machine-Hour Rates*, Prentice Hall, Englewood Cliffs, NJ, 1962, 253.

Vance, Mary A., *Selected List Of Books On Building Cost Estimating*, Vance Bibliographies, Monticello, IL, 1979, 3.

Walker, Frank R., *The Building Estimator's Reference Book,* Frank R. Walker, 1915.

Wass, Alonzo, *Building Construction Estimating*, 2d ed., Prentice-Hall, Englewood Cliffs, NY, 1970, 324.

Westney, Richard E., ed, *The Cost Engineer's Handbook*, Marcel Dekker, Inc., New York, NY, 1997, 749.

Willenbrock, Jack H., Harvey Manbeck, and Michael B. Suchar, *Residential Building: Design and Construction*, Prentice Hall, Englewood Cliffs, NJ, 1997, 700.

SOFTWARE FIRMS HAVING CONSTRUCTION INTEREST

Building Systems Design	Microsoft
Computer Aided Management	Monenco ACRA
Computer Associates International	OnTrack Engineering, Ltd
CSC Artemis	Panorama Software Corporation
Dekker, Ltd.	Primavera Systems, Inc.
Delta Research Corporation	Project Software & Development, Inc.
DNA Consultants, Inc.	Resolution Management Consultants
EPCON International	Richardson Engineering
G2, Inc.	SoftCOST, Inc.
ICARUS Corporation	Timberline Software Corporation
LogiConstruct SA	U.S. Cost, Inc.
Management Computer Controls (MC2)	Welcom Software Technology
Meridian Project Technologies	WinEstimator, Inc.
Micro-Frame Technologies	

Selected Answers

CHAPTER 1

1.2	280,000 Btu	295,400,000 J = 295.4 MJ
	7,500,000 kWh	2.7×10^{13} J = 27×10^6 MJ
1.4	18 ft	5.5 m
	2.0 ft	0.61 m
	10 in.	0.254 m$/10^{-3}$ = 254 mm
	0.01 in.	0.25 mm
	100 in.	2,540 mm
	0.00015 in.	0.0038 mm
1.6	15 lbm/ft^3	240 kg/m^3
	180 kg/m^3	11.24 lbm/ft^3
1.8	200° F	93.3 C
	1000° F	537.8 C
	1500 C	2732° F
	200 C	392° F
	1000 C	1832° F
1.10	0.37 ft^3	0.01 m^3
	125 ft^3	3.54 m^3
	700 ft^3	19.8 m^3
	0.01 in.^{3}	16.4 mm^3
	12 in.^{3}	19,656 mm^3
	150 in.^{3}	245,000 mm^3
	1000 yd^3	765 m^3
1.12	Euros	1607.86 euros
	Pesos	15,681 pesos
	Canada dollars	2,569.00 Canada dollars
1.16	Germany	$197.5365 million
	England	196.5971
	Canada	198.6251

CHAPTER 2

2.2 **(a)** Productivity = Hours/output = 400 hours/50(100) sfca = 8 hr/100 sfca

(b) 5 workers $\times$ 8 hours/10 tons = 4 worker hours per ton.

2.4 **(a)** Duration (days) = 6000/2000 = 3 days.

Also, $(6000\text{ft}^2)(0.008)/8 \times 2 = 3$ days

(b) Duration days = $6000 \times 0.008/(8 \times 3) = 3$ days.

2.6 $465.75. Cost/100 ft^2 = $4.66.

2.8 **(a)**

Task Element	Total Man Hour
Poles in dirt	15.0
Anchors	6.4
Cross arms	21.0
Transformer	4.4
Lightning arrester	1.8
Farm light	1.3
	49.9
Direct labor and equipment	$4491

2.10 352 ft/min.

6.61 min/yd^3

At 90%, the time = 12.2 hr.

2.12 Productivity $4.18 sfca

2.18 Efficiency, $(2436 + 832)/5848 = 56\%$

2.20 Units per hour 4.84

Hr/100 units 20.64

Man hour per ft^2 0.00645

2.24

Job Tasks	I_i	Relative Accuracy, %
A	0.0061	±18.3
B	0.0118	±8.8
C	0.0224	±3.4
D	0.0234	±2.0

Example for A: $I = 2(1.645)\left[(0.0167)(0.9833)/4800\right]^{1/2} = 0.0061$

Relative accuracy = $I/2p' = 0.0061/2(0.0167) = 0.1828 = \pm 18.3\%$

2.26 **(a)** $I = 10\% = 0.1$; $p' = 0.70$; $CI = 95\%, Z = 1.96$.

$N = 4(1.96)^2(0.7)(0.3)/(0.1)^2 = 323$ observations

For 90%, $N = 227$ observations.

2.29 The most significant reason for nonproductivity is labor, i.e., 60%.
Cost of steel per ton placed during study = 6375 × 33.94/84 = $2576.
Expected cost of steel erection for building = 935 × 2576 = $2,408,375.
The bid uses man hour per ton = 62 = 58,000/935
Actual man hour per ton = 6375/84 = 75.89
Labor efficiency = 62.03/75.89 = 82%.

2.32 Job cost $2,366.33

2.34 Total $1872.00
$/ft^2$ $5.85
$/lft $46.80
Wall 95 ft long will cost $4446.

2.37 **(a)** Crew working elsewhere while not assigned to job.
$/yd^3$ $69.37

(b) Crew at job site for fully scheduled work day. Idle time devoted to job.
$/yd^3$ $69.99
Additional cost of nonproductive time = $248.

2.39 Standard crew time = 5.59 hr.
Typical steel member = 2040 lb = 1.02 ton.
Standard crew productivity = 5.59/1.02 = 5.48 man hours/ton.

Case Study: Highway Construction

Cost per mile for equipment and gross hourly direct labor cost = 400,800/9.2 = $43,565

CHAPTER 3

3.2 (a)

Item	
Total bft	66.3
Addition for waste about 5%	70 bft

(b)

Item	
Fire stop	6.9
Studs	48.0
Bottom plate	10.7
Top plate	21.3
Total bft	86.9
Addition for waste about 5%	92.0 bft

3.4 Total cost for standard- length boards $66,650.00
$64,800 for precut lengths. For these numbers, it is cheaper to buy precut.

3.6

Calculation	Result
Volume of excavation: $165 \times 42.75 = 7053.75$ ft^3	261 yd^3
Concrete void: footing: $2 \times 1 \times 165/27$	12.2 yd^3
Concrete void: wall: $1 \times 5.25 \times 165/27$	32.1 yd^3
Net volume for back filling	217 yd^3
Less swell of material: 217×0.85	184.2 yd^3
Removed from the site	33 yd^3

3.8 **(a)** Cost = $1868 (b) Cost = $2805

3.12 Cost = $22,800 and including the transportation = $23,800.

3.14 Total cost of direct material = $945,770

3.16 At 1250 ft^2, ratio = 11/1.

3.18 Cost for shell: C = $29,000

Cost for ends (assuming spherical ends, which costs more than elliptical ends) = $29,500.
Total cost = $58,500.

Case Study: Chemical Process Plant

Total direct and indirect material cost = $2,290,000

CHAPTER 4

4.2

Weichman Contractor
Balance Sheet
March 31

Assets		Liabilities	
Cash	$1200	Acct. payable	$360
Acct rec.	700	Notes payable	3200
Supplies on hand	250	Subtotal	$3560
Equipment	3200	Owners Equity	
		Capital stock	1790
Total	$5350	Total	$5350

4.4

CCM Construction Management
Balance-Sheet Statement
December 31

Assets	
Total	$1,500,000

4.5 Total income $652,000
Total expense $415,000
Retained earnings, Dec 31 $325,000

4.8

Lyell Construction
Profit-and-Loss Statement
December 31

Income		
Sales		$600,000
Total		$600,000
Expenses		
Labor		$100,000
Materials		150,000
Depreciation		50,000
Total		$300,000
Gross profit (to retained earnings)		$300,000

Assets		Liabilities	
Cash	$250,000	Accts. Payable	$50,000
Inventory	100,000		
Subtotal current	$350,000	Net worth	550,000
Fixed	$550,000	Retained earnings	300,000
Total	$900,000	Total	$900,000

4.10

Income			
Total	$185,000		
Expenses			
Total	$140,000		
Net profit	$25,000		
Assets	$695,000.00	Liabilities	$170,000

4.12

Net profit	*$700*		
Assets	$14,160	Liabilities	$3400

4.16

Year	S.L. Dep., $	B.V., $	A.C. Rate, %	A.C. Dep., $	B.V., $
1	47,000	$203,000	20	$50,000	$200,000
2	47,000	156,000	32	80,000	120,000

4.18 Profits (to retained earnings) $4000

4.19

No. of Units	Total Costs	Average Cost, $/Unit
15	$50	$3.33
30	70	2.33

4.22

	C_p	C_o	C_j
Total	$1140.00	$54.00	$28.20

4.27 Total job cost $407,550

      Project overhead costs $79,200

      Project overhead rate   26.5%

CHAPTER 5

5.3

	Mean	Median	Mode	Range	Std. Dev.
a.	4.9	5	5	8	2.38

5.4 $y = 2436$

5.6 $T = 175.9$

5.10 error$=+1.43\%$

    $Y \pm ts_{yi} = 134.5 \pm 4.455$

5.16 Correlation -0.127

5.18 Second-degree polynomial $Y = 5.84 - 8.65 \times 10^{-4}x + 4.47 \times 10^{-7}x^2$

5.24

Year	1999 = 100 Index
1999	100.0

5.28 Cost = $231,404

5.34 $I_3 = 111.93$

CHAPTER 6

6.5 Cost per yd^3 $416.93

6.6 The cost = $69,500

6.8 $13,373,702.00

6.10 **(a)** $T_{u=4} = 13{,}624$ h

   $s = -0.2009$

   $T_{u=6} = 12{,}558$ h

6.14 $27.73 million

6.18 The expected profit becomes $22{,}500(650 - 475) = \$3.9375$ million

6.20 $P(\text{Cost} > 25) = 0.5 + 0.4441 = 94.41\%$

6.24 $/ft^2 6.21

6.28 Total $61,857,711

CHAPTER 7

7.2 For 30 ft, total cost = $1118. For 12 ft, total cost = $4471.

7.4 Total cost $122.00

7.6 Total cost $557

   Cost/yd^3 $492.91

7.8

Work and Material Takeoff	Quantity	Cost Factor	Cost
No. 6 bars, 25 ft, 12 pc	300 ft	$1.277/ft	$383
No. 4 bars, 10 ft 9 in., 12 pairs, or 24 pc	258 ft	0.561/ft	145
Labor for No. 6 bars	450 lb	0.0125 hr/lb × $34/hr = $0.425/lb	191
Labor for No. 4 bars	172 lb	0.023 hr/lb × $34/hr = $0.78/lb	135
14.8 yd^3 × 1.03	15.3 yd^3	$63.50	969
Labor for pour, 15.3 × 0.5	7.7 hr	$90/hr (= 4 × 15 + 30)	693
		Total cost	$2516.00

7.10 Cost per sq. ft = $3.46.

7.12 Annual yearly capital cost consumption = $54,167

   Annual yearly capital cost consumption = $48,750

   Unrecovered capital value = $195,000

7.16 Total cost per hour, excluding labor $76.58/hour

7.18

$$\text{Cost per job} = \$131.12 + 0.822 \times \text{No. of yd}^3 + 15.40 \times \text{No. of miles}$$

$$+ \begin{bmatrix} \text{yd}^3 \text{ pumped} & \$ \\ 15 < 25 & 56 \\ 25 < 60 & 104 \\ 60 \text{ or more} & 144 \end{bmatrix}$$

7.20 Total cost $77,522.00

       Cost per 100 ft = $3876; cost per ft^2 = $1.55.

7.22

$$\text{Cost per job} = \$137.28 + 0.860 \times \text{No. of yd}^3 + 16.13 \times \text{No. of miles}$$

$$+ \begin{bmatrix} \text{yd}^3 \text{ pumped} & \$ \\ 15 < 25 & 58.24 \\ 25 < 60 & 108.16 \\ 60 \text{ or more} & 149.76 \end{bmatrix}$$

CHAPTER 8

8.2 Cost each bridge $2,679,686

       Project cost $13,398,430

8.4 Total cost is 690×10^6.

8.11 t_{ai} = 0.7th month for average and initial cash flow. t_{af} = 3.6th month for average and final point of cash flow.

     t_i = 1st month when the max cash flow is reached. t_f = 3rd month when max cash flow ends.

8.12

Interest, $
Total $174,403

8.14 Bid $33,600,000

Work Package Item
Bid $24,400,000

CHAPTER 9

9.1 Optimum bid is approximately $520,000.

9.2 Peak point occurs at B/C_e of 1.22. Bid value = $884,500.

9.4 Optimum location about 1.25. Bid value is $4,656,250.

9.6 Gain = $2513.

9.8 **(a)** Capture rate = 52.7%.

9.10 The percentage error of the installed system is = ± 13%.

9.12 Bid $1,039,500

9.15 7.2%.

9.20 Optimum location is 1.05 = B/C_e. Bid = $761,250. A range of profit can be determined probabilistically as anywhere there is a positive curve above $0 C_e.

CHAPTER 10

10.2

	Taxable Income	Income Taxes
Total	$515,000.00	$175,100.00

10.4 40%.

10.8

End of Year, n	Actual Dollars, $
4	646.50

10.10 Break-even point in percentage of sales = 85.2%

10.14

Project Units	Total Cost, $	Average Cost, $	Marginal Cost, $
9	38,000	4,220	10,000

10.16 Payback = 6 years

10.19 (a): $I = \$90$

   (b) $i = 4\%$.

   (c) $F = \$90,000$.

10.22 (a) Net PW = $18.

   Net FW = $24.

   Net AEW = $281.10.

   Net AEW = $4.80.

10.30 Cost/unit (A) = $71,250

10.36 $i_c = 10.24\%, \$162,820$

CHAPTER 11

11.1 Bid $9,817,625

11.2 $3,775,753

11.3 $9,455

11.7 Underrun in price $1,600,000

11.8 Overrun in price, variance unfavorable $800,000

11.9 Overrun in price, variance unfavorable   600,000

   Underrun in price, variance favorable   400,000

Index

Symbols and Conversions from U.S. Customary Units to International System of Units

Dimension	To Convert from Customary Units	To International System of Units	Multiply by[1]
Area			
acre (U.S. survey)	acre	m^2 (meter^2)	4.046×10^3
square foot	ft^2	m^2	9.290×10^{-2}
square inch	in.^{2}	m^2	6.451×10^{-4}
square yard	yd^2	m^2	8.361×10^{-1}
Energy			
Btu		J (Joule)	
foot-pound-force	ft-lb-f	J	1.355
kilowatt hour	kWh	J	3.600×10^6
Force			
pound-force	lbf	N (Neuton)	4.448
Length			
foot	ft	m (meter)	3.048×10^{-1}
inch	in.	m	2.54×10^{-2}
mile (U.S. survey)	mile	m	1.609×10^3
yard	yd	m	9.144×10^{-1}
Mass			
pound-mass	lbm	kg (kilogram)	4.535×10^{-1}
sack (94 lb)	sk	kg	4.535×10^{-1}
ton (long, 2240 lbm)	ton, long	kg	1.016×10^3
ton (short 2000 lbm)	ton	kg	9.071×10^2
Mass/Volume (Includes density)			
pound-mass/foot^3	lbm/ft^3	kg/m^3 (kilogram/meter^3)	1.601×10
pound-mass/in.^{3}	lbm/in.^{3}	kg/m^3	2.767×10^4
ton(long, mass)/yard^3		kg/m^3	
Power			
horsepower (550 ft-lbf/s)	hp	w (watt)	7.456×10^2
Stress (Force/Area)			
pound-force/foot^2	lbf/ft^2	Pa (Pascal)	4.788×10
pound/in.^{2}	psi	Pa	6.894×10^3
Temperature			
degree Fahrenheit	°F	C (degree Celsius)	$t°_C = (t°_F - 32)/1.8$
Velocity			
foot/hour	ft/hr	m/s (meter/second)	8.466×10^{-5}
foot/minute	ft/min	m/s	5.080×10^{-3}
foot/second	ft/s	m/s	3.048×10^{-1}
inch/second	in./s	m/s	2.54×10^{-2}
Volume			
board foot	bft	m^3 (meter^3)	2.359×10^{-3}
cubic foot	ft^3	m^3	2.831×10^{-2}
gallon (U.S. dry)	gal	m^3	4.404×10^{-3}
gallon (U.S. liquid)	gal	m^3	3.785×10^{-3}
cubic inch	in.^{3}	m^3	1.638×10^{-5}
cubic yard	yd^3	m^3	7.645×10^{-1}
cubic inch	in.^{3}	mm^3	1.638×10^4
Volume/time			
cubic foot/min	ft^3/min	m^3/s (meter^3/s)	4.719×10^{-4}
cubic foot/second	ft^3/s	m^3/s	2.831×10^{-2}
cubic yard/hour	yd^3/hr	m^3/s	2.123×10^{-4}
cubic yard/min	yd^3/min	m^3/s	1.274×10^{-2}

[1] Conversion values truncated from American Society for Testing and Materials E 380-74 values.